Medicine and Evolution
Current Applications, Future Prospects

Society for the Study of Human Biology Series

Numbers 1–9 were published by Pergamon Press, Headington Hill Hall, Headington, Oxford OX3 0BY. Numbers 10–24 were published by Taylor & Francis Ltd, 10–14 Macklin Street, London WC2B 5NF. Numbers 25–40 were published by Cambridge University Press, The Pitt Building, Trumpington Street Cambridge CB2 1RP. Further details and prices of back-list numbers are available from the Secretary of the Society for the Study of Human Biology.

Medicine and Evolution
Current Applications, Future Prospects

Edited by
Sarah Elton
Paul O'Higgins

CRC Press
Taylor & Francis Group
Boca Raton London New York

CRC Press is an imprint of the
Taylor & Francis Group, an **informa** business

CRC Press
Taylor & Francis Group
6000 Broken Sound Parkway NW, Suite 300
Boca Raton, FL 33487-2742

First issued in paperback 2019

ISBN-13: 978-1-4200-5134-6 (hbk)
ISBN-13: 978-0-367-38725-9 (pbk)

Library of Congress Cataloging-in-Publication Data

Medicine and evolution : current applications, future prospects / editor(s), Sarah
    Elton, Paul O'Higgins.
        p. ; cm. -- (Society for the Study of Human Biology series ; 48)
    "A CRC title."
    Includes bibliographical references and index.
    ISBN 978-1-4200-5134-6 (hardback : alk. paper) 1. Human evolution--Congresses.
    2. Evolution (Biology)--Congresses. 3. Medicine--Congresses. I. Elton, Sarah. II.
    O'Higgins, Paul. III. Title. IV. Series.
    [DNLM: 1. Darwin, Charles, 1809-1882. 2. Hominidae--physiology--Congresses. 3.
    Anthropology, Physical--Congresses. 4. Evolution--Congresses. W1 SO861 v.48 2008 /
    GN 281 M489 2008]

    GN281.4.M43 2008
    599.93'8--dc22

                                                                      2008010668

Visit the Taylor & Francis Web site at
http://www.taylorandfrancis.com

and the CRC Press Web site at
http://www.crcpress.com

Dedicated to the memory of Nick Norgan

Contents

Foreword

The dawn of Darwinian medicine has turned out to be an awkward metaphor. Instead of the start of a long day, however good, the application of evolutionary thinking to medicine is proving to be more like the birth of an organism that is fast growing up. While one could argue about whether we are now at the toddler stage or gangly adolescence, the field clearly remains young. That vigour is on display in this volume. Old ideas, such as the thrifty genotype, are presented not as museum pieces, but active hypotheses that began as intriguing but incorrect hunches and now are maturing to become heuristic perspectives that spur new research important to human health. Common beliefs, such as the role decreased parasite load plays in increasing vulnerability to certain diseases, are examined from novel viewpoints. Generalizations about the ancestral environment are here replaced by anthropologically sophisticated perspectives on the variations in human environments and human adaptability. Long-standing questions about the public health significance of protein requirements, reasons for early pregnancy loss, and optimal mother-infant sleeping arrangements are examined with new data. And fresh ideas about syndromes as diverse as delusions and polycystic ovary syndrome emerge, increasingly with a consideration of the relevant evidence.

This is all very interesting, as doctors so often tell me. However, as doctors predictably ask, is it useful? Should doctors learn the relevant principles of evolutionary biology in the same way they learn biochemistry, anatomy, or physiology? Some answers are here in the chapters on practical applications and medical education. Bentley, in her innovative seminar, discovered what happens when practitioners and researchers are given an opportunity to think more deeply about the implications of evolution for their work; they come up with good ideas and see their own projects in a new light. The chapter on education reviews the rather sorry state of evolutionary teaching in medicine, but from the perspective of insiders who recognize why evolution is not taught. The chapters in this book, taken together, make a strong case for evolution in medical education. The benefits include a better ability to assess advice on specific questions such as diet and recommended daily allowances, and how best to house mothers and neonates in maternity wards. In addition, and perhaps even more importantly, an evolutionary perspective emphasizes that diseases arise not from broken parts in a poorly designed machine, but from the inevitable compromises of an evolved body interacting with novel environments. As it spreads, this more accurate view of the body and disease will change medicine fundamentally.

Darwinian medicine is developing rapidly, in part because of the insights that a new perspective brings to old problems. Its development has also been nurtured by its many interfaces with related disciplines. I hope this book will stimulate readers to want to know more about this field, and I recommend http://EvolutionAndMedicine.org as a useful first port of call to keep abreast of new findings and learn about upcoming meetings.

Randolph M. Nesse
University of Michigan–Ann Arbor

Preface

This volume arose out of the Society for the Study of Human Biology and Biosocial Society Symposium "Medicine and Evolution," held in York, UK, on December 11–12, 2006. As anthropologists working at the recently opened Hull York Medical School (HYMS), one of a number of institutions established in answer to the UK need for more doctors, we were keen to explore how we might link our research and teaching interests in new and exciting ways. The symposium provided an opportunity to expose our colleagues and students to the opportunities offered by an overarching evolutionary view. Given the constraints of modern medical education, we cannot expect to witness a radical infusion of evolutionary thinking in medical schools, but we hope this volume will at least suggest some avenues worthy of exploration and stimulate interest among medical students and practitioners. The York symposium was energised by interaction between the speakers, who for the most part were anthropologists and evolutionary biologists, and the discussants, most of whom were practising clinicians. The resulting conversations were enormously important in shaping this volume.

A great many people have helped us in the preparation of this volume. First and foremost we extend our thanks to those who attended the symposium in York and whose participation made for a highly enjoyable and stimulating meeting. We are especially grateful to the colleagues who contributed chapters for this volume, as well as to those who gave up their time to review the contributions. We gratefully acknowledge the financial assistance of the Society for the Study of Human Biology, the Biosocial Society, and the Hull York Medical School toward symposium costs, and the York Medical Society for generously allowing us to use its superb meeting rooms in the centre of York. Jackie Houlton, Richard Nicholson, and Nadine Webster of HYMS were constant sources of help in arranging and running the symposium as well as preparing the subsequent volume. Barbara Norwitz, Amy Rodriguez, and Pat Roberson of Taylor & Francis have generously and patiently given advice over the past year. Finally, we thank John Russell for his continued support.

Sarah Elton and Paul O'Higgins
York, UK

The Editors

Sarah Elton is senior lecturer in anatomy at Hull York Medical School. She is a biological anthropologist with a research focus on the environmental factors that influence variation and adaptation, with special interest in Old World primates. She uses evolutionary approaches to teach concepts in the core medical curriculum as well as providing elective modules on evolution for medical students.

Paul O'Higgins is chair of anatomy at Hull York Medical School. He is a qualified clinician working in biological anthropology and morphology. His research interests concern human and primate evolution and variation. He has extensive experience of teaching health professionals at undergraduate and postgraduate levels and frequently bases his teaching on the evolutionary principles underlying normal and disordered structure and function.

Contributors

Robert Aunger
Hygiene Centre
London School of Hygiene and
  Tropical Medicine
London, United Kingdom

Jack Baker
Bureau of Business and Economic
  Research
University of New Mexico
Albuquerque, New Mexico

Helen Ball
Parent Infant Sleep Lab and
  Medical Anthropology Research
  Group
Department of Anthropology
Durham University
Durham, United Kingdom

Gillian R. Bentley
Department of Anthropology and
  Wolfson Research Institute
Durham University
Durham, United Kingdom

M. Christopher Dean
Department of Cell Anatomy and
  Developmental Biology
University College, London
London, United Kingdom

Sarah Elton
Functional Morphology and
  Evolution Unit
Hull York Medical School
University of Hull
Hull, United Kingdom

M. Anderson Frey
Department of Anthropology
University of New Mexico
Albuquerque, New Mexico

Edward H. Hagen
Department of Anthropology
Washington State University
Vancouver, Washington

Kim Hill
School of Human Evolution and
  Social Change
Arizona State University
Tempe, Arizona

A. Magdalena Hurtado
School of Human Evolution and
  Social Change
Arizona State University
Tempe, Arizona

Inés Hurtado
Department of Anthropology
University of New Mexico
Albuquerque, New Mexico
and
Instituto Venezolana de
  Investigaciones Cientificas
Venezuela

Jonathan Marks
Department of Sociology and
  Anthropology
University of North Carolina at
  Charlottle
Charlotte, North Carolina

Alejandra Núñez-de la Mora
Medical Anthropology Research Group
Department of Anthropology
Durham University
Durham, United Kingdom

Paul O'Higgins
Functional Morphology and
  Evolution Unit
Hull York Medical School
University of York
York, United Kingdom

Tessa M. Pollard
Medical Anthropology Research Group
Department of Anthropology
Durham University
Durham, United Kingdom

Laurence Shaw
London Bridge Fertility
  Gynaecology and Genetics Centre
London, United Kingdom

Stanley J. Ulijaszek
Institute of Social and Cultural
  Anthropology
University of Oxford
Oxford, United Kingdom

Nigel C. Unwin
Institute of Health and Society
University of Newcastle
Newcastle-upon-Tyne, United Kingdom

Virginia J. Vitzthum
Anthropology Department and
  the Kinsey Institute
Indiana University
Bloomington, Indiana
and
Institute for Primary and
  Preventative Health Care
Binghamton University
Binghamton, New York

1

Introduction

Sarah Elton
Functional Morphology and Evolution Unit, Hull
York Medical School, University of Hull

Paul O'Higgins
Functional Morphology and Evolution Unit, Hull
York Medical School, University of York

There is increasing interest in examining aspects of health and disease in the context of evolutionary theory,[1-3] and the past decade has witnessed the rise of evolutionary or Darwinian medicine as an entity distinct from anthropology, evolutionary biology, or evolutionary psychology. Nonetheless, evolutionary concepts are often viewed as being tangential to medical teaching and practise, which tend to emphasise technical and proximate factors together with the treatment of the individual. Williams and Nesse[4] recognised early on that it would not be easy to persuade clinicians of the relevance of evolutionary medicine, regardless of the benefits brought by the approach. To this end, one major aim of this volume is that contributors should, where possible, indicate how their research and scholarship informs practical applications in clinical settings, health promotion, or medical education.

The next eight chapters of this volume address topics that will be familiar to most students of evolutionary medicine: nutrition (Chapters 2, 3), Type 2 diabetes (Chapter 4), fertility and childbirth (Chapters 5, 6, 7), immune regulation (Chapter 8), and psychiatry (Chapter 9). In contrast to Chapters 2 to 9, which focus on evolutionary insights into particular aspects of health and disease and that have a long history of anthropological investigation, Chapters 10 to 13 consider in much broader ways how evolutionary medicine might be "useful" for medical practise or education.

In Chapters 2 and 3 Elton and Ulijaszek, respectively, examine Stone Age diets from different perspectives. Elton, in Chapter 2, draws on examples from palaeoenvironmental studies, environmental archaeology, and modern human and primate ecology to critique the concept of the environment of evolutionary adaptedness. In doing so, she concludes that given the inherent flexibility of the human diet, there is no obvious benefit to employing a dietary regimen based on Stone Age principles, and points to alternative health promotion messages. Another issue to come out of this chapter as well as others (Chapter 4, Chapter 13) is the importance of tackling the impact of social inequality and marginalisation, which are often much greater risk factors for ill-health and disease than any mismatch between genetic (or cultural) heritage and contemporary environments. Thus, one opportunity for evolutionary medicine in the future is to show how it complements other approaches to the study of health and diseases. As Alan Goodman[5] has argued, long and hard, researchers working within traditional "four field" anthropology are in a unique position to integrate biological and social studies of past populations with those conducted on contemporary human populations. Certainly, in the past few years, anthropologists—particularly those engaged in applied anthropology—have become increasingly interested in issues relating to socioeconomic development, migration, social inequality, and marginalisation. And since anthropology is one of the disciplines from which evolutionary medicine emerged, it should provide a context in which it is possible to integrate different perspectives—including those based on evolution—in studies of modern human health.

Ulijaszek focuses on a single aspect of human diet—protein consumption—and examines how protein intake might have varied as a result of the transition to agriculture. In turn, he considers how this, on its own and through feedback with infectious disease, might explain the reduction of stature evident in several Neolithic populations on the transition from hunting to farming. These observations reinforce recent calls for increases in the recommended daily intakes of protein, especially in places where infectious disease is prevalent. Ulijaszek's chapter also illustrates the breadth of approaches that can be classified under the heading "evolutionary medicine," drawing as it does on palaeopathology, observations from the archaeological record, and models based on modern human ecology.

The burgeoning literature shows that evolutionary medicine comprises a number of diverse strands, beyond those that could be included here and encompassing anthropological, biological, biomedical, and psychological approaches. Although not discussed explicitly in this volume, the potential breadth of evolutionary studies within medicine is brought further into relief by consideration of the facets of contemporary evolutionary theory—rather than natural selection alone—that might be brought to bear in the study and practise of medicine. Important concepts such as genetic drift, Neutral Theory, Evo-Devo, and epigenetics can all be applied to research in health and medicine (see Chapters 10, 11, 12 and 13). Since all these emerged after Darwin, we preferentially use the term "evolutionary" rather than "Darwinian" medicine in this volume, to better reflect the complexities of modern evolutionary thinking. The use of contemporary evolutionary principles and ideas has already been shown to be highly profitable in at least one prominent contribution on evolutionary medicine,[2] and also helps to counter the suggestion that evolutionary perspectives within medicine are all tied to simple adaptationist interpretations. Indeed, shifting the focus from "adaptationism" is likely to provide opportunities not yet realised for evolutionary perspectives within medicine.

Chapters 4 and 5 explore another mainstay of evolutionary medicine, insulin resistance, again from two different viewpoints. In Chapter 4, Pollard and colleagues discuss type 2 diabetes in Asian populations and assess the genetic and developmental contributions to the condition. They recommend that until potential thrifty or nonthrifty genes have been identified with certainty, preventive strategies should be focused on the better-understood developmental rather than genetic factors that predispose to type 2 diabetes. Shaw and Elton, in Chapter 5, also discuss the developmental influences on another condition of insulin resistance, polycystic ovary syndrome (PCOS), and present various evolutionary scenarios relating to climatic instability, seasonality, and food availability that may explain the high incidence of the syndrome in populations throughout the world. The authors draw on Shaw's experience as a practicing clinician to suggest when and how it might be appropriate to use evolutionary explanations about PCOS in patient consultation. In particular, they argue that providing an evolutionary viewpoint may encourage general lifestyle modification rather than short-term dieting in women with the syndrome who are seeking fertility treatment.

In Chapter 6, Vitzthum's life history approach to understanding the regulation of pregnancy outcomes demonstrates the role of maternal resource availability, in a view that resonates with the ideas discussed in Chapters 4 and 5. Her insights, based on many years of fieldwork on the Bolivian *altiplano*, underline the potential importance of viewing early pregnancy loss in the context of evolution and life history, not least by helping to provide explanations that may reduce the feelings of inadequacy in those who have experienced it. Examining conditions or events that are commonly seen as dysfunctions through an evolutionary lens is one clear and stated aim of evolutionary medicine,[1] yet translating such observations into health care policy and practise has proven to be far from straightforward, notwithstanding the promise of recent work, such as that described in Chapters 4 to 6. Evolutionary and ethno-paediatrics is one area within evolutionary medicine that has arguably had the most success in translating the results of research findings and clinical trials into medical practise. In Chapter 7, Ball uses an evolutionary and comparative approach to evaluate how best the needs of mothers and babies should be accommodated in hospital maternity wards, with particular reference to breast feeding. Her study, undertaken as a randomised controlled trial, demonstrates the importance of maintaining physical contact in the neonatal period. This and similar contributions reinforce the practical, political, and research efforts of midwives and are helping to shift attitudes in the UK away from the medicalisation of pregnancy, birth, and postnatal care. However, as Ball points out, there is far from universal adoption of the recommendations that have arisen from evolutionary paediatrics.[6] Nonetheless, encouraging such dialogue between anthropologists, life scientists, and health professionals interested in evolutionary approaches is one important aspect of the development of evolutionary medicine.

Another is the need to move from ideas to hypotheses, and from there to experimental and empirical research. One criticism levelled at evolutionary medicine is that much of it is speculative and remains untested or even untestable.[7,8] It has been suggested that generating and testing specific hypotheses is an important way of demonstrating the utility of an evolutionary framework within medicine.[7,8] In this volume we endeavour to make at least a step in this direction by presenting studies grounded in empirical observation, of which Chapters 6 and 7 are excellent examples. In a time when randomised controlled trials are seen as the gold standard for clinical research, developing such studies is one of the next big challenges for evolutionary approaches within the medical arena. Only by doing this can evolutionary research within medicine truly move from academic theory to clinical practise.

Chapters 8 and 9 examine two further topics, immune regulation and psychiatry, that have received considerable attention from those interested in evolutionary approaches to health and disease. Hurtado and colleagues, in Chapter 8, examine human parasite load in comparative, temporal, and global perspectives and argue that living in modern, sanitized environments away from a diverse parasite fauna may result in several chronic conditions, including asthma. Although this observation is not new,[2] the authors draw on their previous work on life history in modern indigenous populations to present a novel interpretation of human-parasite co-occurrence, consequent adaptation, and implicaitons for global health patterns. In Chapter 9, Hagen discusses the potential strategic value of delusions, behavioural phenomena that may be "adaptive" rather than "insane." Since psychiatry primarily focuses on illness, the tendency is to medicate people experiencing delusions. However, Hagen suggests that modification of social networks might prove to be a more effective "treatment" than pharmacological alteration of thought processes.

The use of the term "medicine" implies application to patient care, and some prominent advocates of evolutionary medicine are indeed practising clinicians. However, much research and scholarship into evolutionary medicine is conducted by academics in anthropology and the life sciences who are not health care professionals. One way in which the field could be moved nearer to mainstream medicine is through increasing the interaction between those generating the concepts, ideas, and case studies and those who are actively engaged in providing health care. The challenges inherent to this are exemplified in this volume: Only one contributor is currently a practicing clinician, although others are qualified in medicine or dentistry. However, many of the chapters have been reviewed by researchers who are also engaged in the delivery of health care, and their input has been especially important in reinforcing that contributors should indicate how their research and scholarship informs practical applications in clinical settings, health promotion, or medical education. Bentley and Aunger, in Chapter 10, describe how Bentley's interactions with medical and academic colleagues as well as her own students allowed them to survey opinions about the utility (or potential utility) of evolutionary medicine. Looking at five topics (racial medicine, epigenetics, emergency medicine, public health, and obstetrics) that link evolutionary theory and medical practise, they argue that far from being without practical application, several areas of evolutionary medicine have the scope to improve clinical practise and treatment.

In a similar vein, Dean (Chapter 11) suggests that evolutionary biology is a vital part of the educational framework of prospective health care practitioners. He provides many examples of how adding evolution to topics in the curriculum not only brings new perspectives but also acts to energise the student experience. Dean also argues that the increasing marginalisation of evolution in medical education will have serious implications for the way in which disease is defined and treated in the future. Unfortunately, as Elton and O'Higgins describe in Chapter 12, and as discussed by others,[4] the opportunities to organize curricula around an evolutionary paradigm are very limited. Nonetheless, it is possible—especially through elective programmes—to bring evolutionary medicine to the attention of students, with consequent educational benefits as well as future awareness in medical practise and research.

Most of the chapters in this volume bring out the positive aspects of evolutionary medicine, but the approach also has its detractors. It is vital to consider when and how an evolutionary perspective might not make an important contribution to medical science and practise. We feel that it is necessary to address this issue head-on, and to this end, in a thoughtful and occasionally provocative contribution, Marks (Chapter 13) seriously questions the practical benefits for the patient of an evolutionary approach to medicine and rightly raises concerns about some uncomfortable associations between Darwinism and "Big Pharma," as well as critically evaluating adaptationism. His chapter draws attention to the fact that medical research and clinical practise cannot be organised around a single set of theories. Medicine is an applied discipline which draws on, and quite properly supersedes, theoretical and empirical knowledge from a diverse range of natural and social science disciplines—including evolutionary biology. To overemphasise any one of these runs the risk of ignoring others and failing to identify where and how best to intervene.

The maturation of evolutionary medicine is dependent on critical, reflexive thought from within the field itself, and other contributors also question whether evolutionary approaches are universally appropriate or applicable. Elton (Chapter 2) highlights that human evolution has taken place over many millions of years and in many places, so fixing the origin (and therefore adaptations) of humans in the African Pleistocene is problematic. This is reinforced by research into epigenetics (Chapters 4, 5, and 10) and studies adopting life history and ecological perspectives (Chapters 6 and

8) which stress variation rather than universals. Another complex idea is that human health may suffer because of "mismatches" between the way human bodies were crafted by natural selection during our evolutionary history and the very different challenges posed by contemporary human environments. Identifying such mismatches could offer exciting insights for advancing clinical practise. One way would be though distinguishing between "natural" and "pathological" variation and thereby avoiding the unnecessary medicalisation of apparent dysfunction—be it polycystic ovary syndrome (Chapter 5) or delusional behaviour (Chapter 9)—that actually reflect beneficial or neutral adaptations for coping with (past) environmental challenges. Another would be by focusing attention on those aspects of contemporary environments (be they affluent or poor) that might be modified to better fit the short- and long-term needs of human bodies at the extremes of their developmental and physiological flexibility. Such flexibility, which encompasses cultural and behavioural adaptation as well as genetic and physiological responses, makes the study of adaptation in humans especially challenging. Thus, much future work will be required to identify which traits were selected for in which environments and to assess whether they were actually optimal—if, in fact, there ever exists an optimum fit between organism and environment.

Contemporary environments and lifestyles are not necessarily less healthy overall than those prevailing during much of human evolutionary history,[9] even if they are different. High income countries may well be facing an epidemic of obesity, but the "quality" and "quantity" of health therein has, arguably, never been better (albeit when measured by such crude measures as the prevalence of stunting and life expectancy, respectively). That is not to say that human health cannot be improved, by adopting a more sensitive and naturalistic approach to health and disease (which considers the biological heritage of humans as extraordinary, imperfect and natural organisms) has no contribution to make to any improvements we can make. Vitzthum, in Chapter 6, argues that environments could be modified to reduce early pregnancy loss. Thus, there is an obvious role for evolutionary medicine in informing decisions about how we can improve the fit between human biology and environment—in short, by developing even better environments to live in than those we see today and those prevailing during our evolutionary history. To do this, as demonstrated in Chapter 6, we need to appreciate the range of human adaptability and how and under what circumstances humans—and their diseases—evolved.

An in-depth understanding of human evolutionary history is also vital if we are to discourage biological determinism or legitimising therapeutic programmes such as "racialised" medicine that represent flawed and potentially dangerous interpretations of population histories and biological variation (see discussions in Chapters 4, 5, 10 and 13). Pollard and colleagues (Chapter 4) and Shaw and Elton (Chapter 5) caution against genetic determinism when considering differential susceptibilities to insulin resistance, and also highlight the important developmental aspects of type 2 diabetes and PCOS. Thus, those working within evolutionary medicine have a crucial role to play in advocating the appropriate uses of evolutionary arguments and perspectives. However, as MacCallum[10] points out, we should be wary of dismissing all evolutionary viewpoints within medicine on the basis of a few inappropriate studies or clinical applications.

Evolutionary medicine has the potential to contribute a number of interesting insights to healthcare, not only through health promotion and counselling but also in developing novel therapies. However, it also faces substantial challenges, noted by Williams and Nesse[4] at the outset. The chapters in this volume demonstrate how compelling evolutionary perspectives on health, illness, and medicine can be. Singly and collectively, they shed light on health and wellbeing in past, present, and future humans, at the level of the individual as well as the population. In the process, they also explicitly or implicitly highlight areas in which evolutionary research in medicine might be improved, by developing more sophisticated models of past human ecologies, embedding evolutionary ideas within existing academic or clinical frameworks, and designing studies that can be tested using accepted research tools, such as controlled trials. Given that one recent study[11] has suggested that the term "evolution" is actively avoided in the medical literature surrounding antibiotic resistance (even though many consider this to be the biomedical phenomenon which best reflects the relevance of evolutionary perspectives to contemporary medical practise), evolutionary medicine also plays a vital role in supporting the continuing presence of evolutionary theory in the medical curriculum and research arena.

Pollard and colleagues (Chapter 4) and Shaw and Elton (Chapter 5) caution against genetic determinism when considering differential susceptibilities to insulin resistance, and also highlight the important developmental aspects of type 2 diabetes and PCOS.

References

1. Nesse, R. M., and Williams, G. C. 1995. *Evolution and healing: The new science of Darwinian medicine.* London: Weidenfeld and Nicholson.
2. Stearns, S. C. 1998. *Evolution in health and disease.* New York: Oxford University Press.
3. Trevathan, W. R., Smith, E. O., and McKenna, J. J. 1999. *Evolutionary medicine.* Oxford: Oxford University Press.
4. Williams, G. W., and Nesse, R. M. 1991. The dawn of Darwinian medicine. *Q. Rev. Biol.* 66:1–22.
5. Goodman, A.H. Seeing culture in biology. In: Ellison, G.T.H. and Goodman, A.H. (Eds). *The Nature of Difference. Science, Society and Human Biology.* Taylor & Francis: Boca Raton; 2006, pp. 225–242.
6. Clift-Matthews, V. 2007. How much change in 50 years? *Br. J. Midwif.* 15:64.
7. Bull, J. J. 1995. (R)evolutionary medicine. *Evolution* 49:1296–98.

8. Ebert, D., and Sokolova, N. V. 2001. Morning has broken: Ten years after the dawn of evolutionary medicine. *J. Evol. Biol.* 14:194–96.

9. Gage, T.B., Are modern environments really bad for us? Revisiting the demographic and epidemiologic transitions. *Am. J. Phys. Anthropol.*, 128, 96–117, 2005.

10. MacCallum, C.J., Does medicine without evolution make sense? *PloS Biology*, 5, e112, 2007.

11. Antonovics, J. et al., Evolution by any other name: antibiotic resistance and avoidance of the E-word, *PloS Biology*, 5, e30, 2007.

2

Environments, Adaptation, and Evolutionary Medicine
Should We Be Eating a Stone Age Diet?

Sarah Elton
Functional Morphology and Evolution Unit, Hull
York Medical School, University of Hull

Contents

Introduction

A central tenet of evolutionary or Darwinian medicine is that many chronic diseases and degenerative conditions evident in modern Western populations have arisen because of a mismatch between Stone Age genes and recently adopted lifestyles.[1–5] In a nutshell, genes or traits that may have been selectively advantageous or neutral in the past are argued to be potentially deleterious within the context of industrialisation and modernisation. Some suggest that this mismatch can be extended even further back in time, to the widespread adoption of agriculture.[5] It is believed that chronic and degenerative conditions persist at such high levels in many populations because the rate at which selection operates is not sufficient to respond to the current pace of cultural and environmental change.[1,5] In other words, it is thought by many advocates of evolutionary medicine that our environments are evolving faster than we are.

Environments consist of biotic components such as animals, plants, and microorganisms as well as abiotic factors like temperature and topographic features. They exert numerous and varied selective pressures, and because human environments differ, environmentally induced polymorphisms exist. In humans a good example is the high frequency of the haemoglobin S (sickle cell) allele in areas of Africa with endemic *Plasmodium falciparum* malaria.[6] Another often quoted example is the causal link between intensity of UV radiation and skin pigmentation.[7] An albeit more controversial yet increasingly plausible instance of environmental adaptation is the strong selective pressure apparently exerted by climate on several genes involved in regulating blood pressure.[8] Culture also plays an important part in determining human environments—it is possible for different populations to coexist in the same physical area but nonetheless experience different environments and therefore selection pressures because of cultural practises. A classic example is lactase persistence in pastoralist populations,[9] which tends to be significantly higher than that seen in their nonpastoralist neighbours: the cattle-herding Fulani and the agriculturalist Yoruba of Nigeria are a case in point.[10]

There is thus no question that there are some clear examples of environmental adaptation in human populations, and many of these must have appeared relatively recently, and therefore rapidly, since the origin of modern *Homo sapiens* or even since the adoption of specific cultural practises. It is also evident that certain adaptations may no longer be beneficial outside the context in which they evolved: the loss of heterozygote advantage for the sickle cell gene in nonmalarial regions, melanomas in light-skinned people in low latitudes, vitamin D deficiencies in dark-skinned people in high latitudes, and differential interpopulation susceptibilies to hypertension and lactose intolerance are but a few examples. The role of relatively recent environments in shaping human variation is recognised in evolutionary medicine. Nonetheless, it is commonly argued that adaptations that have arisen in our recent evolutionary past and which are dependent on a small number of allelic variants do not reflect the true nature and speed of human adaptation to the environment, particularly for complex conditions, such as degenerative diseases, that involve many genes.[5] In these cases, the Stone Age is seen as being the environment of evolutionary adaptedness (EEA).

The Stone Age, as used in evolutionary medicine, is often loosely defined. It sometimes appears to be synonymous with the Pleistocene (the geological epoch that lasted from c. 1.75 million years ago to c. 11 thousand years ago) and at others the Palaeolithic (a more amorphous era that extends from the first appearance of stone tools c. 2.5 million years ago to around 10,000 years B.P.). In fact, the Stone Age archaeologically encompasses the Palaeolithic, Mesolithic, and Neolithic, and so technically should include the period in which plants were first domesticated. In this chapter, the terms *Stone Age* and *Palaeolithic* as descriptors of time are avoided where possible, and instead used only with reference to the general concept of Stone Age or Palaeolithic adaptations.

Many discussions of Stone Age adaptations use reconstructions of past human behaviour based on observations of modern gatherer-hunter groups that inhabit open and arid tropical or subtropical environments (for example, the !Kung of southwestern Africa or aboriginal Australians), an assumption that is critiqued below. Stone Age diets inferred from such observations apparently show the importance of a variety of plant foods (including nuts, seeds, fruits, tubers, and berries) as well as wild (i.e., nondomesticated) animal products, primarily bushmeat or game.[2,3,11] Products like honey, fish, and shellfish may have been available but are argued to be minor components of the diet.[2] The significance of domesticated foodstuffs, including milk and cultivated grains, is downplayed.[2,3] Descriptions of Stone Age diets stress high-fibre foods that are low in salt and high in essential nutrients, as well as having a large protein component.[2,3,11] The energy expended in subsistence activities is also highlighted,[2,3] asserting the need for a balance between intake and expenditure.

The concept of adaptation to such a Stone Age diet is undeniably seductive. At first glance, environmental mismatch is a plausible theory that explains why we get fat, suffer from heart disease, or develop type 2 diabetes. The links between these conditions and modern Western lifestyles are well documented, even if the precise mechanisms are yet to be elucidated fully.[12] But is it really accurate to view their occurrence in terms of a conflict between our Palaeolithic bodies and rapidly changing modern environments? In this chapter I will consider and critically evaluate notions of the EEA and Stone Age adaptation and their utility in understanding human evolutionary history and current human health, focusing particularly on diet. It is not my intention to set them up as straw men. Instead, I am interested in addressing a number of interrelated questions: What EEA? What Stone Age? What Palaeolithic diet? Which populations? Which members of the population? The variation and variability that is not only inherent to humans but also to many successful primates, probably including several hominin species, goes to the heart of these questions. Those engaged in evolutionary medicine should recognise that just as the modern world is complex, the world in which humans evolved was also heterogeneous and not confined to the Pleistocene. The ideas that I synthesise in this chapter are not new; evolutionary biologists, ecologists, palaeontologists, and anthropologists have been using concepts of variability and variation to critique notions of the EEA and Stone Age adaptations for the best part of two decades.[13-21] However, prominent advocates of evolutionary medicine repeatedly fall back on explanations for some chronic diseases that rely on a homogeneous reconstruction of Stone Age environments,[4,5] despite acknowledging the probability of greater complexity.[1-3,22] I argue that refocusing hypotheses away from a one-size-fits-all idea that humans are imperfectly adapted to their environments toward a greater appreciation of human variation and variability would strengthen evolutionary medicine and its applications. To these ends, I conclude the chapter by outlining a more flexible, strategic model of human nutrition and suggesting an alternative to the Stone Age diet.

The Human EEA and Stone Age Adaptations

The environment of evolutionary adaptedness (EEA) is the environment or environments in which the current characteristics of a species evolved.[23,24] Recently, the concept of the EEA has most commonly been applied to humans, adopted particularly by evolutionary psychologists (e.g., Tooby and Cosmides[25]), but also used in evolutionary medicine (e.g., Nesse and Williams[1]). However, the theoretical basis of the EEA has been criticised and its practical utility questioned.[14,17] Of all the criticisms, those most pertinent to the topics covered in this chapter, and which certainly apply to the way the EEA has been used in evolutionary medicine, are that the concept stresses similarities and universals rather than the variation that is necessary for selection,[14] and that the mosaic nature of evolution is often disregarded.[17] In addition, the idea does not adequately address the fact that selective pressures still operate, leading to the assumption that all evolution has occurred in the past.[14]

Tooby and Cosmides[25] (p. 388) assert that the EEA is adaptation specific and comprises a "statistical composite of environments." However, they and many others still depend largely on the Pleistocene as the human EEA. From a general perspective this may be justifiable, given that many of the anatomical and behavioural features that make us human—big brains, obligate terrestrial bipedalism, and complex language and culture—evolved in the Pleistocene. However, as discussed below, the Pleistocene is far from being a uniform temporal and spatial entity. If it is useful to think of a human EEA in the Pleistocene at all, which chronological subdivisions of the Pleistocene are particularly relevant, and which localities? On a more specific level, the Pleistocene is only one period in time that could be classed within the human EEA, and a significant weakness in the way the concept has been applied within evolutionary medicine is that the evolution that occurred before or after the Stone Age is rarely taken into consideration.[17] If the EEA is adaptation specific, different time periods will encompass the EEAs of particular traits,[14,17] and Foley[14] has argued convincingly that if the concept of the EEA is to be used at all, it should be within this context. Thus, to take one example, the emergence of high levels of sociality in human evolutionary history would be linked not to the selective pressures of the Pleistocene, but to the much earlier origin of the catarrhine primates.[14]

It is therefore obvious that the data used to reconstruct human EEAs, and by extension Stone Age adaptations, are crucial to their ultimate validity. Even if the specific concept of the EEA is rejected as being theoretically flawed, accurate and detailed evidence for the environmental, ecological, and behavioural background to human evolution is essential to the understanding of processes that may have impacted on the biology and health of extant human populations. Such evidence can be direct, taken from studies of the fossil and archaeological records, or it can be analogical. Although methodological developments such as advanced morphometric techniques, tooth microwear, and stable isotope analyses have broadened and deepened the ways in which the adaptations of past humans can be examined, it is still necessary to use analogy, which provides a framework for studying human evolution.[26] A range of animal groups (see, for example, Arcadi[27]) and human populations have been used as analogues, with modern gatherer-hunters and nonhuman primates being the most common.[26,28] Of the nonhuman primates, chimpanzees and Old World monkeys such as baboons are used most often.[26,28] Models based on analogy have usefully been divided into two main types: those that are simple or referential and those that are strategic or conceptual.[29] Referential models are common in palaeoanthropology and palaeontology,[29] and include direct comparison of one anatomical trait, species, or population with another. Conceptual or strategic models, in which the emphasis is on identifying the underlying processes that influence a behaviour or trait in the observable world, then extending this to the past,[29] are also used, but even now are less common. A conceptual approach relies on gathering detailed data on the past from a range of sources, including the archaeological and fossil records and palaeoenvironmental reconstruction, which then form the basis for strategic inference.[29]

Work into Stone Age adaptations would benefit enormously from a more extensive application of strategic modelling. An apparently pervasive belief in popular discussions of Stone Age adaptations is that the lifestyles and ecologies of some modern gatherer-hunter groups provide a window onto the EEA. A recent UK television programme, for example, sent people to the Kalahari to experience a "Stone Age Fat Camp" in the company of the !Kung. Unfortunately, this simplistic perception is also evident in the scientific literature,[2,3] with insufficient attention paid to the variation observed in modern populations and the archaeological record, despite recent research[11] that makes more of an attempt to address dietary differences. It is rarely acknowledged that there are numerous types of gatherer-hunter lifestyles or that such populations today show the products of evolution and interactions with other human groups in their ecologies, cultures, and behaviours.[14,21,30]

The subsistence strategies observed in modern gathering and hunting populations have evolved alongside agriculture and other economies, and are far from being relic, ancestral behaviours.[31] On this basis, there is no compelling reason why modern foraging lifeways are better examples of the EEA and its products than pastoralist or agricultural populations using traditional, preindustrialised subsistence practises. Equally, the diets of gatherer-hunters have not been preserved in aspic since the Mesolithic, with food procurement activities in many areas responding to prevailing ecological or cultural conditions.[32,33] In addition, relatively recent political and social interventions may have marginalised populations engaged in subsistence economies, with the result that observed "traditional" lifestyles could well be the product of behavioural and cultural changes that have occurred over the past couple of centuries or even a few decades. This is evident, for example, in the Hill Kharia of northern Bengal, whose fluctuations between agriculture and foraging have in part been determined by external social pressures,[33] but is not confined to gatherer-hunter populations, as shown by the effects of Soviet collectivisation on the Evenki herders of Siberia.[34] These factors notwithstanding, a highly selective set of modern foraging groups (primarily those living in the Kalahari or Australian desert environments) are still generally used as referential models for past populations. A strategic approach, on the other hand, would allow the construction of more complex models, based on underlying ecological principles rather than simple analogy, and facilitate predictions of past adaptations and behaviours in a variety of environments and circumstances.

Variation and Variability in Human Evolutionary History

The use of the EEA in evolutionary medicine has moved it away from its roots in evolutionary psychology and led to increasingly wide applications. However, since the concept of the EEA and how it is often employed have some serious limitations, it would arguably be better for those interested in the significant and substantial links among evolution, environment, health, and disease to approach questions from ecological perspectives. These stress the relationships between the organism and its biotic and abiotic environments, and do not rely on assumptions of a uniform environment for all populations or subpopulations within a species, or indeed for a single population through time. Human environments are inevitably complex, with a significant cultural dimension. It is beyond the scope of this chapter to examine in detail all elements of human environments, so as my major aim is to examine the validity and utility of Stone Age adaptations when thinking about modern human health and disease, I will concentrate on aspects that are frequently and relatively reliably reconstructed from evidence in the geological, ethnographic, fossil, and archaeological records. In particular, I will discuss variation and variability in physical environments and habitats from the Pliocene to the Holocene, and diets and foraging behaviours in living and extinct humans and hominins, including intraspecific and intrapopulation dietary variation. Taken together, these form the basis for consideration of human behaviour and lifestyles from an ecological rather than adaptationist perspective.

Physical Environments and Habitats

Today, humans exploit all the major biomes of the world (freshwater, marine, desert, forest, grassland, and tundra), a pattern that was probably established during the Late Pleistocene.[35] The ability of hominins to inhabit or exploit a relatively wide range of environments appears to date from at least the Late Pliocene and Early Pleistocene.[16,35] Fossil hominins have been recovered from Plio-Pleistocene deposits in East, southern, and central Africa. This extensive geographic distribution alone suggests that early hominins would have experienced different types of environments. The potential variability of hominin environments is supported by palaeoenvironmental data that indicate the availability of a wide range of habitats at many Plio-Pleistocene localities, including woodlands, grasslands, and tropical forests.[28,36-41] African environments altered considerably over time, partly in response to global climatic trends,[16,42] but also due to regional processes, like tectonic activity.[43,44] A single species, such as *Paranthropus boisei*, with a relatively extensive geographic range and a long tenure in the fossil record would have been required to survive in different habitats and "ride out" numerous environmental fluctuations.[35,45] At Koobi Fora in East Africa, for example, where *P. boisei* fossils have been recovered from horizons dated to between 2 and 1.4 million years ago, woodland-dominated habitats in the Upper Burgi Member gave way to more open environments in the succeeding KBS Member and edaphic grassland, with much less tree coverage in the more recent Okote Member.[37]

Dispersal, although likely to have been determined partially by habitat corridors, would have exposed hominins to novel and varied environments, both within and outside Africa.[35] The 1.8-million-year-old *Homo* fossils from Dmanisi in Georgia demonstrate that extra-African dispersal occurred early in the Pleistocene,[46] if not before.[47] Reconstruction of past global biomes indicates that hominins dispersing out of Africa in the Early Pleistocene would have needed to exploit diverse habitats, including grassland and temperate forest, during their expansion into Asia and particularly Europe.[48] Archaeological and geological data from Early Pleistocene deposits in China suggest that hominins were subject to significant climate (and thus environmental) variability driven by both regional and global processes.[49] The extreme climatic fluctuations of the Pleistocene would have made a significant impact on hominin environments. Although there was an overall trend to more open, arid environments associated with global climatic cooling during the Pleistocene, this was not a straightforward progression.[35] In Africa, data from fossil lakes indicate that the climate made rapid switches between arid and humid periods during the early to mid Pleistocene,[50] a pattern that apparently persisted throughout the Pleistocene.[51,52] There were also substantial habitat differences in East and southern Africa, the two regions that have yielded most of the African hominin fossil record, with East Africa apparently sampling a much wider array of open and closed habitats than southern Africa during the Pleistocene.[53] Ice sheet incursion and contraction periodically altered the extent of the major vegatational biomes in Eurasia,[54] and changes to the intensity of the annual monsoon and other frequent climatic events like El Niño would have heavily influenced the distribution of different types of vegetation in Asia,[54,55] with concomitant effects on habitats throughout the Pleistocene.

Throughout much of their evolution, hominins would have been subject to climatic and environmental fluctuation. Some of this variation would have happened at scales that corresponded to glacial cycles, acting over periods measured in tens of thousands of years.[56] However, some climate change was much more rapid, such as the short-lived warming events that occurred within glacials.[57,58] The transition to these stadial periods might have occurred over periods as short as a few decades,[57] within one or two generations of hominins, and would have made a major impact on high-latitude environments. Superimposed on these climatic events were annual shifts caused by seasonality, which very probably affected even the earliest hominins.[59]

Neanderthals are the classic example of hominins that were able to inhabit high-latitude regions during rapid climatic and environmental change. Their body proportions (such as relatively short distal limb lengths and large bi-iliac breadths) appear to have evolved, at least in part, from the influence of cold climates.[60–62] However, it was largely their behaviour that allowed them to respond to variations in climate and environment.[58] Like other organisms faced with environmental variability, hominins had the option to evolve morphologically to new conditions, innovate behaviourally, or disperse. Neanderthals probably employed the latter two options, surviving some of the most extreme climatic fluctuations of the Pleistocene, although ultimately their relatively specialised morphology may have prevented them from reacting effectively to the environmental changes of the very late Pleistocene.[58]

Modern humans, on the other hand, with more gracile morphology and highly complex social behaviours, were very well suited to tracking the vagaries of the environment and were able to exploit marginal habitats.[58] It is evident that modern humans have responded in many ways to different environments. Morphological evolution is demonstrated by, for instance, differences in skin pigmentation[7] and body proportions.[63] Other biological adaptations include lactase persistence[9,10] and haemoglobin polymorphisms.[6] Significant cultural and social diversity is also observed, some of which might be mediated or driven by the demands of different environments or the ways in which resources are exploited.[64] An excellent example of the importance of culture to human survival in marginal areas is seen in Arctic populations, in which cultural traditions promote social integrity in regions that require a high level of cooperation, and alterations to technology and material culture help accommodate sometimes rapidly changing environments.[65] Pastoralism is another example of a cultural mechanism that enables habitation and use of marginal environments, as low-quality plant food, often not suitable as a dietary staple for humans, is converted into high-quality animal food.[66] High-latitude gatherer-hunter populations also have a high dependency on meat for this reason.[30] In addition, pastoralists and some gatherer-hunters use movement as a means of living with short-term environmental shifts. Seasonal movement is evident in many parts of the world, and there is evidence of a long history of this in some areas. In the Arctic, for instance, it is indicated by archaeological evidence dating to 4000–5000 years B.P.[67] and has also been inferred from the earlier Palaeolithic record.[68]

Data from palaeoenvironments and the fossil and archaeological records clearly show that responding to environmental variability is the norm rather than the exception in modern humans and was even evident in the congeneric Neanderthals. The adaptive flexibility of hominins may predate the emergence of our own genus, *Homo*. Indeed, as part of the variability selection hypothesis it has been argued that important adaptive features of early hominins evolved in response to variable environments.[16] Such features could plausibly include foraging behaviours and dietary strategies.[16] Although most of the hominin species currently recognized were probably not part of the lineage that led to modern humans, investigating their habitat preferences helps to establish that in human evolutionary history, environmental variation (and therefore probably behavioural and biological variation) was common and has ancient roots. Even with a narrow focus on the candidates most likely to have been part of the modern human lineage, it is clear that their environments varied in time and space. Just after 2 million years ago, the appearance in the fossil record of *Homo ergaster/Homo erectus*, the first unequivocal member of the genus *Homo*,[69] broadly corresponds to an increase in open environments in Africa.[42,56] However, the apparently swift dispersal of *Homo* out of Africa into Asia, reaching latitudes as high as 40° N in the Caucasus by 1.8 million years ago[46] and northeast China by 1.3 million years ago (Ma),[49] demonstrates that *H. erectus* grade hominins were not confined to arid savanna grasslands in the tropics. Working on the assumption that these dispersed Eurasian populations did not contribute to the gene pool of modern humans does not significantly limit the importance of environmental variability and potential for adaptive flexibility in the human lineage—*H. ergaster* and descendents in Africa would have been subject to the climatic cyclicity described above, as well as experiencing marked spatial differences in environments, which may well have included dispersal into and exploitation of tropical forest well before the development of agriculture.[70] It is therefore erroneous to assume that modern gatherer-hunters who live in open, arid environments are automatically the best representatives of Stone Age populations, and that African savanna grasslands represent the physical component of the human EEA.

Diets and Foraging Behaviours

The marked differences in environments exploited by modern humans, both today and in the past, and the likelihood that hominins from the Pliocene onward occupied varied and variable habitats suggest that diets and foraging behaviours were similarly diverse. Indeed, intraspecific dietary flexibility appears to be a hallmark of at least some Plio-Pleistocene hominins.[45] To date, the links between flexibility of diet and habitat variation in early hominins are largely circumstantial. However, several nonhuman primate species inhabit varied environments and exhibit considerable intraspecific behavioural and ecological variation, which includes dietary flexibility. Stable carbon isotope analysis of rhesus macaque hair suggests clear dietary differences between populations living in different regions, which appear to be related to the seasonal availability (or lack thereof) of favoured foods such as ripe fruit.[71] Baboons show dietary variation,[72] linked to ecological differences between sites,[73] seasonality,[74] and factors such as age associated with individual life stages and preferences.[75] Dietary heterogeneity, in time and space, is also seen in other geographically dispersed Old World primates, including the vervet monkey[76] and the black-and-white colobus monkey.[77] These observations indicate that dietary variability is the norm rather than the exception in many modern primates, especially those that are widely dispersed or exist in fluctuating environments. Using a strategic approach, this would suggest that environmental variability often results in dietary heterogeneity in primates, an association that could be predicted for early hominins.

In modern gatherer-hunter populations there is a close relationship between diet and environment.[30] Temperature, precipitation, and solar radiation are key variables that influence productivity and therefore food availability.[30] Habitat productivity, measured either directly or via rainfall as a proxy, is increasingly being recognised as an important determinant of intraspecific variations in body size—an indicator of dietary quality—in tropical nonhuman primates.[78-80] In humans, which have a much wider geographic distribution, the relationships between productivity and diet in foraging populations are potentially even more important. At higher latitudes and lower temperatures, productivity tends to be less and the number and range of plant foods included in gatherer-hunter diets decrease.[30] One of the reasons that humans, unlike other primates, are able to inhabit these environments is their ability to procure and consume large quantities of meat and fish, exploiting marine/aquatic resources and the specialist herbivores that are adapted to eating low-quality plant food. It has been argued that this was a crucial factor in the dispersals and success of Pleistocene hominins into temperate Eurasia,[81] and later into extreme high-latitude environments in the northern hemisphere, which have been inhabited by humans for at least 30,000 years.[82] Polar Inuits, who live in areas with exceptionally low primary productivity (around 45 $g/m^2/yr$), only spend around 10% of their time gathering plant foods directly, depending instead on hunting (40%) and fishing (50%) for subsistence.[30] A similar subsistence pattern is seen in at least forty other gatherer-hunter groups.[30] In fact, the high dependence on gathered plant foods (>80%) reported for the archetypal gatherer-hunters, the !Kung and the G/wi from the Kalahari,[30] is exceptional in modern terms. This does not mean that some past gatherer-hunter populations were not equally dependent on plant foods. Rather, the crucial point is that dependence on particular food groups may be driven by the environment, and as environments vary, so will diets.

The archaeological record supports the notion that human diets have varied in time and space. Mesolithic gatherer-hunters exhibited significant dietary heterogeneity, even within a relatively small area like Britain and Ireland.[83] Stable isotope analysis suggests that individuals, grouped by site, had differential dependence on terrestrial and marine resources, with some having mixed diets and others appearing to subsist almost entirely on terrestrial foods.[83] At the well-known British site of Star Carr, over thirty potential plant and animal foodstuffs (birds and mammals) have been identified.[83] Potential dietary variability is also emphasised by the recovery of wild mammal (cattle, deer, and boar) bones from Danish shell mounds.[84] Interestingly, fish and seafood, which have a relatively low profile in many reconstructions of Stone Age diets, appear to have been important components of Mesolithic and Neolithic diets at coastal sites in northern Europe, with plentiful evidence for fish consumption at sites in Denmark.[84] At least one Late Stone Age site, in Lesotho, also has strong evidence for extensive seasonal exploitation of freshwater fish, despite other data suggesting that routine use of river fish in Africa did not become common until around 5,000 years ago.[85] This underlines the links between environmental variation, including seasonality, and subsistence patterns, and again demonstrates that there was no single or typical dietary strategy in humans during the Pleistocene and Early Holocene.

Human dietary variation is not restricted to between-population differences. Significant within-population variation in subsistence activity is also evident, based on sex, social position, or age. For example, those who hunt, often adult males, may have preferential access to game compared to other members of the population,[84] although a simple "man the hunter, woman the gatherer" sexual division of labour is certainly not observed in all populations.[30,86] Children's roles in modern gatherer-hunter food procurement are not universally defined. !Kung children do very little foraging, mainly because of the patchy nature of their food resources and the need to travel considerable distances in search of food.[87] They therefore have little control over what they eat, whereas Hadza youngsters are able to forage very near their homes, and thus have greater opportunity to determine their food consumption.[87] Children in foraging populations may also chose to procure and eat foods that adults may reject or search for foods in locations normally avoided by adults, as demonstrated by the behaviour of Meriam juveniles in the Torres Strait.[88] It has been argued that perceptions of diet derived from ethnographic research may be skewed because the mainly male ethnographers may not have interacted routinely with women,[19] and by extension those dependent on women, like children. This has implications not only for the accurate reconstruction of diet across a population, but also for the ways in which dietary variability and variation are perceived. It is illogical to develop a theory of Stone Age adaptation based on the diets of one sex, social class, or life stage, but failure to acknowledge intrapopulation variation leads to such a risk. This again shows the need to consider the importance of human dietary, environmental, and behavioural diversity when working within the framework of evolutionary medicine.

Adaptation and Maladaptation

The brief review of environmental variation and variability as indicated by habitats, diets, and intrapopulation differences presented in this chapter only scratches the surface of the diversity evident in modern, past, and extinct human populations. Nonetheless, it is clear that reconstructing Stone Age adaptations and a typical human EEA is very difficult. The utility of these concepts as they are currently applied in evolutionary medicine is questionable on this basis alone. In addition, much of the work on Stone Age adaptations relies on identifying the transition point at which adaptation turns into maladaptation. The boundary between the Mesolithic and the Neolithic is often assumed in theories of Stone Age adaptation to represent a switch from a natural mode of subsistence to one increasingly dominated by cultivation—the beginnings of the supposed gene-environment mismatch. However, the evidence that the move to agriculture represented a shift from adaptation to maladaptation is far from conclusive,[17,19] and transitions between one mode of subsistence to another appear to be, and have been, fluid and dependent on localised conditions.

Multiple lines of archaeological evidence show the probable importance of a range of gathered and hunted foods, a pattern that does not necessarily change abruptly at the boundary between the Mesolithic and Neolithic.[84] When transitions to domestication occurred, it is unlikely that there were straightforward replacements of one way of life with another. This general principle is demonstrated very well by some modern gatherer-hunter populations in which subsistence strategies are highly responsive to external social or environmental pressures, switching, for example, from cultivation to foraging and back again (e.g., Reddy[33]). Modern populations described as foragers may engage in long-term cultivation alongside gathering, hunting, or both, as shown by the sago extraction practised by the Nuaulu.[32] Along the same spectrum, agrarian or pastoral populations may also forage, pastoralists may cultivate, and groups largely dependent on cultivated crops may also keep beasts. Interaction with those engaged in different subsistence economies is also a reality in many regions. In past populations, true demarcation of different subsistence strategies and assessment of major transitions that could cause environmental mismatch may thus be very difficult. This indicates that the use of terms such as Stone Age diet fail to capture the complexity of human dietary transitions and thus evolution.

The potentially fuzzy boundaries between different modes of subsistence in the past have implications for the way in which health status is interpreted in relation to changes in diet. Since *Man the Hunter*[89] and the introduction of the idea that gatherer-hunters were the "original affluent society,"[90] foraging populations have commonly been viewed as having adequate or even ample leisure time and sufficient food to meet their needs. Closely related to this is the argument that the health of past gatherer-hunters was favourable in comparison with those who had adopted agriculture (cf. Cohen and Armelagos[91]), one of the fundamental ideas that underlie discussions of Stone Age adaptation. This picture of well-adapted gatherer-hunters and maladapted agriculturalists can be critiqued using several different lines of evidence. The first, that subsistence strategies might not fall neatly into one category or another, is discussed above. A second is that observations of modern gatherer-hunters indicate that foraging societies are not necessarily affluent, experiencing food shortage and resultant insults to both physical and evolutionary fitness. The Ache have been reported as being hungry, preferring more food than they are able to procure, and it is possible that fertility and child survivorship would increase if they consumed more.[92] The usual caveats about inferring past behaviours from modern observations apply, but by placing the evidence from the Ache within a general ecological framework, it is plausible to suggest that a similar situation may have been experienced by some past foragers. It has already been argued in this chapter that seasonality and other environmental fluctuations would have made an impact upon many early human populations, and it is likely that resource fluctuation would have caused periods of undernutrition. This might have been exacerbated by lack of food storage, which can act as a buffer against seasonality. It is also possible that, regardless of seasonal shifts, time was a limiting factor on resource procurement in past foragers. Time is a significant constraint in primate behavioural ecology, and even if resources are plentiful, the time budgeted for foraging could only be increased at the expense of other essential activities, including resting, travelling, and socialising.[93]

Another line of critique comes from skeletal data. Some skeletons of gatherer-hunters recovered from the archaeological record show growth perturbations associated with episodic seasonal stress.[94] Such markers appear to be found frequently in individuals from populations, both agrarian and foraging, that live in areas with marked seasonality and which have restricted resource diversity.[94] Thus, it is not always accurate to distinguish the nutritional and health status of groups based on a straightforward assessment of their subsistence economies. Equally, it is a gross oversimplification to assume that ancient gatherer-hunter groups had better overall health status than agriculturalists or pastoralists. Major diseases and health problems often shift as social organization changes. However, such shifts cannot be attributed solely to the adoption of a new mode of subsistence such as agriculture, and emerging health problems are not necessarily induced by diet. For example, the increase of gastrointestinal infections is often attributed to sedentarisation,[17] but it is well known by now that sedentary living is not exclusive to agriculturalists, and has been observed in some modern and past gatherer-hunters, with arguably the most famous archaeological example being the Mesolithic Ertebølle people.[95] Thus, although social organization is often linked to mode of subsistence, there are relatively few unique associations between particular types of group structure and methods of food procurement.

There is also no strong evidence that people dependent on food groups excluded from Stone Age diets, such as milk or domesticated grains and starches, automatically have poorer health than gatherer-hunters or suffer disproportionately from dietary intolerances. Many populations have adopted agriculture, and in nonindustrialised economies, including those that are agrarian, chronic diseases of affluence in people following traditional (i.e., non-Westernised) diets are rare.[20,96] Maladaptation is therefore not an automatic consequence of a move away from foraging. In some, possibly many, instances, gathering and hunting may be a less than optimal way to exploit a particular environment. For example, the Turkana, a well-studied group of modern East African pastoralists, live in an arid environment that is relatively poor in edible and easily processed plant resources.[97] Their reliance on the meat, blood, and particularly milk of domesticated animals (primarily cattle) allows them to exploit an environment that would otherwise be difficult to inhabit.[97] In addition, they have an armoury of cultural and behavioural responses to environmental insults, including regulation of fertility, outmigration, and activity patterns.[97] Although detailed studies of lactase persistence in the Turkana, a biological response to their chosen environment, are yet to be undertaken, it is likely that they are adapted to the consumption of cow, sheep, and goat milk into adulthood.[98] Data on the health status of transhumant Turkana populations are limited, but despite high rates of infectious diseases and relatively high rates of infant mortality, low rates of malnutrition and other nutritional deficiencies have been observed.[97]

These low rates of malnutrition in combination with data from other populations with a history of milk-based pastoralism[9] illustrate that long-term milk consumption is not universally maladaptive. It is true that from a global perspective lactase persistence into adulthood is the exception rather than the rule. Nonetheless, given that pastoralism might be the most appropriate strategy in unproductive environments, it could have significant nutritional and fitness benefits for those able to digest milk sugar.[9] This environmental adaptation, dismissed by many commentators on Darwinian nutrition as being a relatively unimportant single allele mutation,[5] is a vital element of the ability of some humans to exploit a range of unproductive habitats that would normally be inaccessible to hominoids. The advantages of milk consumption under certain ecological conditions would be hard to identify through a homogeneous Palaeolithic view of human nutrition.

Compared to modern diets, starches and carbohydrates, especially from grains, are argued to have been much less important as Stone Age energy sources, for which animal protein and fat are seen to be paramount.[11] This is despite strong evidence that starchy foods (including grains), often from a small number of plants, are or were vital dietary components for many modern and ancient human populations, agrarian as well as foraging, in low-, mid-, and high-latitude regions.[19,99,100] It has also been suggested that starchy underground tubers were vital fallback foods for australopiths in the Pliocene and Early Pleistocene.[101] In addition to downplaying the overall importance of many sources of carbohydrates, there is an apparent assumption in Stone Age dietary reconstructions that carbohydrates from fruits and vegetables are superior to those from cereal grains (e.g., Eaton and Eaton,[2] p. 253), the refined products of which are often seen as being empty calories. However, cereal grain refinement of the type evident in Westernised diets is a product of increased industrialisation rather than of grain cultivation per se. Far from being deleterious, the consumption of whole grains has been linked to lower incidences of type 2 diabetes and cardiovascular disease.[102,103]

Another common charge levelled at non-Stone Age foods is that they promote food allergies and intolerances, and are thus maladaptive and a further indication of environmental mismatch. Grains, along with dairy products, are often causally implicated by sufferers of the common gut disorder irritable bowel syndrome,[104] even though there is no conclusive evidence that these foods promote or exacerbate the condition.[105] In many cultures, regardless of whether or not there is a long history of cereal grain consumption, true intolerance to gluten is relatively uncommon. In the UK, the prevalence of coeliac disease as diagnosed by serological tests is estimated to be in the order of 1%,[106] a pattern that is replicated in many other populations.[107] Interestingly, prevalence is higher in some isolated groups, including the Finnish, the Sardinians, and the Saharawis.[107,108]

Although it has been argued that populations with a longer history of grain cultivation and consumption show lower prevalence of coeliac disease,[109] current data do not support this theory, with no obvious cline of increasing susceptibility (indicated by disease prevalence and frequency of the candidate HLA-D genes) from the Fertile Crescent region to northern Europe.[107] It is suggested that in the population with the highest reported incidence of coeliac disease, displaced Saharawis in Algeria (in which 5.6% of the children studied tested positive for the disease), high prevalence indicates positive selection for the coeliac disease genotype due to its protective effects against parasites.[107,108] However, this hypothesis is yet to be adequately tested, and it is also possible (especially given the higher prevalence in other genetic isolates) that stochastic factors, such as founder effect, might also be at work. Thus, although coeliac disease is precipitated by diet and is undoubtedly debilitating for many sufferers, who may experience reduced fertility, malnutrition, and increased risk of certain cancers,[106] it is not a particularly common response to the consumption of cultivated grains, and its etiology in the context of the development of human subsistence practises requires further clarification. On the basis of current evidence, therefore, general maladaptation to cultivated cereal consumption is unlikely, a conclusion reinforced by studies indicating the health benefits of consuming whole-grain cereals.

So where does this leave the debate on dietary adaptation versus maladaptation? In its evolutionary rather than developmental sense, adaptation is genetic, yet few genetic adaptations to diet have been discovered in humans.[19] This indicates that true maladaptations to specific diets or ranges of foodstuffs should be rare. In the case of the best-documented example of dietary adaptation, lactase persistence, the ability to digest lactose appears to have tracked milk-based diets, rather than the reverse,[9] and (at least until the creeping Westernisation of diets across the globe) adult individuals without this adaptation tended to exist under dietary regimens with no history of milk consumption. It is thus difficult to argue, from an ecological perspective, that the consumption of milk in humans is universally adaptive or maladaptive—it is a product of varying environments and thus selection pressures in different populations, leading to polymorphism.

A major implication of the diversity and flexibility evident in human diets is that far from being adapted to a Stone Age diet and maladapted to Holocene postagricultural subsistence, modern humans are suited to exploiting and consuming a vast range of foodstuffs. It has been suggested that the inherent flexibility of human dietary behaviour is a result of our hominoid or even anthropoid origins.[19] This is supported by the evidence reviewed in this chapter, both dietary and environmental. Furthermore, analogy with modern eurytopic primates[71,75,76] alongside data from the Old World monkey and early hominin fossil records[110–113] indicate that the broad, plant-based diet (in some cases supplemented by faunal material) characteristic of Old World primates, including hominins, was established at least by the Pliocene (c. 5 to 2 million years ago), and probably earlier, in the Miocene. Such a diet has been argued to be beneficial to modern humans, albeit with modifications that include increased meat eating facilitated by technology and decreased dietary bulk.[19,20,114]

Translated into nutritional advice for humans with Westernised diets high in refined and processed foods, this means eating fewer refined and processed foods and replacing them with fruits and vegetables (and other plants with relatively low digestible energy) as well as opting where possible for grass-fed rather than fatter grain-fed meat.[20] This should be coupled with increased energy expenditure, so that individuals remain in energy balance.[20] This approach to human diet, although based in comparative primatological study (and therefore to an extent drawing on our evolutionary heritage) does not proscribe particular food groups because their widespread adoption occurred in the Holocene rather than the Pleistocene. Instead, it encompasses the variability and flexibility that is a hallmark of human subsistence behaviour while identifying the major underlying factors—specifically a limited dependence on highly processed foods coupled with high physical activity—that appear to prevent chronic diseases of affluence in many populations with traditional diets, whether they be foragers, agriculturalists, or pastoralists.

Conclusion: Implications for Medicine

Given the factors outlined in this chapter—environmental variation and consequent dietary flexibility that can be traced back to the earliest hominins and possibly to the radiation of Old World primates, indistinct transitions between different modes of subsistence, and few actual dietary maladaptations—it is unlikely that Western diseases of affluence can be explained simply with reference to Stone Age adaptations and environmental changes that began with the origins of agriculture. Rather than there being a mismatch between our Palaeolithic bodies and modern environments, it is much more likely that diseases of affluence are the result of decoupling intake and expenditure, mechanisation of the modes of production, with consequent industrial processing of food, and living longer. Our bodies are not poorly adapted to cultivated foods en masse—in fact, humans have few true genetic adaptations to diet compared to many other animals.[19] Instead, highly processing food makes it more energy dense, so more calories are consumed before satiety is reached. This leads to weight gain if not offset by activity, and it is also possible that foods with a high glycaemic index (such as refined flours), from which glucose is liberated easily, promote fluctuations in blood sugar levels, and therefore an increased risk of insulin resistance.[115] Processing food is not a new phenomenon in humans or even hominins: technologies such as stone tools that were designed to aid food procurement have been around for at least 2.5 million years, and gut proportions in modern humans hint at the importance of food processing—both mechanical and chemical—prior to ingestion.[19] Many starch staples in traditional diets, such as sago, need extensive processing before they are fit to consume.[32] However, industrial-style processing has led to much food being more extensively pre-digested than ever before.

Industrialisation (through both the industrial and the agricultural revolutions) was the driving force of the demographic transition in Europe and other Western nations and led to a decline in mortality and fertility, in addition to massive social change in which food production and supply were concentrated in the hands of a relatively small proportion of the population. Such demographic changes, with attendant alterations to diet, are increasingly being seen in less developed nations, with consequent shifts in disease profiles,[96] although the European model of the demographic transition is not exactly replicated in currently developing nations, since chronic diseases of affluence are often coupled with a high prevalence of infectious diseases, leading to a double burden on health. This notwithstanding, alterations to diet caused by industrialisation and urbanisation throughout the world have resulted in a massive rise in obesity, heart disease, type 2 diabetes, and other chronic conditions associated with diet. It has been argued that rather than the emergence of agriculture representing the transition in human diet from adaptation to maladaptation, it is much more likely to be the very recent demographic transition that has caused a widespread shift in disease burdens, including those that are diet related.[17]

Adopting a diet relatively low in highly processed foods plus increasing daily activity may be beneficial to health.[20] Although health advice is notoriously complex and often contradictory, this message is currently being promoted by some public health agencies. In particular, the Harvard School of Public Health advocates a nutrition pyramid with physical activity at its base, followed by whole-grain foods and plant oils, then fruits and vegetables. Highly processed foods, animal fats, and red meats are at the top of the pyramid.[116] Despite its apparent general popularity, there appears to have been little adoption of the Stone Age diet by agencies promoting health and well-being. In fact, some clinicians (Vickers and Zollman,[117] p. 1420) have argued that exclusion diets, specifically the Stone Age diet, can be "highly restrictive, socially disruptive and expensive." This is not a trivial matter. In industrialised and industrialising nations, socioeconomic status is a powerful determinant of environment, in that it influences exposure to certain pollutants, infrastructure (including education and health care), and access to food resources. Thus, promoting diets, like the Stone Age diet, that involve avoidance of easily available, cheap foods in favour of those that tend to be harder to find or more expensive may in fact reinforce chronic disease burdens in certain parts of the community. Health stratification of this type is already evident in the UK: between 1970–1972 and 1991–1993 there was a marked decline in deaths from heart disease in males from social class I, whereas the heart disease death rate in social class V, which was only slightly higher than that in social class I in the early 1970s, remained more or less static into the 1990s.[118] Given these figures, it is likely that social inequality is a much greater risk factor in chronic disease than mismatched genes and environments.

A proximate and sociological perspective on very recent social factors (for example, migrant populations and their health problems) may well be more valuable in the understanding of chronic diet-related diseases than one that stresses deeper population histories. This is supported by increasing awareness that developmental insults *in utero* caused by poor maternal nutrition can significantly increase the chances of developing chronic diseases in adulthood (Barker[119]; see also Chapters 4 and 5). So does this mean that there is no place for human nutritional advice based on evolutionary or comparative data? It is worth stressing that some of the concepts outlined in Stone Age diets are beneficial, in particular the emphasis on physical activity. Other suggestions, like reducing sodium intake, also align well with mainstream nutritional advice and evidence-based studies of chronic disease risk. The adoption of these measures, although probably easier for those of higher socioeconomic status, does not necessarily rely on access to specific resources or facilities or require major modifications to normal lifestyles. As already indicated, the main difficulties that emerge when promoting good nutrition on the basis of Stone Age reconstructions are the emphasis on animal protein, since the (cheap) meat available to most people is much higher in fat than wild game,[19] the exclusion of milk (even in populations where lactase persistence is common), and the exclusion of grain, which in its whole form may actually be beneficial.

Appreciating the variation and variability in human diet and its evolution, rather than holding fast to Stone Age criteria, would strengthen understanding about the foods that are best for humans to eat, and possibly facilitate mainstream applications of evolutionary medicine in this area. Although the suggestions of Milton[19,20,114] about the importance of a plant-based diet low in processed foods draw on evolutionary principles and comparative behavioural ecology, they do not restrict dietary adaptation to a single point in time, and there is no blanket exclusion of food groups on the basis that they are not Palaeolithic. This approach is thus an excellent model for the ways in which evolutionary perspectives can be integrated into achievable public health strategies. Advocating Milton's model rather than the more fashionable and better publicised Stone Age approach not only would improve the theoretical basis of evolutionary perspectives on nutrition, but also may be useful in supporting movements calling for social interventions that improve health (such as better urban planning) and uncomplicated, basic nutritional advice (see, for example, Goldacre[120]).

In contrast to a strongly adaptationist perspective that stresses the universal nature of human biology, an ecological perspective on human diets highlights behavioural flexibility and adaptability in varied and varying environments alongside a very small number of rapid genetic adaptations as being crucial to our understanding of human nutrition. Does this in turn mean that the nutritional advice given to individuals should be tailored to their population histories? To an extent, this already occurs with the widespread recognition that not all people have the gene that enables lactose digestion into adulthood. Official U.S. nutritional advice currently plays on the fact that individuals of different body sizes have different energy requirements, and it has been suggested that a next step may be dietary advice based on genotype.[121] This suggestion was very much tongue in cheek, but dietary therapeutics are already used not only for those suffering from certain chronic conditions, but also for people at risk of developing them (for example, those with a family history of heart disease). Nonetheless, given the inherent flexibility of human diets and the very few genetic adaptations we have to them, as long as people follow the straightforward and basic rules of keeping highly processed foods to a minimum and staying in energy balance, it is difficult to see many general advantages in eating right for your genotype, Stone Age or not.

Acknowledgements

I am grateful to Paul O'Higgins, Randy Nesse, and two reviewers for their comments and suggestions.

References

1. Nesse, R. M., and Williams, G. C. 1999. On Darwinian medicine. *Life Sci. Res.* 3:1–17, 79–91.
2. Eaton, S. B., and Eaton, S. B. 1998. The evolutionary context of chronic degenerative diseases. In *Evolution in health and disease*, ed. S. C. Stearns, 251–59. Oxford: Oxford University Press.
3. Eaton, S. B., Eaton, S. B., and Konner, M. J. 1999. Palaeolithic nutrition revisited. In *Evolutionary medicine*, ed. W. R. Trevathan, E. O. Smith, and J. J. McKenna, 313–32. Oxford: Oxford University Press.
4. Nesse, R. M. 2001. How is Darwinian medicine useful? *West. J. Med.* 174:358–60.
5. Eaton, S. B., et al. 2002. Evolutionary health promotion. *Prev. Med.* 34:109–18.
6. Carter, R., and Mendis, K. N. 2002. Evolutionary and historical aspects of the burden of malaria. *Clin. Microbiol. Rev.* 15:564–94.
7. Jablonski, N. G., and Chaplin, G. 2000. The evolution of human skin coloration. *J. Hum. Evol.* 39:57–106.
8. Hunter Young, J., et al. 2005. Differential susceptibility to hypertension is due to selection during the Out-of-Africa expansion. *PLoS Genet.* 1 (e82):0730–0738.
9. Holden, C., and Mace, R. 2002. Pastoralism and the evolution of lactase persistence. In *Human biology of pastoral populations*, ed. W. R. Leonard and M. H. Crawford, 280–307. Cambridge: Cambridge University Press.
10. Cook, G. C. 1978. Did persistence of intestinal lactase into adult life originate on the Arabian Peninsula? *Man* 13:418–27.
11. Cordain, L., et al. 2000. Plant-animal subsistence ratios and macronutrient energy estimations in worldwide hunter-gatherer diets. *Am. J. Clin. Nutr.* 71:682–92.
12. Elliot, R., and Ong, T. J. 2002. Science, medicine and the future: Nutritional genomics. *BMJ* 324:1438–42.
13. Turke, P. W. 1990. Which humans behave adaptively, and why does it matter? *Ethol. Sociobiol.* 11:305–39.
14. Foley, R. A. 1995. The adaptive legacy of human evolution: A search for the environment of evolutionary adaptedness. *Evol. Anthropol.* 4:194–203.
15. Irons, W. 1998. Adaptively relevant environments versus the environment of evolutionary adaptedness. *Evol. Anthropol.* 6:194–204.
16. Potts, R. 1998. Variability selection in hominid evolution. *Evol. Anthropol.* 7:81–96.
17. Strassmann, B. I., and Dunbar, R. I. M. 1998. Human evolution and disease: Putting the Stone Age in perspective. In *Evolution in health and disease*, ed. S. C. Stearns, 91–101. Oxford: Oxford University Press.
18. Stearns, S. C., and Ebert, D. 2001. Evolution in health and disease: Work in progress. *Q. Rev. Biol.* 76:417–32.
19. Milton, K. 2000. Hunter-gatherer diets—A different perspective. *Am. J. Clin. Nutr.* 71:665–67.
20. Milton, K. 2002. Hunter-gatherer diets: Wild foods signal relief from diseases of affluence. In *Human diet: Its origin and evolution*, ed. P. S. Ungar and M. F. Teaford, 111–22. Westport, CT: Bengin and Garvey Press.
21. Tattersall, I. 2001. Evolution, genes and behaviour. *Zygon* 36:657–66.
22. Nesse, R. M., and Williams, G. C. 1995. *Evolution and healing: The new science of Darwinian medicine.* London: Weidenfeld and Nicholson.
23. Bowlby, J. 1969. *Attachment and loss.* Vol. 1. New York: Basic Books.

24. Bowlby, J. 1973. *Attachment and loss.* Vol. 2. New York: Basic Books.
25. Tooby, J., and Cosmides, L. 1990. The past explains the present: Emotional adaptations and the structure of ancestral environments. *Ethol. Sociobiol.* 11:375–424.
26. Foley, R. A. 1992. Analogue models in palaeoanthropology. In *The Cambridge encyclopaedia of human evolution,* ed. S. Jones, R. D. Martin, and D. Pilbeam, 335–41. Cambridge: Cambridge University Press.
27. Arcadi, A. C. 2006. Species resilience in Pleistocene hominids that traveled far and ate widely: An analogy to the wold-like canids. *J. Hum. Evol.* 51:383–94.
28. Elton, S. 2006. Forty years on and still going strong: The use of the hominin-cercopithecid comparison in palaeoanthropology. *J. R. Anthropol. Inst.* 12:19–38.
29. Tooby, J., and DeVore, I. 1987. The reconstruction of hominid behavioural evolution through strategic modelling. In *The evolution of human behaviour. Primate models,* ed. W. Kinzey, 183–238. New York: State University of New York Press.
30. Kelly, R. L. 1995. *The foraging spectrum: Diversity in hunter-gatherer lifeways.* Washington, DC: Smithsonian Institution Press.
31. Foley, R. 1991. Hominids, humans and hunter-gatherers: An evolutionary perspective. In *Hunters and gatherers 1: History, evolution and social change,* ed. T. Ingold, D. Riches, and J. Woodburn, 207–21. Oxford: Berg.
32. Ellen, R. 1991. Foraging, starch extraction and the sedentary lifestyle in the lowland rainforest of central Seram. In *Hunters and gatherers 1: History, evolution and social change,* ed. T. Ingold, D. Riches, and J. Woodburn, 117–34. Oxford: Berg.
33. Reddy, G. P. 1994. Hunter-gatherers and the politics of environment and development in India. In *Key issues in hunter-gatherer research,* ed. E. S. Burch and L. J. Ellanna, 357–76. Oxford: Berg.
34. Leonard, W. R., et al. 2002. Ecology, health and lifestyle change among the Evenki herders of Siberia. In *Human biology of pastoral populations,* ed. W. R. Leonard and M. H. Crawford, 206–35. Cambridge: Cambridge University Press.
35. Elton, S. The environmental context of human evolutionary history in Eurasia and Africa. *J. Anat.,* in press.
36. Plummer, T. W., and Bishop, L. C. 1994. Hominid palaeoecology at Olduvai Gorge, Tanzania as indicated by antelope remains. *J. Hum. Evol.* 27:47–75.
37. Reed, K. E. 1997. Early hominid evolution and ecological change through the African Plio-Pleistocene. *J. Hum. Evol.* 32:289–322.
38. Kappelman, J., et al. 1997. Bovids as indicators of Plio-Pleistocene palaeoenvironments in East Africa. *J. Hum. Evol.* 32:229–56.
39. Sikes, N. 1999. Plio-Pleistocene floral context and habitat preferences of sympatric hominid species in East Africa. In *African biogeography, climate change and early hominid evolution,* ed. T. G. Bromage and F. Schrenk, 301–15. New York: Oxford University Press.
40. Elton, S. 2001. Locomotor and habitat classification of cercopithecoid postcranial material from Sterkfontein Member 4, Bolt's Farm and Swartkrans Members 1 and 2, South Africa. *Palaeont. Afr.* 37:115–26.
41. Kovarovic, K., and Andrews, P. 2007. A bovid postcranial ecomorphological survey of the Laetoli palaeoenvironment. *J. Hum. Evol.* 52:663–80.
42. DeMenocal, P. B. 1995. Plio-Pleistocene African climate. *Science* 270:53–59.
43. Partridge, T. C., et al. 1995. Climatic effects of Late Neogene tectonism and volcanism. In *Paleoclimate and evolution, with emphasis on human origins,* ed. E. S. Vrba, G. H. Denton, T. C. Partridge, and L. H. Burckle, 8–23. New Haven, CT: Yale University Press.

44. King, G., and Bailey, G. 2006. Tectonics and human evolution. *Antiquity* 80:265–86.
45. Wood, B., and Strait, D. 2004. Patterns of resource use in early *Homo* and *Paranthropus*. *J. Hum. Evol.* 46:119–62.
46. Gabunia, L., et al. 2000. Earliest Pleistocene hominid cranial remains from Dmanisi, Republic of Georgia: Taxonomy, geological setting, and age. *Science* 288:1019–25.
47. Dennell, R. 2003. Dispersal and colonisation, long and short chronologies: How continuous is the Early Pleistocene record for hominids outside East Africa? *J. Hum. Evol.* 45:421–40.
48. Hughes, J. K., Elton, S., and O'Regan, H. J. 2008. *Theropithecus* and "Out of Africa" dispersal in the Plio-Pleistocene. *J. Hum. Evol.* 54:43–77.
49. Zhu, R. X., et al. 2001. Earliest presence of humans in northeast Asia. *Nature* 413:413–17.
50. Trauth, M. H., Maslin, M. A., Deino, A., and Strecker, M. R. 2005. Late Cenozoic moisture history of East Africa. *Science* 309:2051–53.
51. Street, F. A., and Grove, A. T. 1976. Environmental and climatic implications of Late Quaternary lake-level fluctuations in Africa. *Nature* 261:385–90.
52. DeMenocal, P., et al. 2000. Abrupt onset and termination of the African Humid Period: Rapid climate responses to gradual insolation forcing. *Quat. Sci. Rev.* 19:347–61.
53. Elton, S. 2007. Environmental correlates of the cercopithecoid radiations. *Fol. Primatol.* 78:344–64.
54. Hope, G., et al. 2004. History of vegetation and habitat change in the Austral-Asian region. *Quat. Int.* 118/119:103–126.
55. Quade, J., and Cerling, T. E. 1995. Expansion of C_4 grasses in the Late Miocene of northern Pakistan: Evidence from stable isotopes in paleosols. *Paleogeog. Paleoclimatol. Paleoecol.* 115:91–116.
56. DeMenocal, P. B. 2004. African climate change and faunal evolution during the Pliocene-Pleistocene. *Earth Planet. Sci. Lett.* 220:3–24.
57. Dowdeswell, J. A., and White, J. W. C. 1995. Greenland ice core records and rapid climate change. *Phil. Trans. R. Soc. Phys. Sci. Eng.* 352:359–71.
58. Finlayson, C. 2004. *Neanderthals and modern humans: An ecological and evolutionary perspective.* Cambridge: Cambridge University Press.
59. Foley, R. A. 1993. The influence of seasonality on hominid evolution. In *Seasonality and human ecology*, ed. S. J. Ulijaszek and S. S. Strickland, 17–37. Cambridge: Cambridge Univesity Press.
60. Holliday, T. W. 1999. Brachial and crural indices of European Late Upper Palaeolithic and Mesolithic humans. *J. Hum. Evol.* 36:549–66.
61. Ruff, C. B. 1994. Morphological adaptation to climate in modern and fossil hominids. *Ybk. Phys. Anthropol.* 37:65–107.
62. Ruff, C. 2002. Variation in human body size and shape. *Annu. Rev. Anthropol.* 31:211–32.
63. Pearson, O. M. 2000. Activity, climate and postcranial robusticity: Implications for modern human origins and scenarios of adaptive change. *Curr. Anthropol.* 41:569–607.
64. Collard, I. F., and Foley, R. A. 2002. Latitudinal patterns and environmental determinants of recent human cultural diversity: Do humans follow biogeographical rules? *Evol. Ecol. Res.* 4:371–83.
65. Fitzhugh, W. W. 1997. Biogeographical archaeology in the Eastern North American Arctic. *Hum. Ecol.* 25:385–418.

66. Crawford, M. H., and Leonard, W. R. 2002. The biological diversity of herding populations: An introduction. In *Human biology of pastoral populations*, ed. W. R. Leonard and M. H. Crawford, 1–9. Cambridge: Cambridge University Press.

67. Powers, W. R., and Jordan, R. H. 1990. Human biogeography and climate change in Siberia and Arctic North America in the fourth and fifth millennia BP. *Phil. Trans. R. Soc. Lond. A* 330:665–70.

68. Goebel, T. 1999. Pleistocene human colonization of Siberia and peopling of the Americas: An ecological approach. *Evol. Anthropol.* 8:208–27.

69. Wood, B., and Collard, M. 1999. The human genus. *Science* 284:65–71.

70. Mercader, J. 2002. Forest people: The role of African rainforests in human evolution and dispersal. *Evol. Anthropol.* 11:117–24.

71. O'Regan, H. J., et al. Diets of modern and fossil macaques assessed using stable isotope analysis of hair, bone and tooth enamel. *J. Hum. Evol.*, submitted.

72. Barton, R. A., Whiten, A., Byrne, R. W., and English, M. 1993. Chemical composition of baboon plant foods: Implications for the interpretation of intra- and interspecific differences in diet. *Folia Primatol.* 61:1–20.

73. Hill, R. A., and Dunbar, R. I. M. 2002. Climatic determinants of diets and foraging behaviour in baboons. *Evol. Ecol.* 16:579–93.

74. Codron, D., Lee-Thorp, J. A., Sponheimer, M., De Ruiter, D., and Codron, J. 2006. Inter- and intra-habitat dietary variability of **chacma** baboons (*Papio ursinus*) in South African savannas based on fecal $\delta^{13}C$ and %N. *Am. J. Phys. Anthropol.* 129:204–14.

75. Altmann, S. A. 1997. *Foraging for survival: Yearling baboons in Africa.* Chicago: University of Chicago Press.

76. Fedigan, L., and Fedigan, L. M. 1988. *Cercopithecus aethiops:* A review of field studies. In *A primate radiation: Evolutionary biology of the African guenons*, ed. A. Gautier-Hion, F. Bourlière, J. P. Gautier, and J. Kingdon, 387–411. Cambridge: Cambridge University Press.

77. Richard, A. F. 1985. *Primates in nature.* New York: W. H. Freeman.

78. Dunbar, R. I. M. 1990. Environmental determinants of intraspecific variation in body weight in baboons (*Papio* spp.). *J. Zool.* 220:157–69.

79. Lehman, S. M., Mayor, M., and Wright, P. C. 2005. Ecogeographic size variation in sifakas: A test of the resource seasonality and resource quality hypotheses. *Am. J. Phys. Anthropol.* 126:318–28.

80. Cardini, A., Jansson, A.-U., and Elton, S. 2007. A geometric morphometric approach to the study of ecogeographical and clinal variation in vervet monkeys. *J. Biogeogr.* 34:1663–78.

81. Turner, A. 1992. Large carnivores and earliest European hominids—Changing determinants of resource availability during the Lower and Middle Pleistocene. *J. Hum. Evol.* 22:109–26.

82. Pitulko, V. V., et al. 2004. The Yana RHS site: Humans in the arctic before the Last Glacial Maximum. *Science* 303:52–56.

83. Milner, N. 2006. Subsistence. In *Mesolithic Britain and Ireland*, ed. C. Conneller and G. Warren, 61–82. Stroud: Tempus.

84. Milner, N., et al. 2004. Something fishy in the Neolithic? A re-evaluation of stable isotope analysis of Mesolithic and Neolithic coastal populations. *Antiquity* 78:9–22.

85. Plug, I. 2002. The exploitation of freshwater fish during the Later Stone Age of Lesotho: Preliminary results. In *Beyond affluent foragers*, ed. C. Grier, J. Kim, and J. Uchiyama, 24–33. Oxford: Oxbow.

86. Bird, R. 1999. Co-operation and conflict: The behavioural ecology of the sexual division of labor. *Evol. Anthropol.* 8:65–75.
87. Blurton Jones, N., Hawkes, K., and Draper, P. 1994. Differences between Hadza and !Kung children's work: Affluence or practical reasons? In *Key issues in hunter-gatherer research*, ed. E. S. Burch and L. J. Ellanna, 188–215. Oxford: Berg.
88. Bird, D. W., and Bliege Bird, R. 2000. The ethnoarchaeology of juvenile foragers: Shellfishing strategies among Meriam children. *J. Anthropol. Archaeol.* 19:461–76.
89. Lee, R. B., and DeVore, I. 1968. *Man the hunter*. New York: Aldine de Gruyter.
90. Sahlins, M. 1972. *Stone Age economics*. New York: Aldine de Gruyter.
91. Cohen, M. N., and Armelagos, G. L. 1984. *Paleopathology at the origins of agriculture*. New York: Academic Press.
92. Hill, K. A., and Hurtado, A. M. 1996. *Ache life history: The ecology and demography of a foraging people*. New York: Aldine de Gruyter.
93. Dunbar, R. I. M. 1988. *Primate social systems*. London: Crook Helm.
94. Yesner, D. R. 1994. Seasonality and resource "stress" among hunter-gatherers: Archaeological signatures. In *Key issues in hunter-gatherer research*, ed. E. S. Burch and L. J. Ellanna, 151–68. Oxford: Berg.
95. Rowley-Conway, P. 1983. Sedentary hunters: The Ertebolle example. In *Hunter-gatherer economy in prehistory: A European perspective*, ed. G. Bailey, 111–26. Cambridge: Cambridge University Press.
96. McKeown, T. 1983. A basis for health strategies: A classification of disease. *BMJ* 287:594–96.
97. Little, M. A. 2002. Human biology, health and ecology of nomadic Turkana pastoralists. In *Human biology of pastoral populations*, ed. W. R. Leonard and M. H. Crawford, 151–82. Cambridge: Cambridge University Press.
98. Little, M. A. 1989. Human biology of pastoralist populations. *Ybk. Phys. Anthropol.* 32:215–47.
99. Kubiak-Martens, L. 1996. Evidence for possible use of plant foods in Palaeolithic and Mesolithic diet from the site of Całowanie in the central part of the Polish Plain. *Veg. Hist. Archaeobotan.* 5:33–38.
100. Lamb, J., and Loy, T. 2005. Seeing red: The use of Congo Red dye to identify cooked and damaged starch grains in archaeological residues. *J. Archaeol. Sci.* 32:1433–40.
101. Laden, G., and Wrangham, R. 2005. The rise of the hominids as an adaptive shift in fallback foods: Plant underground storage organs (USOs) and australopith origins. *J. Hum. Evol.* 49:482–98.
102. Montonen, J., et al. 2003. Whole grain and fiber intake and the incidence of type 2 diabetes. *Am. J. Clin. Nutr.* 77:622–29.
103. Lui, S., et al. 2003. Is intake of breakfast cereals related to total and cause-specific mortality in men? *Am. J. Clin. Nutr.* 77:594–99.
104. Nanda, R., et al. 1989. Food intolerance and the irritable bowel syndrome. *Gut* 30:1099–1104.
105. Shanahan, F., and Whorwell, P. J. 2005. IgG-mediated food intolerance in irritable bowel syndrome: A real phenomenon or an epiphenomenon? *Am. J. Gastroent.* 100:1558–59.
106. Van Heel, D. A., and West, J. 2006. Recent advances in coeliac disease. *Gut* 55:1037–46.
107. Catassi, C. 2005. Where is celiac disease coming from and why? *J. Pediat. Gastroent. Nutr.* 40:279–82.

108. Catassi, C., et al. 1999. Why is coeliac disease endemic in the people of the Sahara? *Lancet* 354:647–48.
109. Simoons, F. J. 1981. Celiac disease as a geographic problem. In *Food, nutrition and evolution*, ed. D. N. Walcher and N. Kretchmer, 179–99. New York: Masson.
110. Sponheimer, M., and Lee-Thorp, J. A. 1999. Isotopic evidence for the diet of an early hominid, *Australopithecus africanus. Science* 283:368–69.
111. Teaford, M. F., and Ungar, P. S. 2000. Diet and the evolution of the earliest human ancestors. *Proc. Natl. Acad. Sci. U.S.A.* 97:13506–11.
112. Codron, D., Luyt, J., Lee-Thorp, J. A., Sponheimer, M., de Ruiter, D., and Codron, J. 2005. Utilizations of savanna-based resources by Plio-Pleistocene baboons. *S. Afr. J. Sci.* 101:245–48.
113. El-Zaatari, S., Grine, F. E., Teaford, M. F., and Smith, H. F. 2005. Molar microwear and dietary reconstructions of fossil Cercopithecoidea from the Plio-Pleistocene deposits of South Africa. *J. Hum. Evol.* 49:180–205.
114. Milton, K. 1999. Nutritional characteristics of wild primate foods: Do the diets of our closest living relatives have lessons for us? *Nutrition* 15:488–98.
115. Willett, W., Manson, J. A., and Liu, S. 2002. Glycemic index, glycemic load, and risk of type 2 diabetes. *Am. J. Clin. Nutr.* 76:274S–280S.
116. Harvard School of Public Health. 2007. http://www.hsph.harvard.edu/nutritionsource/pyramids.html (accessed October 23, 2007).
117. Vickers, A., and Zollman, C. 1999. ABC of complementary medicine: Unconventional approaches to nutritional medicine. *BMJ* 319:1419–22.
118. Drever, F., and Bunting, J. 1997. Patterns and trends in male mortality. In *Health inequalities*, ed. F. Drever and M. Whitehead, 95–107. London: Stationary Office.
119. Barker, D. J. P. 1998. The fetal origins of coronary heart disease and stroke: Evolutionary implications. In *Evolution in health and disease*, ed. S. C. Stearns, 246–50. Oxford: Oxford University Press.
120. Goldacre, B. 2007. Tell us the truth about nutritionists. *BMJ* 334:292.
121. McGinty, S. A. 2007. Pyramids galore: Rapid response to Goldacre 2007 "Tell us the truth about nutritionists." *BMJ*, February 12. http://www.bmj.com/cgi/eletters/334/7588/292#159250 (accessed October 23, 2007).

3

Human Protein Requirements and Infection Stress among Young Children at the Origins of Agriculture

Stanley J. Ulijaszek
Institute of Social and Cultural Anthropology, University of Oxford

Contents

Introduction

The human need to meet the energy and protein needs of everyday life is fundamental to population health and survivorship, and is therefore an appropriate topic for Darwinian medicine to consider. The transition to agriculture involved major shifts in subsistence, resulting in change in diet from foods that humans may have been adapted to, to new ones that often challenged physiological adaptability. This change contributed to changes in patterns of human health that persist to the present day among rural agricultural populations of the developing world. Relationships among energetics, survivorship, and human adaptation and evolution have been considered extensively elsewhere[1-3] and are not elaborated further here. The metabolic interchangeability of the macronutrients carbohydrate, fat, and protein by way of intermediary metabolism links energetics to all the macronutrients. Protein is the most satiating of the macronutrients, and humans and other vertebrates prioritize the intake of protein relative to other macronutrients when seeking to achieve satiety.[4]

Human protein intakes are extremely variable within populations, while protein metabolism at the level of the individual is plastic in response to intake,[5] and low, per unit of body size, relative to that of most other primates. This is largely a function of the greater body size of humans relative to that of most other primate species. With the advent of increased animal capture and meat eating, *Homo erectus* is viewed to have consumed a diet rich in meat,[6] obtaining large amounts of high-quality protein from this source. As prehistoric hunter-gatherers, humans acquired very high proportions of their dietary energy from animal sources. With the advent of agriculture, novel foods were introduced as staples for which the hominin genome had little evolutionary experience.[7] One consequence of the transition to agriculture and animal husbandry was a decline in the proportion of dietary energy and protein from meat, and an increase in dependence on a more limited range of seeds and grains. Without the low protein requirements of humans relative to those of other primates,[8] the transition to agriculture could not have taken place, since cereal-based diets supply far less protein than most hunter-gatherer diets. While it is unlikely that adults suffered protein deficiency after the origins of agriculture, at least in the Near East and Mediterranean,[9] it is not clear whether there was protein deficiency (in either quantity or quality) among young children. Although breast feeding infants are unlikely to suffer protein deficiency, at least for the first 6 months of life,[10] this is much less so thereafter. The protein quality of cereal foods is much lower than that of meat, and the transition to cereal-based diets may have involved a reduction in protein quality. Density-dependent infectious diseases emerged as significant stressors at this time.[11] Given the relative immaturity of the human immune system in early childhood and the increased protein needs for the immune system to fight infection,[12] it is plausible that infectious disease could have imposed a burden on protein nutritional status of young children, particularly since physiological protein needs are highest at this stage of life.

The transition from hunting and gathering to agriculture and animal husbandry in the Near East and Mediterranean region occurred some 13,000 years ago (ya), with the deliberate growing of wild cereal crops and the taming of small mammals being a means of expanding the food supply in response to population increase.[13] This transition saw a negative secular trend in adult stature[14] at the time that infant mortality and life expectancy declined.[15] The most plausible explanation is that of combined stresses of nutrition and infection on child growth,[16] particularly in the first two years of life, when the potential effects of these stressors are greatest.[17] The extent to which young children at the origins of agriculture may have suffered protein deficiency is examined here by modelling of possible dietary, protein, and essential amino intakes of children aged 6 to 36 months consuming constructed Neolithic diets with and without the additional protein costs of infection. Specifically, three questions are addressed in relation to stress and survivorship of young children in the Near East and Mediterranean at the origins of agriculture: (1) Was the likelihood of protein deficiency low? (2) To what extent might the negative secular trend have been associated with possible protein deficiency? (3) Might infections emerging with the origins of agriculture have increased protein requirements to a level that made protein deficiency, in either quantity or quality, possible?

Research into physiological protein requirements has been central to the understanding of nutritional health from Berthollet's discovery of nitrogen-containing "animal substance" in 1785.[18] The history of protein requirement across the twentieth century has been one of progressive decline, with new research identifying physiological protein needs as being lower than previously thought. In the 1950s the clinical condition of kwashiorkor, then common in much of the developing world, was considered to have been caused by simple protein deficiency.[18] By the 1980s, it was considered to be due to a range of causes of which low protein intake was one[19] and infection another.[20] More recently, new knowledge concerning the protein costs of infection have led to calls for increases in recommended daily intakes of protein, particularly for developing countries where exposure to infection is high. While humans have evolved to have low physiological protein requirements relative to those of other primate species, and to be able to adapt to even lower intakes, this does not eliminate the need to consider its importance to nutritional health or clinical practise. The clinical area that is best informed by the present analysis is paediatric care among populations experiencing high infectious disease burdens.

Human Protein Needs

Estimates of protein requirements for good public health nutrition acknowledge that humans are exposed to a very wide range of protein intakes, and vary in their adaptation to low protein intake. As it is very difficult to establish the lower limits of protein needs for humans,[21] dietary requirements set recommendations that provide safe levels of intake, which are generally two standard deviations above physiologically determined values for protein balance, according to life stage.[22] Requirements incorporate the protein needs for child growth, as well as for pregnancy and lactation. Although the need to identify the impact of infections on the protein requirements of infants and children was clearly identified in the 1980s,[22] there have remained few quantitative estimates of this.[23] As with most human nutritional requirements, protein requirements vary with age, sex, physical activity, and health. In public health nutrition, protein requirements are expressed in two major ways. The physiological need is expressed by the estimated average requirement (EAR), which is an estimate of the average requirement for protein for a nutritionally healthy population. In this construct, approximately half of the group will require less and half will require more protein than this value. For a group obtaining adequate protein, the range of intakes varies around the EAR. The level of protein intake for a population that accommodates an adequate intake for the vast majority of the population is represented by the reference nutrient intake (RNI), which is the amount of protein that is enough to ensure that the needs of nearly all the group (97.5%) are met. This value is usually the EAR plus two standard deviations. In this construction, the majority of the population has protein intakes in excess of their physiological need. Speth[24] has warned of the possible adaptive stresses associated with consumption of very high protein intakes. However, detrimental effects of high levels of protein consumption on health and well-being remain questionable; excess consumption does not result in physiological overload, except possibly at the very extremes of intake.[25] Conversely, a number of positive attributes have been placed on high protein intake. It confers satiety[26] and is useful in weight loss diets.[27] Furthermore, high-protein diets stimulate insulin-like growth factor 1 production, which promotes skeletal development and bone formation.[28] They also allow adequate intake of essential amino acids across a range of different foods, including ones with predominantly low protein quantity or quality, and are permissive of the protein costs of infection.[12,29]

The average additional protein loss during infection has been estimated to be around 0.6 g/kg/day for adults, while the protein costs of specific infections have been estimated to be around 0.9 g/kg/day for diarrhoeal infection and 1.2 g/kg/day for typhoid.[29] Furthermore, Tontisirin et al.[30] estimated the total protein needs of children aged 8 to 11 and 12 to 18 months to be raised by 20 and 14%, respectively, during infection. Waterlow[12] estimated that the additional acute phase reactant protein synthesis of infants at the height of infection was as high as 1.2 g/kg body weight. This protein demand is met from diet and, where this is limited, from the catabolism of skeletal muscle and the metabolic diversion of amino acids. There is also increased demand for some essential amino acids during infection. The tripeptide glutathione, which in its reduced state plays a central role in cellular defence against oxygen-derived free radicals generated in infection, is rich in the essential amino acid cysteine.[20] Most of the major acute phase proteins, central to the activation of most arms of the immune system in response to infection, are rich in the essential amino acids phenylalanine and tryptophan.[31]

The slow rate and extended duration of childhood and juvenile growth relative to other primate species resulted in lower daily energy costs, despite higher overall need[2]; this also applies to daily protein needs. Both energy and protein metabolism scale allometrically with body size,[32] the relationships across species obeying functions of the following sort:

$$\text{basal metabolic rate (kcal/day)} = 70 \times \text{body weight}^{0.75}$$
$$\text{obligatory nitrogen loss (mg nitrogen/day)} = 140 \times \text{Wt}^{0.75}$$

Larger primates need to consume less meat to obtain their protein needs than smaller ones.[3,33] As large primates, humans have low protein needs that can be met in a range of ways, and which are permissive of great diet breadth.

Across the evolution of the hominins, it is extremely unlikely that diet would have been limited by protein availability, since dietary choices would have been necessarily limited to minimally processed, wild plant and animal foods.[7] According to Speth,[24] protein limitation probably did not exert selective pressure in hominin evolution; one consequence of this is the very wide range of protein intakes that human populations can subsist on, without obvious signs of ill-health. At the lower level, there are no unequivocal biochemical or physiological deficiency symptoms, apart from growth failure and tissue wasting.[21] In industrialized society, clinically manifest protein deficiency, in the form of kwashiorkor, is extremely rare now, and occurred in the past hundred years or so only under near-famine conditions.[34] Some very low protein intakes have been recorded in some traditional societies of New Guinea, in absence of signs of protein deficiency.[35] Subsequent investigations in the New Guinea Highlands identified the utilization of urea nitrogen for protein synthesis[36] as a physiological adaptation to the very low protein content of the traditional diet there, which comprised largely of *Colocasia* taro until around 400 years ago, and of sweet potato since then. Elsewhere in the developing world, kwashiorkor usually occurs in association with infection, especially diarrhoea.[20] Another possible route for protein deficiency to act as a selective pressure is in pregnancy and pregnancy outcome. However, this is again unlikely, since the most usual measure of pregnancy outcome, that of low birthweight, is most usually limited by energy or iron deficiency than by protein deficiency.

Understanding Macronutrient Needs and Infection at the Origins of Agriculture

The regular exploitation of cereal grains in the Near East and Mediterranean region arose with the emergence of the Natufian culture in the Levant around 13,000 years ago (ya),[37] with the domestication of emmer and einkorn wheat having taken place by 10,000 to 11,000 ya, from strains of wild wheat localized to southeastern Turkey.[38] The shift from hunting and gathering to cultivation and domestication was not an immediate and direct one; during the transition human groups probably varied in the extent to which they practised each strategy. In addition, the time at which farming and herding took precedence over hunting and gathering varied from site to site, although by 7,000 ya the transition from one mode of subsistence to another was more or less complete.[9] Although sheep were domesticated by around 11,000 years

B.P.[39] and goats and cows by about 10,000 ya,[40,41] early direct chemical evidence for dairying, from residues of dairy fats found on pottery in Britain, dates only to 6,100 to 5,500 ya.[42]

Angel[14] compiled data on estimated stature of populations in the Eastern Mediterranean across a time frame that saw changes in subsistence economies from hunting to developed farming. In his analysis, he identified declining stature (table 3.1) to be associated with declining health across the transition from hunting to farming. Cohen[15] associated shorter adult stature in the Neolithic with patterns of village settlement based on subsistence farming and stockbreeding and greater physiological stress due to undernutrition and infectious disease in childhood. Increased infant mortality and reduced life span, noted by Cohen,[15] are likely to be outcomes of these stresses. Sedentism, the clearing of land for agriculture and animal husbandry, and the increased potential of human contact with human and animal faeces provided ideal conditions for the transmission of helminthic and protozoal parasites, with substantially increased infection rates.[15] Increased population density would have also increased the likelihood of the spread of density-dependent infectious diseases. New diseases may have included intestinal parasites, tuberculosis, leprosy, schistosomiasis, and malaria.[43] These would have had their greatest impacts on children.

When poor growth or small stature is observed in the archaeological record, the complexity of nutrition-infection relationships makes any interpretation of growth faltering difficult;[16] one way forward is to study the environmental stresses impacting on young child growth in developing countries in the present. In such populations, exclusive breast feeding usually provides adequate energy and protein to support good child growth until 6 months of age.[10] Interactions between undernutrition and infection lead to growth faltering relative to growth potential from about the age of supplementation of breast feeding with other foods. Such faltering may continue for months or years, depending on the severity of the disease environment, and the abun-

Table 3.1   Mean Stature of Eastern Mediterranean Populations, from the Paleolithic to the Neolithic Periods

Period	Years before Present	Stature			
		Males		Females	
		n	Mean	n	Mean
Paleolithic	32–11,000	35	177.1	28	166.6
Mesolithic	11–9,000	61	172.5	35	159.7
Early Neolithic	9–7,000	39	169.6	31	155.5
Late Neolithic	7–5,000	6	161.3	13	154.3

Source: Adapted from Ulijaszek.[11]

dance and quality of the nutritional environment. In most populations, the process of growth faltering is complete by the age of 2 years. Undernutrition-infection interactions can be initiated in either of two ways. The first involves poor nutritional status leading to impaired immune system competence and reduced resistance to infection,[44] while the second involves an exposure to infectious disease that leads to a combination of the following: loss of appetite, malabsorption of nutrients in the gut, elevated maintenance metabolism as a consequence of fever,[45] and the breakdown of bodily protein to partially or totally fuel the production of acute phase proteins needed in the immune response.[46] In addition, the balance between possible immune system depression and adaptation as a consequence of one bout of infection is important in determining the immune system response to subsequent exposure to infectious agents and disease experience, and the extent of anorexia, fever, and malabsorption during infectious episodes.[47] Usually, children susceptible to infectious disease in this way undergo repeated bouts of infection across their earliest years, physical growth being inhibited directly by low nutrient intake, the diversion of nutrients to fuel the immune response, and changed levels of immunological and endocrine factors that control growth.

The positive secular trend observed in many nations across the twentieth century has been attributed to the decline in stresses associated with poor growth in the first 2 years of life.[17,48] Nutrition-infection interactions have their greatest potential impact on child growth in this age range.[49,50] It is possible that the negative secular trend represented by the decline in stature in the Near East and Mediterranean region could have been due to the emergence of these combined stresses.

Methods

There are many qualitative descriptions of dietary change during the Neolithic in the Near East[9,51]; quantification is more problematic. The relative contribution of different animal and plant species to the diet has been estimated for Paleolithic and Neolithic populations represented in archaeological excavations in the Near East and Mediterranean. These include sites in the Jordan Rift Valley, western Turkey, Israel, Iran, Syria, Iraq, and Greece.[14,52–56] Evidence for dietary usage comes from study of assemblages of animal bones, plant food, and faecal remains at, or close to, sites of human habitation.[57,58]

In the present analysis it is assumed that animal foods contributed the following percentages, by weight, to the diet: gazelle, 30; goats and sheep, 30; pig, 15; cattle, 10; fish, 7.5; and molluscs, 7.5. It is also assumed that plant foods contributed the following percentages by weight, to the diet: domesticated cereals, 63 (wheat, rye, barley in equal balance); legumes, 24 (peas and lentils in equal measure); leafy vegetables, 8; soft fruit, 4; and roots, 1. The energy and protein composition of these foods is determined from food composition tables for the Middle East[59] and the UK,[60] while the essential amino acid composition is calculated from Amino-Acid Content of Foods and Biological Data on Proteins.[61] These values are combined to create a constructed Neolithic diet, to be used in modeling the protein and amino acid adequacy of the possible food intakes of young children. The protein, energy, and essential amino acid content of the animal and plant food components of this constructed diet are given in table 3.2. The composition of breast milk used in this analysis is also given here, and is derived from estimates given in Butte et al.[10] and Dewey et al.[23]

From these measures, the protein and amino acid adequacy of the infant and young child diet at the origins of agriculture is estimated in the following way, and using the following assumptions. Exclusive breast feeding usually provides adequate protein to support good child growth until 6 months of age,[10] and in most traditional societies observed in the twentieth century, breast feeding continues to beyond 2 years of age.[62] In this analysis, it is assumed that infants in the Neolithic did likewise. After 6 months of age, however, exclusive breast feeding is inadequate to sustain sufficient child growth.[10] Therefore, the extent to which a diet in which breast milk is supplemented by the constructed Neolithic diet (supplementation diet) meets the protein and essential amino acid needs of a child aged 6 to 11 months is estimated from energy intake values equivalent to the revised FAO/WHO/UNU[22] recommendations.[23] This is compared to the protein and amino acid contributions of constructed Neolithic diets that supply energy requirements to the total exclusion of breast milk; for the purposes of this analysis, this is called the weaning diet. In all cases, it is assumed that energy needs are met. This is therefore an analysis of the best case. The amino acid profiles of the constructed diets are compared to reference patterns for essential amino acids put forward by Dewey et al.[23] Two protein requirement constructs are used to determine whether protein needs are met: the physiological one of EAR, and the safe intake of protein by populations, as represented by the RNI.

The proportion of the diet that came from animal sources is likely to have varied across time and place at the origins of agriculture. Therefore, the proportion of the Neolithic diet that comes from animal sources is varied in the analysis between 0 and 15%, to represent different levels of diet quality, and to estimate the lower limits of animal sources in the diet that allow adequacy

Table 3.2 Energy and Protein Composition of Constructed Neolithic Diet, per 100 g Edible Portion

		Energy (kcal)	Protein (g)	Val	Tryp	Thre	Phen + Tyr	Meth + Cys	Lys	Leu	Iso leu
						Essential Amino Acids (mg)					
Breast milk		67	0.9	43	21	37	71	31	60	88	45
Animal foods	Mean	132	16.2	313	28	287	500	249	556	507	301
Plant foods	Mean	95	3.3	315	15	207	515	246	216	417	224

of protein and essential amino acid intake. The process is repeated for a diet that is comprised totally of the constructed Neolithic diet (the weaning diet), to determine the extent to which such a diet, if consumed in the absence of breast milk, could be associated with protein or amino acid deficiency. This process is repeated for children aged 12 to 36 months. Table 3.3 shows energy and protein requirements, and the energy supplied by breast milk and constructed Neolithic diets to the two groups of young children.

The protein adequacy of the supplementation and weaning diets is also examined when the protein costs of infection are factored in. To this end, the daily protein needs of children aged 6 to 11 and 12 to 36 months are raised by 20 and 14%, respectively, values in line with the observations of Tontisirin et al.,[30] in relation to the protein costs of infection. It would be useful to know whether essential amino acid adequacy might be challenged by increased need during infection. There are currently inadequate data on which to base a calculation of amino acid requirements in acute disease.[63]

Results

Figure 3.1 shows the protein intake, in grams per day, of a child aged 6 to 11 and 12 to 36 months, respectively, consuming (1) the supplementation diet and (2) the weaning diet, with constructed Neolithic diet at 15, 10, 5, and 0% from animal sources, in both cases, relative to adjusted FAO/WHO/UNU[22] reference values.[23] Protein intake of the child aged 6 to 11 months consuming the supplementation diet is well above both EAR and RNI, regardless of how much dietary energy comes from meat in the constructed Neolithic diet. The protein intake of the child aged 12 to 36 months consuming the supplementation diet is above both the EAR and RNI regardless of whether the constructed Neolithic diet contains animal food sources. The same pattern is exhibited for the child consuming the weaning diet. Figure 3.2 shows amino acid profiles of a child aged 6 to 11 months, consuming supplementation and weaning diets, respectively, with the constructed Neolithic diet at 15, 10, 5, and 0% from animal sources, relative to estimates of essential amino acid needs of infants.[23] On both diets, the child is lysine deficient when 5% or less of the constructed Neolithic diet comes from animal sources. Figure 3.3 shows amino acid profiles of a child aged 12 to 36 months, consuming the same supplementation and weaning diets. In this case, there is lysine deficiency when the supplementation diet has no energy from animal sources, and when animal sources contribute either no energy or 5% dietary energy to the weaning diet.

Figure 3.4 shows protein intake, in grams per day, of a child aged 6 to 11 and 12 to 36 months, consuming supplementation and weaning diets, respectively, with the constructed Neolithic diet at 15, 10, 5, and 0% from animal sources, relative to adjusted FAO/WHO/UNU[22] reference values[23] plus the protein cost of infection.[30] The child aged 6 to 11 months is protein deficient relative to the RNI on the supplementation diet when 5% or less of the con-

Table 3.3 Protein and Essential Amino Acid Adequacy Calculated from Infant/Young Child Diet Type and Composition

Age Group (months)	Requirement		Infant/Young Child Diet Type	Energy (kcal) from:	
	Energy (kcal)	Protein (g) (EAR)		Breast Milk	Constructed Neolithic Diet
6–11	844	11.1	Supplementation	539	305
			Weaning	0	844
12–36	1197	11.7	Supplementation	539	358
			Weaning	0	1197

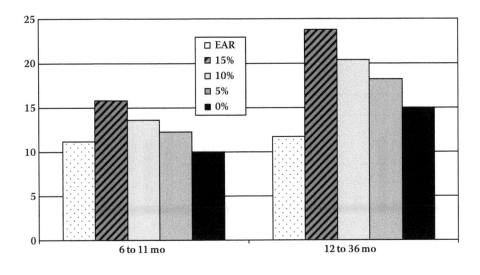

a

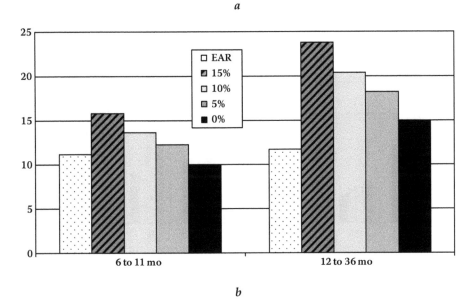

b

Figure 3.1   (a) Protein intake (g/day) of child aged 6 to 11 and 12 to 36 months consuming supplementation diet (with constructed Neolithic diet at 15, 10, 5, and 0% from animal sources), relative to adjusted FAO/WHO/UNU[22] reference values.[23] (b) Protein intake (g/day) of child aged 6 to 11 and 12 to 36 months consuming weaning diet at 15, 10, 5, and 0% animal sources, relative to adjusted FAO/WHO/UNU[22] reference values.[23]

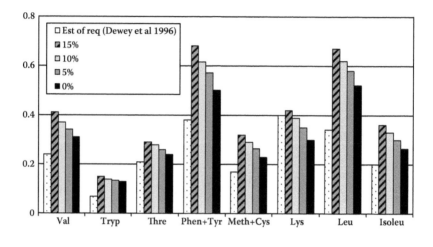

a

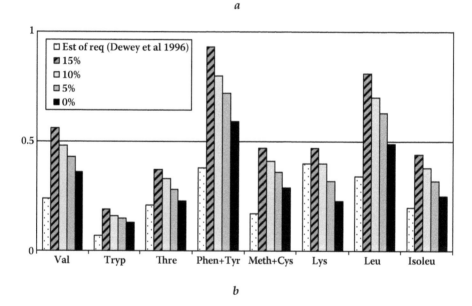

b

Figure 3.2   (a) Amino acid profile of child aged 6 to 11 months consuming supplementation diet (with constructed Neolithic diet at 15, 10, 5, and 0% from animal sources), relative to estimates of essential amino acid needs of infants.[23] (b) Amino acid profile of child aged 6 to 11 months consuming weaning diet at 15, 10, 5, and 0% animal sources, relative to estimates of essential amino acid needs of infants.[23]

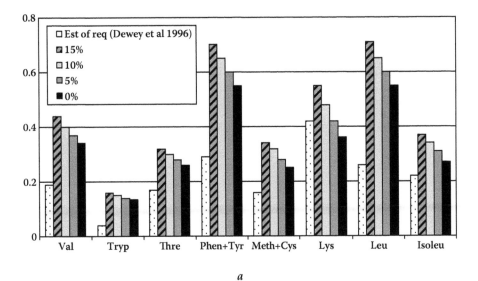

a

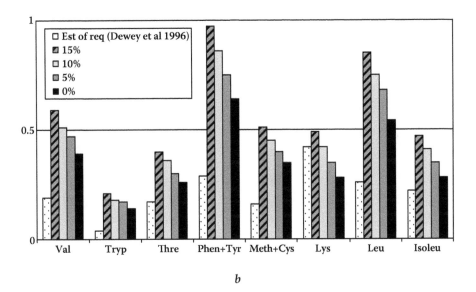

b

Figure 3.3 (a) Amino acid profile of child aged 12 to 36 months consuming supplementation diet (with constructed Neolithic diet at 15, 10, 5, and 0% from animal sources), relative to estimates of essential amino acid needs of infants.[23] (b) Amino acid profile of child aged 12 to 36 months consuming weaning diet at 15, 10, 5, and 0% animal sources, relative to estimates of essential amino acid needs of infants.[23]

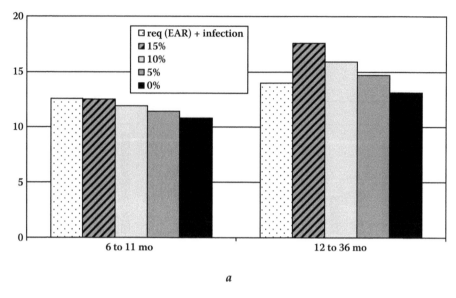

a

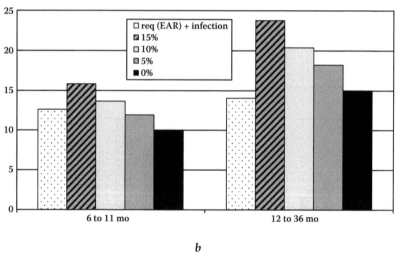

b

Figure 3.4   (a) Protein intake (g/day) of child aged 6 to 11 and 12 to 36 months consuming supplementation diet (with constructed Neolithic diet at 15, 10, 5, and 0% from animal sources), relative to adjusted FAO/WHO/UNU[22] reference values[23] plus protein cost of infection.[30] (b) Protein intake (g/day) of child aged 6 to 11 and 12 to 36 months consuming weaning diet at 15, 10, 5, and 0% animal sources, relative to adjusted FAO/WHO/UNU[22] reference values[23] plus protein cost of infection.[30]

structed Neolithic diet comes from animal sources. In the child aged 12 to 36 months, protein deficiency is only likely on the supplementation diet when the constructed Neolithic diet has no dietary energy from animal sources.

Discussion

In the presence of breast feeding, there is little likelihood of there having been protein deficiency among young children at the origins of agriculture if physiological energy needs were met, even when the diet included no animal sources. However, if the diet had no energy from animal sources, children aged 6 to 11 months on a weaning diet may have experienced deficiency in protein quality. In this age group, lysine deficiency was identified for the supplementation diet when 10% or less of the constructed Neolithic diet came from animal sources, and on the weaning diet when 5% or less of the dietary energy in the constructed Neolithic diet came from animal sources, despite both supplementation and weaning diets being adequate in total protein. However, the measures used to determine adequacy, the EAR and RNI, are population constructs, and even though the model shows that the group on average meets its protein needs, there may be individuals within it that do not. While there is no quantitative deficiency of protein intake for any of the diets for the child aged 12 to 36 months, the supplementation and weaning diets may be deficient in lysine when there are no animal foods in the constructed Neolithic diet for the former, and when animal foods comprise 5% or less of the dietary energy from the constructed Neolithic diet, for the latter. This observation must be treated with caution, since dietary requirements for lysine remain uncertain,[64] and may be lower than those used in this analysis.

Protein deficit is unlikely to have been a primary factor limiting the growth of young children beyond 6 months of age, even in infants receiving only plant-based supplementary foods, with the possible exception of diets based on sago or cassava.[65] In an examination of protein intakes of children aged 18 to 30 months in Egypt, Mexico, and Kenya, Beaton et al.[66] concluded that protein intakes, even in the absence of breast feeding, were unlikely to limit child growth, although lysine deficiency was seen as having a minor contribution to the growth faltering of the Kenyan children. Even if meat consumption was low among young children in the Neolithic Near East and Mediterranean, deficiency of protein quantity or quality is unlikely to have been a major limiting factor in growth and development. On the basis of this evidence, the negative secular trend toward decreased adult body stature across the transition to agriculture in the Near East is unlikely to have been directly due to protein deficiency, as long as infants were breast fed for the first 2 years of life. There are three other possible explanations for this negative secular trend. The first is that of zinc deficiency, associated with transition to a wheat-based diet, which can impact on child growth. The second is that of low energy density of the weaning diet, making it difficult for young children to meet their energy needs. The third is the additional protein requirement for infection.

Raising the physiological protein requirement bar by adding possible protein costs of infection increases the likelihood of protein deficiency of the 6- to 11-month-old child consuming a supplementation diet with 5% or less of the constructed Neolithic diet from animal sources. It increases the possibility of protein deficiency in the child aged 12 to 36 months, but only with the supplementation diet when animal sources contribute none of the dietary energy. Although it was not possible to model it in this analysis, it is likely that such diets would also be deficient in lysine, to varying extents. Protein deficiency as an explanation for the negative secular trend toward decreased adult body stature becomes more plausible in the context of increased prevalence of infectious diseases, to which young children are particularly susceptible. Although the possibility of adaptation to low protein intake cannot be discounted, the human biological cost of such adaptation might have been increased mortality due to impaired immunological resistance to infectious disease. Angel[14] identified declining stature to be associated with declining health across the transition from hunting to farming. Cohen[15] and Ulijaszek[11] associated shorter adult stature in the Neolithic with greater physiological stress due to undernutrition and infectious disease; such stress is likely to be the mechanism underlying increased infant mortality, reduced life span, and increased incidence of infectious disease[15] across this period. The present study points to possible quantitative and qualitative deficiencies in protein intake among young children as a specific factor in this relationship.

Paediatric practitioners in societies with high infectious disease burdens may be so underresourced that the consideration of protein needs at the transition to agriculture may be a luxury they do not have. However, it is worth placing the debate concerning the protein needs of infants exposed to both undernutrition and infection in a longer-term context. Humans have evolved to have low physiological protein requirements relative to other primate species, and to be able to adapt to even lower intakes. This fine balance was upset with the changed ecologies that were created by humans with the origins of agriculture, particularly with the emergence of a range of density-dependent infectious diseases. While controlled or eliminated in most of the industrialised world, most of the developing world continues to suffer with high infectious disease burdens. Recent calls for increases in recommended daily intakes of protein, particularly for developing countries where exposure to infection is high, tacitly acknowledge the low likelihood that infectious disease burdens will return to the low levels that existed before the origins of agriculture.

References

1. Ulijaszek, S. J. 1995. *Human energetics in biological anthropology.* Cambridge: Cambridge University Press.

2. Aiello, L. C., and Wells, J. C. K. 2002. Enegretics and the evolution of the genus. *Homo. Annu. Rev. Anthropol.* 31:325.
3. Leonard, W. R., Robertson, M. L., Snodgrass, J. J., and Kuzawa, C. W. 2003. Metabolic correlates of hominid brain evolution. *Comp. Biochem. Physiol. A Mol. Integrative Physiol.* 136:5.
4. Raubenheimer, D., and Simpson, S. 2005. Obesity: The protein leverage hypothesis. *Obes. Rev.* 6:133.
5. Millward, D. J. 2003. An adaptive metabolic demand model for protein and amino acid requirements. *Br. J. Nutr.* 90:249.
6. Leonard, W. R. 2000. Human nutritional evolution. In *Human biology. An evolutionary and biocultural perspective*, ed. S. Stinson, B. Bogin, R. Huss-Ashmore, and D. O'Rourke, chap. 9. New York: Wiley-Liss.
7. Cordain, L., Eaton, S. B., Sebastian, A., et al. 2005. Origins and evolution of the Western diet: Health implications for the 21st century. *Am. J. Clin. Nutr.* 81:341.
8. Oftedal, O. T. 1991. The nutritional consequences of foraging in primates: The relationship of nutrient intakes to nutrient requirements. *Phil. Trans. R. Soc. Lond. B* 334:161.
9. Ulijaszek, S. J. 1991. Human dietary change. *Phil. Trans. R. Soc. Lond. B* 334:271.
10. Butte, N. F., Lopez-Alarcon, M. G., and Garza, C. 2002. *Nutrient adequacy of exclusive breastfeeding for the term infant during the first six months of life.* Geneva: World Health Organization.
11. Ulijaszek, S. J. 2000. Interpreting patterns of growth and development among past populations. *Perspect. Hum. Biol.* 5:1.
12. Waterlow, J. C. 1991. Protein-energy malnutrition: Challenges and controversies. *Proc. Nutr. Soc. India* 37:59.
13. Hillman, G. C., Colledge, S. M., and Harris, D. R. 1989. Plant food economy during the Epipalaeolithic period at Tell Abu Hureya, Syria: Dietary diversity, seasonality, and modes of exploitation. In *Foraging and farming. The evolution of plant exploitation*, ed. D. R. Harris and G. C. Hillman, 240–68. London: Unwin Hyman.
14. Angel, J. L. 1984. Health as a crucial factor in the changes from hunting to developed farming in the Eastern Mediterranean. In *Paleopathology at the origins of agriculture*, ed. M. N. Cohen and G. J. Armelagos, chap. 3. New York: Academic Press.
15. Cohen, M. N. 1989. *Health and the rise of civilisation.* New Haven, CT: Yale University Press.
16. King, S. E., and Ulijaszek, S. J. 1999. Invisible insults during growth and development: Contemporary theories and past populations. In *Human growth in the past*, ed. R. Hoppa and C. Fitzgerald, 161. Cambridge: Cambridge University Press.
17. Cole, T. J. 2003. The secular trend in human physical growth: A biological view. *Econ. Hum. Biol.* 1:161.
18. Carpenter, K. J. 1994. *Protein and energy. A study of changing ideas in nutrition.* Cambridge: Cambridge University Press.
19. Golden, M. H. N. 1985. The consequences of protein deficiency in man and its relationship to the features of kwashiorkor. In *Nutritional adaptation in man*, ed. K. Blaxter and J. C. Waterlow, 169. London: John Libbey.
20. Jackson, A. A. 1990. The aetiology of kwashiorkor. In *Diet and disease*, ed. G. A. Harrison and J. C. Waterlow, chap. 5. Cambridge: Cambridge University Press.
21. Millward, D. J. 2005. Protein and amino acid requirements. In *Human nutrition*, ed. C. Geissler and H. Powers, chap. 8. Edinburgh: Elsevier Churchill Livingstone.
22. FAO/WHO/UNU. 1985. *Energy and protein requirements.* World Health Organization Technical Report Series 724. Geneva: WHO.

23. Dewey, K. G., Beaton, G., Fjeld, C., Lonnerdal, B., et al. 1996. Protein requirements of infants and children. *Eur. J. Clin. Nutr.* 50 (Suppl. 1):S119.
24. Speth, J. 1989. Early hominid hunting and scavenging. The role of meat as an energy source. *J. Hum. Evol.* 18:329.
25. Ulijaszek, S. J., and Strickland, S. S. 1993. *Nutritional anthropology. Prospects and perspectives*. London: Smith-Gordon.
26. Batterham, R. L., Heffron, H., Kapoor, S., et al. 2006. Critical role for peptise YY in protein-mediated satiation and body-weight regulation. *Cell Metabol.* 4:223.
27. Abete, I., Parra, M. D., Zulet, M. A., and Martinez, J. A. 2006. Different dietary strategies for weight loss in obesity: Role of energy and macronutrient content. *Nutr. Res. Rev.* 19:5.
28. Bonjour, J. P. 2005. Dietary protein: An essential nutrient for bone health. *J. Am. Coll. Nutr.* 24 (Suppl. S):526S.
29. Scrimshaw, N. S. 1992. Effect of infection on nutritional status. *Proc. Natl. Sci. Council* 16:46.
30. Tontisirin, K., Aimanwra, N., and Valyasevi, A. 1984. Long-term study on the adequacy of usual Thai weaning food for young children. *Food Nutr. Bull.* 10:265.
31. Reeds, P. J., Fjeld, C. R., and Jahoor, F. 1994. Do the differences in the amino acid compositions of acute phase and muscle proteins have a bearing on nitrogen loss in traumatic states? *J. Nutr.* 124:906.
32. Brody, S. 1945. *Bioenergetics and growth*. New York: Reinhold Publishing Corporation.
33. Hamilton, W. J., and Busse, C. 1978. Primate carnivory and its significance to human diets. *Bioscience* 28:761.
34. Waterlow, J. C. 1992. *Protein energy malnutrition*. London: Edward Arnold.
35. Oomen, H. A. P. C. 1970. Interrelationship of the human intestinal flora and protein utilization. *Proc. Nutr. Soc.* 29:197.
36. Koishi, H. 1990. Nutritional adaptation of Papua New Guinea highlanders. *Eur. J. Clin. Nutr.* 44:853.
37. Bar-Yosef, O. 1998. The Natufian culture in the Levant, threshold to the origins of agriculture. *Evol. Anthropol.* 6:159.
38. Salami, F., Ozkan, H., Brandolini, A., Schafer-Pregl, R., and Martin, W. 2003. Genetics and geography of wild cereal domestication in the near east. *Nat. Rev. Genet.* 3:429.
39. Hiendleder, S., Kaupe, B., Wassmuth, R., and Janke, A. 2002. Molecular analysis of wild and domestic sheep questions current nomenclature and provides evidence for domestication from two different subspecies. *Proc. R. Soc. Lond. B* 269:893.
40. Luikart, G., Gielly, L., Excoffier, L., et al. 2001. Multiple maternal origins and weak phylogeographic structure in domestic goats. *Proc. Natl. Acad. Sci. U.S.A.* 98:5927.
41. Loftus, R. T., Ertugrul, O., Harba, A. H., et al. 1999. A microsatellite survey of cattle from a centre of origin: The Near East. *Mol. Ecol.* 8:2015.
42. Copley, M. S., Berstan, R., Dudd, S. N., et al. 2003. Direct chemical evidence for widespread dairying in prehistoric Britain. *Proc. Natl. Acad. Sci. U.S.A.* 100:1524.
43. Ortner, D. J., and Theobald, G. 1993. Diseases in the pre-Roman world. In *The Cambridge world history of human disease*, ed. K. F. Kiple, 247–61. Cambridge: Cambridge University Press.
44. Chandra, R. K. 1972. Immunocompetence in undernutrition. *J. Pediatr.* 81:1194.
45. Ulijaszek, S. J. 1997. Transdisciplinarity in the study of undernutrition-infection interactions. *Coll. Antropologicum* 21:3.
46. Grimble, R. F. 1992. Dietary manipulation of the inflammatory response. *Proc. Nutr. Soc.* 51:285.

47. Tomkins, A. M. 1986. Protein-energy malnutrition and risk of infection. *Proc. Nutr. Soc.* 45:289.
48. Takaishi, M. 1994. Secular changes in growth of Japanese children. *J. Pediatr. Endocrinol.* 7:163.
49. Ulijaszek, S. J. 1990. Nutritional status and susceptibility to infectious disease. In *Diet and disease,* ed. G. A. Harrison and J. C. Waterlow, chap 7. Cambridge: Cambridge University Press.
50. Neumann, C. G., and Harrison, G. G. 1994. Onset and evolution of stunting in infants and children. Examples from the Human Nutrition Collaborative Research Support Program. Kenya and Egypt studies. *Eur. J. Clin. Nutr.* 48 (Suppl. 1):S90.
51. Cordain, L. 1999. Cereal grains: Humanity's double-edged sword. In *Evolutionary aspects of nutrition and health, diet, exercise, genetics and chronic disease,* ed. A. P. Simopoulos, 19. *World Rev. Nutr. Diet.* 84. Basel: Karger.
52. Cutting, M. V. 2005. *The Neolithic and Early Chalcolithic farmers of Central and Southwest Anatolia: Household, community and the changing use of space.* Oxford: Archaeopress.
53. Perles, C. 2001. *The Early Neolithic in Greece. The first farming communities in Europe.* Cambridge: Cambridge University Press.
54. Renfrew, C., and Bahn, P. 1996. *Archaeology: Theories, methods and practise.* London: Thames and Hudson.
55. Flannery, K. V. 1969. Origins and ecological effects of early domestication in Iran and the Near East. In *The domestication and exploitation of plants and animals,* ed. P. J. Ucko and G. W. Dimbleby, 73. London: Gerald Duckworth and Company Limited.
56. Reed, C. A. 1977. A model for the origin of agriculture in the Near East. In *Origins of agriculture,* ed. C. A. Reed, 543–67. The Hague: Mouton.
57. Harris, D. R., and Hillman, G. C. 1989. *Foraging and farming. The evolution of plant exploitation.* London: Unwin Hyman.
58. Clutton-Brock, J. 1989. *The walking larder. Patterns of domestication, pastoralism, and predation.* Boston: Unwin Hyman.
59. FAO. 1982. *Food composition tables for the Near East.* FAO Food and Nutrition Paper 26. Rome: FAO.
60. McCance, R. A., and Widdowson, E. 2002. *The composition of foods.* 6th ed. London: Royal Society of Chemistry and the Food Standards Agency.
61. FAO. 1981. *Amino-acid content of foods and biological data on proteins.* FAO Nutritional Studies 24. Rome: FAO.
62. Sellen, D. W. 2001. Comparison of infant feeding patterns reported for nonindustrial populations with current recommendations. *J. Nutr.* 131:2707–2715.
63. Obled, C., Papet, I., and Breuille, D. 2002. Metabolic bases of amino requirements in acute diseases. *Curr. Opin. Clin. Nutr. Metabol. Care* 5:189.
64. Millward, D. J., and Jackson, A. 2004. Protein/energy ratios of current diets in developed and developing countries compared with a safe protein/energy ration: Implications for recommended protein and amino acid intakes. *Public Health Nutr.* 7:387–405.
65. Gibson, J. 2001. Food demand in the rural and urban sectors of PNG. In *Food security for PNG,* ed. R. M. Bourke, M. G. Allen, and J. G. Salisbury, 45. Canberra: Australian Centre for International Agricultural Research.
66. Beaton, G. H., Calloway, D. H., and Murphy, S. P. 1992. Estimated protein intakes of toddlers: Predicted prevalence of inadequate intakes in village populations in Egypt, Kenya, and Mexico. *Am. J. Clin. Nutr.* 55:902.

4

Evolutionary Perspectives on Type 2 Diabetes in Asia

Tessa M. Pollard
*Medical Anthropology Research Group, Department
of Anthropology, Durham University*

Alejandra Núñez-de la Mora
*Medical Anthropology Research Group, Department
of Anthropology, Durham University*

Nigel C. Unwin
Institute of Health and Society, University of Newcastle

Contents

Introduction

Type 2 diabetes is one of the most important health problems faced by the world today. It accounts for 85 to 95% of diabetes in populations worldwide,[1] arising from a combination of resistance to the action of insulin, which is strongly associated with obesity and physical inactivity, and the inability of the insulin-producing cells of the pancreas to secrete enough insulin to overcome that resistance. This leads to rising blood glucose levels and, eventually, diagnosis of diabetes, based on glucose rising above defined levels.[2]

Diabetes leads to a number of disabling and life-threatening complications. Largely through the damage it causes to small blood vessels and nerves (so-called microvascular complications), it is one of the most common causes of blindness,[3] renal failure,[4] and lower limb amputation.[5] Through the damage it causes to larger blood vessels (macrovascular complications) it markedly increases the risk of coronary heart disease, stroke, and peripheral vascular disease. Indeed, at least in the West, the majority of people with diabetes will die from cardiovascular disease.[6,7] Not surprisingly, type 2 diabetes significantly reduces life expectancy.[8] The onset of the disease is usually gradual and insidious, and it may go undiagnosed for years. In many populations, particularly in poorer parts of the world, 50% or more of people with type 2 diabetes have not been diagnosed, and even in those diagnosed, the disease is often poorly controlled.[1,9,10]

Type 2 diabetes is strongly associated with overweight and obesity,[11–13] and until recently it was considered a disease of greater importance to rich, Western populations than to populations in low- and middle-income countries. However, in 2000 the countries with the largest numbers of estimated cases of type 2 diabetes were India (32 million) and China (21 million), with the United States in third place (18 million).[9] These numbers can be explained partly by the fact that India and China have very large populations, but surveys have also suggested that in some urban areas of Asia the prevalence of the disease is higher than in the United States.[14] For example, the prevalence of type 2 diabetes in adults was 8.2% in the United States in 1999–2000, but the National Urban Diabetes Survey of six major cities in India conducted at the same time found a prevalence of 12.1% (both studies having adjusted for the age structure of the populations surveyed).[12,15] The largest increases in prevalence between 1995 and 2025 are predicted for China (68%) and India (59%),[16] and projections suggest that by 2030 the gap between India, China, and the United States in the number of people with diabetes will have widened[9] (figure 4.1). These projections are based on ageing of the populations (the risk of type 2 diabetes increases sharply with age), and trends toward an increasing proportion of the population living in urban centres. However, the projections do not specifically take into account increasing levels of obesity, and are thus likely to be conservative.

The prevalence of overweight and obesity in Asia, as indicated by the number of people with a BMI (body mass index, calculated as weight/height^2) equal to or greater than 25 for overweight and equal to or greater than 30 for obesity, has been increasing rapidly. For example, the prevalence of overweight and obesity in China, as assessed by BMI greater than or equal to 25, has increased by 400% over the last 20 years, in comparison with an increase of 20% in Australia.[17] The National Urban Diabetes Survey in India showed that 31% of the urban population was overweight or obese as assessed by BMI.[12] These changes have been driven partly by the so-called nutrition transition, a term describing consumption of food higher in fat, more animal

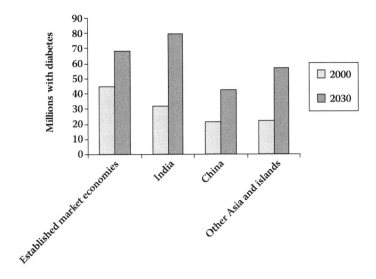

Figure 4.1 Estimated numbers of people with diabetes by region for 2000 and 2030 (figures from Wild et al.[9]). "Other Asia and islands" includes the low- and middle-income economies of Asia (apart from India and China) and the islands of the Indian and Pacific oceans, except Madagascar.

produce, more sugar, and more processed foods,[18,19] caused by changes in farming methods, industrialisation, and the rapid spread of global super-market chains and fast food restaurants.[20,21] At the same time there has been a reduction in manual labour, partly because large numbers of farmers have left the land to work and live in urban areas, most of them in occupations that pay better and need less energy expenditure than farm work.[22] Urbani-sation has also been associated with dramatic increases in private car owner-ship and television watching.[23]

Nevertheless, levels of obesity as assessed by BMI[11] are still much lower in Asia than in most Western countries,[14,17,24] and are not enough in themselves to explain the high rates of type 2 diabetes. The key here appears to be differ-ences in body composition between Asian and Western populations; people in Asia tend to have a higher proportion of body fat for a given level of BMI than do people of European origin.[14,25] Thus, adiposity is greater at lower BMIs in Asian populations. For this reason, there has been some consider-ation of implementing lower BMI cutoff points to indicate overweight and obesity in Asian populations.[26] In addition, fat patterning in Asians tends to favour central or abdominal obesity.[14] This pattern is evident at birth and con-tinues into adulthood. For example, studies have shown that Indian babies are significantly smaller by every measure than European babies but are rel-atively more obese with larger amounts of central fat.[25,27] Central obesity is associated with higher levels of intra-abdominal, visceral fat deposition, and this is a particularly strong risk factor for type 2 diabetes. For example, high waist circumference, a surrogate measure of visceral obesity, increases the risk of diabetes independently of the risk reflected by high BMI.[13] It is clear that any explanation of high rates of type 2 diabetes in Asia as compared to those seen in populations of European origin must also address the origin of this differential body composition.

In this chapter we consider the evolutionary rationales for two hypotheses that have been applied to explain differential susceptibility to type 2 diabetes at the population level, the thrifty genotype and thrifty phenotype hypotheses, and how they may be relevant to the worsening epidemic of type 2 diabetes in Asia. Both address the question of why some people in some populations seem to be particularly good at storing fat or likely to develop insulin resistance. We also use an evolutionary perspective to consider the relevance of recent research making important links between infection, inflammation, abdominal obesity, and several noncommunicable diseases, including type 2 diabetes.

*Thriftiness: The Contribution of Genes**

In 1962 James Neel published the suggestion that the basic defect in diabetes mellitus was a quick insulin trigger in response to hyperglycaemia (high levels of blood glucose).[28] Neel considered that the quick insulin trigger was under genetic control and coined the term *thrifty genotype*. Rapid insulin release in response to a rise in blood glucose was hypothesised to lead to the rapid utilisation of glucose, minimizing its loss in urine. Neel proposed that the thrifty genotype would have been selected under conditions of feast or famine, because of a heightened need to make good use of glucose available during times of plenty to see people through times when food was scarce. He believed that hunter-gatherers during the Palaeolithic, who were ancestral to all modern humans, would have experienced such feast or famine conditions because of cycling of food availability, with periods of both abundance and scarcity. Neel considered that in a contemporary Western environment in which there is more than enough food continuously available, a thrifty genotype would make people vulnerable to developing high levels of obesity and diabetes. He did not suggest any difference between populations in the prevalence of the thrifty genotype, but rather that because of the evolutionary background of the species, people in general would be vulnerable to the overproduction of insulin in the Western environment, with some between-individual variation in genetic vulnerability to diabetes.

---

* This section is based partly on material first published in Pollard, T. M., *Western Diseases: An Evolutionary Perspective* (Cambridge, Cambridge University Press, 2008); reproduced by permission of Cambridge University Press.

Understanding of glucose and insulin metabolism and of the pathologies underlying diabetes was at a rudimentary stage when Neel put forward the thrifty genotype concept, and it later became clear that the specific physiological mechanisms he invoked were not plausible. As a result, he attempted to describe alternative pathways,[29] but he and colleagues later wrote that these modifications to the original idea were not "intellectually satisfactory."[30] They also wrote that "the term 'thrifty genotype' has served its purpose, overtaken by the growing complexity of modern genetic medicine"[30] (p. 60). They preferred to conceptualise obesity and type 2 diabetes as resulting from previously adaptive multifactorial genotypes that had evolved to work together in an integrated manner in the Palaeolithic environment, but were no longer adaptive in a Western environment, where an overabundant diet and sedentary behaviour act together to disrupt this integrated functioning. Specific proposals about the physiology underlying this adaptation to Palaeolithic conditions and how it is disrupted in obese, sedentary populations have since been suggested.[31]

Neel et al.[30] noted "the suspicion that there might be a particular predisposition to the disease in some tribal groups," (p. 45) but suggested that there is no evidence for a simple "ethnic predisposition." (p. 45) Nevertheless, others went on to propose the existence of particularly thrifty genotypes in a number of different populations that developed high rates of type 2 diabetes in the second half of the twentieth century, including Native Americans, Australian Aborigines, and Pacific Islanders. These populations were posited to have evolved thrifty genotypes in response to previous environments in which, for various reasons, energy supplies were limited.[32,33] The notion has also been applied to people from South Asia (the Indian subcontinent) to explain the high levels of type 2 diabetes found in migrant South Asians in the UK.[34] In 1995 an editorial in *Lancet* provided a list of "genetically acquired risk factors that are potentiated and supplemented by westernization" (p. 402) considered to be found in people of South Asian descent.[35] These included low function in the cells in the pancreas that secrete insulin and high levels of insulin resistance. More recently, Wells[36] has noted that the Indian subcontinent was vulnerable to famine as a result of the regular failure of the monsoon, and argued that this created selective pressure for increased visceral adiposity. He suggests that visceral adipose tissue is sensitive to fluctuations in energy balance and is favoured in situations of chronic energy deficiency, such as famine. Kagawa et al.[37] focus on the possibility of a different kind of thrifty genotype in the Japanese, noting that over many hundreds of years Japanese people consumed less fat than Europeans and positing that Japanese people therefore "require efficient lipid absorption."

Despite the attractiveness of these ideas, few candidate thrifty genes have been identified. In fact, only a very few genes have been identified as being associated with the risk of type 2 diabetes, and they explain to no more than a minimal extent differences among populations in susceptibility to disease.[38] There is, however, evidence that migrant South Asians in the United

States have a higher frequency of a variant of the gene (PC-1) K121Q than do U.S. whites. This gene appears to be involved in cellular insulin signaling, and the allele that was more frequent in South Asians was associated with greater levels of insulin resistance.[39] It has been suggested that this allele "confers a biological advantage, such as that postulated in the 'thrifty gene hypothesis.'"[40] While this idea is undeveloped, the gene appears to be a possible candidate for explaining part of the elevated risk of type 2 diabetes in people of South Asian origin living in Westernised environments, including in urban South Asia. Similarly, Kagawa et al.[37] list a number of single nucleotide polymorphisms (SNPs) that they suggest have thrifty effects, emphasising contrasts between Asians and Europeans. In particular, they cite a SNP in fatty acid binding protein 2 that is more frequent in Japanese than in Europeans. However, the existence of these thrifty alleles is not clearly established, and Kagawa et al. acknowledge that limited sample sizes have "precluded rigorous statistical analysis in many studies of 'thrifty' SNPs."

Kagawa et al. also note that of all the major types of early agriculture, the European form was the only one to include livestock, and that the European diet was therefore higher in protein and fat than other diets. Thus, rather than seeking explanations for a supposed thrifty genotype in Asians, we might do better to seek an explanation for a nonthrifty genotype in Europeans. Others have taken this approach, for example, focusing on the relatively high consumption of lactose from milk in the European diet, and proposing that constant exposure to this simple sugar would have led to selective pressure for a nonthrifty genotype in Europeans.[41] The suggestion that the ancestral condition for humans was thrifty, and that this condition persists in most populations, but is less common in Europeans, is certainly logically more appealing than the idea that many different populations independently evolved thrifty genotypes.

Since the origin of *Homo sapiens* human populations have been exposed to environments that differ physically (for example, in climate and geography) and in relation to ecology and subsistence methods, and it is plausible that different selective pressures have led to genetic differentiation[42] that now affects vulnerability to chronic disease. For example, differences in the need to conserve sodium in hot and cold climates are thought to have led to a cline in the frequency of alleles involved in the regulation of blood pressure.[43] Thus, it is plausible that some populations will be genetically more vulnerable to type 2 diabetes than others, but it seems clear that the thrifty gene hypothesis was initially applied too enthusiastically to one population after another. It has also been too closely identified with racial and ethnic groupings, neither of which are genetically meaningful methods of categorisation.[44] It is possible, therefore, that, on average, Asian populations are genetically more susceptible to type 2 diabetes than European populations, when exposed to environmental risk factors associated with Westernisation, but the evidence to support this notion is not, so far, strong.

Developmental Origins?

A surge of interest in early life influences on the development of chronic disease in later life began with the publication of studies showing that areas of Britain with the highest rates of neonatal mortality early in the 20th century tended to have the highest rates of coronary heart disease later in the century.[45] David Barker's group then found evidence of links between birth weight, as well as other measurements at birth, and later risk of cardiovascular disease and related diseases, including type 2 diabetes, with smaller babies having higher risk.[46]

There is now strong evidence linking low birth weight with later risk of type 2 diabetes in those living in Western environments.[47] A systematic review of studies investigating this link found that 13 out of 16 reported that babies with lower birth weight had a greater risk of type 2 diabetes.[48] A later meta-analysis confirmed the link between birth weight and type 2 diabetes.[49] These studies have generally focused on babies who are born small for gestational age, but Harder et al.[49] found that babies born early (and therefore small) are also at a greater risk of type 2 diabetes in later life. They suggest that small babies tend to be overfed and note that rapid growth in infancy and early childhood has also been implicated as a risk factor for type 2 diabetes. However, it is difficult to separate the effects of low birth weight and rapid infant growth, since small babies tend to grow particularly fast.[47,50]

There is also a consistent inverse association between birth weight and central or abdominal obesity, such that smaller babies have the more harmful type of fat patterning.[51] Similarly, there is evidence that those born small for gestational age have a higher percentage of body fat than larger babies.[52] Finally, those born small show higher fasting plasma insulin and glucose concentrations and higher levels of insulin resistance in later life than normal-weight babies.[48] All these characteristics can be considered thrifty.

The effect of low birth weight on later diabetes risk appears to be exacerbated by adiposity in later life. Thus, people who are born small, who have a more central type of fat patterning throughout life, and then put on weight and become heavy, have been shown to have the greatest risk of type 2 diabetes.[53] A study in India found that the greatest degree of insulin resistance in men and women was found in those who were born small for gestational age, but who had a high fat mass between the ages of 2 and 12 years.[54] In contrast, in the Gambia small babies maintain excellent cardiovascular health into adulthood, with a complete absence of the metabolic syndrome, so long as they retain their "lean, fit and frugal" lifestyle.[55]

We normally associate metabolic disease with large body size, so how can we explain what appear to be counterintuitive effects of being small in early life? Why should those who are born small or exhibit catch-up growth in early life be at greater risk of type 2 diabetes if exposed to risk factors in later life? Gluckman et al.[56] have distinguished three possible types of answers to this question. The first is mechanistic and does not invoke an evolutionary perspective, but the second two possibilities draw on evolutionary theory.

First, it is possible that developmental changes that impair health in later life result simply from disruptive effects on normal development that cause long-term dysfunction in affected organs or tissues. In the case of type 2 diabetes, nonadaptive associations between markers of poor fetal growth and later disease may arise because the fetus receives insufficient resources, resulting in impaired development of the pancreatic beta-cells that secrete insulin. In those who develop insulin resistance later in life, as is common in Western environments, an impaired ability to secrete insulin will hasten progression to type 2 diabetes.[57] The original thrifty phenotype hypothesis focused on this mechanism[58] but later incorporated more adaptive explanations for the development of thriftiness.[59]

Second, developmental changes that impair health in later life may derive from plasticity that has evolved because it improves short-term survival in early life.[56] Given that the neonatal and weaning periods are high-risk times for humans, selective pressures favouring adjustments that improve survival through these early stages of life will be strong. Hales and Barker emphasise that "nutritional thrift" selectively protects brain growth at critical stages in early development, by maintaining high levels of plasma glucose, and is therefore adaptive.[59] It is also possible that being a small baby with large central fat reserves (abdominal fat is mobilised more easily than subcutaneous fat) increases chances of survival during the first year or so of life.[60]

The possibility that has attracted most interest, however, is that an individual may make changes to his or her developmental trajectory in order to improve long-term reproductive fitness. Here the developing fetus is considered to derive signals about the external environment from the mother and to use these to set developmental pathways that will be beneficial in that particular environment. These have been termed predictive adaptive responses.[56] In energetically stressed environments such predictive adaptive responses may increase thriftiness.[59] Here the suggestion is that the developing fetus is prepared for a difficult environment in which it makes sense to be small, metabolically thrifty, and have a reserve of abdominal fat. It is only if there is a mismatch between the fetal environment and that encountered in later life that this may result in a greater susceptibility to metabolic disease, including type 2 diabetes[61] (p. 173).

In marked contrast, Wells has argued that the development of thriftiness results not as an adaptive response on the part of the fetus, but rather as a result of the mother limiting her investment in each offspring in order to increase her lifelong reproductive success.[42,62] He notes that given the genetic differences between mother and child, a mother with limited resources will apportion them among all her offspring in ways that increase her own fitness even to the detriment of each individual child. In the case of humans, the long period of parental investment means that it may be adaptive for a mother to restrict the energy requirements of each of her offspring by restricting their early growth, setting them on a low growth trajectory, and increasing the thriftiness of their metabolism. She can do this most effectively during the period of early life when she can exert physiological control over these characteristics, when offspring are *in utero* or breastfeeding. Importantly, this idea contrasts with the expectation, implicit in Gluckman and Hanson's proposal, that maternal-infant physiological interaction is always adjusted in favour of the infant[63] (pp. 37–40).

It has also been pointed out that the association between low birth weight and type 2 diabetes could arise from a genetic source, not as a result of developmental plasticity. Thus, the genotype responsible for type 2 diabetes might itself cause retarded growth *in utero*.[47] Hattersley and Tooke proposed the fetal-insulin hypothesis, which suggests that genetically determined insulin resistance may result in impaired growth in the fetus, as well as insulin resistance in adult life.[64] This is plausible because insulin plays an important role in promoting growth in the fetus. This, then, is a variant of the thrifty genotype hypothesis that also accounts for apparent thrifty phenotype effects, as observed in the correlation between small birth weight and risk of type 2 diabetes. However, as with older versions of the thrifty genotype hypothesis, allelic variation that could underpin such a mechanism has not been identified.[47]

Thus, a number of evolutionary explanations for developmental effects on later type 2 diabetes risk in obesogenic Western or Westernising environments have been proposed, and this remains a controversial area. Assessments of these hypotheses must consider that the mechanisms invoked did not evolve in the modern affluent environment. For example, environments in which fetal growth and development are constrained by lack of energy are often implicitly considered as abnormal, but in evolutionary terms it is modern affluent environments that are abnormal.

We should expect thrifty phenotype effects to be most marked in areas of the world experiencing particularly rapid socioeconomic development, such as Asia, where for current cohorts of adults the environment now contrasts enormously with that experienced in early life.[65,66] Researchers are now pursuing the thrifty phenotype hypothesis in India in particular. It makes good sense to do this work in India, since birth weights there are among the lowest in the world,[67] socioeconomic development is rapid, and, as we have seen, rates of type 2 diabetes and heart disease in South Asians living in the West and in urban areas of South Asia are very high. Results have shown that in comparison to babies born in the UK, Indian babies are small, but have a relatively high proportion of central body fat.[24] This thin-fat body composition is maintained in later life.[68] Further, birth weight is inversely associated with insulin resistance in children, and it is those children who were the smallest at birth and the heaviest at the time of testing who have the highest levels of insulin resistance.[69] Similar studies in other areas of Asia have not, as yet, been published. However, since socioeconomic change is also occurring at a very rapid rate in other areas of Asia, we might expect to see similar effects.

The thrifty phenotype hypothesis offers a relatively parsimonious solution to the puzzle posed by the high rates of type 2 diabetes seen in urban Asian populations.

Intergenerational Effects

The thrifty phenotype approach, as formulated above, suggests that in populations in transition, such as those in Asia, disease risk will decline in generations born into more Westernised environments because they will receive the "correct" signals about their later environment during gestation and early life. Birth weight and other features of the newborn are not, however, determined only by the environment experienced by the mother during her pregnancy. They are affected by a number of factors, some of which relate to the mother's experiences earlier in life and even the experiences of women further back in the maternal line. Thus, maternal nutrition, maternal infection (particularly malaria), and cigarette smoking all have important effects during gestation, but so does maternal body size, including both height and weight.[67] This means that the development of the fetus, as crudely reflected by birth weight, is affected by what happens during gestation, but also by whatever has previously determined the mother's body size. Kuzawa[70] has suggested that this is adaptive because the environment experienced during gestation may not accurately reflect the longer-term environment, for example, because of seasonal effects. He suggests that messages are passed down the maternal line from generation to generation to allow the developing baby a longer-term measure of the external environment, creating what he calls intergenerational phenotypic inertia. The implication of this long-term cueing of developmental processes *in utero* is that babies born to women who may themselves live in a Western or Westernising environment, but whose recent maternal ancestors lived in relatively poor environments, are likely to follow developmental pathways that make them vulnerable to the development of type 2 diabetes and cardiovascular disease in later life.

This is exactly the situation in much of Asia. In China, for example, the contrast between the nutritional and wider environmental experiences of recent cohorts of children and their parents and grandparents is enormous. Most of those who are currently grandparents grew up in a time of severe economic hardship and war, including the Sino-Japanese war (1937–1945) and the Chinese civil war (1946–1949), while most of those who are currently parents of children, who grew up under Mao, also experienced extreme food shortages, including the terrible famine of 1959 and the chronic shortages of basic necessities during the Cultural Revolution (1966–1976).[71] During this period people ate mainly grains and vegetables, so far as they were available. Then, during the 1970s, farming systems changed and food supplies increased, and in the 1980s there was a huge increase in the availability and consumption of cooking oil, meat, poultry, eggs, and seafood.[72] Dietary change has continued at a fast pace, with nutrient composition changing to become more like that seen in affluent Western countries. For example, energy intake from fat increased from 22% to 30% between 1992 and 2002 in the country as a whole, while for urban residents fat contributed 35% of energy intake in 2002, more than the World Health Organisation's suggested upper limit of 30%.[21] In China the intergenerational contrast is perhaps particularly marked because of the drastic decline in fertility in recent decades, resulting partly from the one-child policy, in operation since 1979.[73] It is often said in China that the resulting only children are treated as "little emperors" by their parents and grandparents.[72]

Unfortunately, even the slow washing-out process implied by these ideas is likely to be impeded by the effects of high levels of impaired glucose tolerance or even diabetes during pregnancy in mothers in Westernising Asian environments, which also have effects on the developing fetus. Women with impaired glucose tolerance or diabetes give birth to large babies, who have an increased risk of type 2 diabetes in later life.[74] For this reason, the association between birth weight and later risk of type 2 diabetes is not linear as implied by some studies, but U shaped, with both the smallest and largest babies having increased risk.[49] For example, the offspring of Pima women from Arizona who were diabetic during pregnancy have a particularly high prevalence of young-onset type 2 diabetes.[75] The risk is increased compared with offspring of diabetic fathers and siblings born before the onset of maternal diabetes,[75,76] suggesting that it is caused at least partially by the uterine environment.[77] Thus, Fall[65] suggests that the rise in type 2 diabetes occurring in Westernising countries may be self-perpetuating, predicting that the epidemic will be sustained in those born to the generation who are experiencing rapid lifestyle change. This effect is expected whether thriftiness is genetically or developmentally acquired, as shown in figure 4.2.

Such effects are being investigated in an ongoing study in South India, which is following a cohort of children born in Mysore in 1997–1998.[78] This study found that newborns of diabetic mothers were larger in all body measurements than babies with nondiabetic parents. There was also evidence that milder levels of maternal hyperglycaemia, in nondiabetic women, stimulate insulin secretion and growth in the foetus.[78] At 5 years of age, daughters of diabetic mothers had greater adiposity and higher glucose and insulin concentrations than daughters of nondiabetic parents.[77] It is not clear why the effect was not seen in boys, but the small sample size (forty-one) of children born to diabetic parents may be to blame.

Recent work suggests that transgenerational effects on risk of metabolic syndrome may also be transmitted via epigenetic mechanisms.[79] Here, the environment experienced by an individual, especially in early life, is thought to affect the expression of his or her genes in ways that are heritable by future generations. It seems likely that future work to understand these mechanisms will be fruitful in establishing inherited and early life effects on vulnerability to type 2 diabetes.

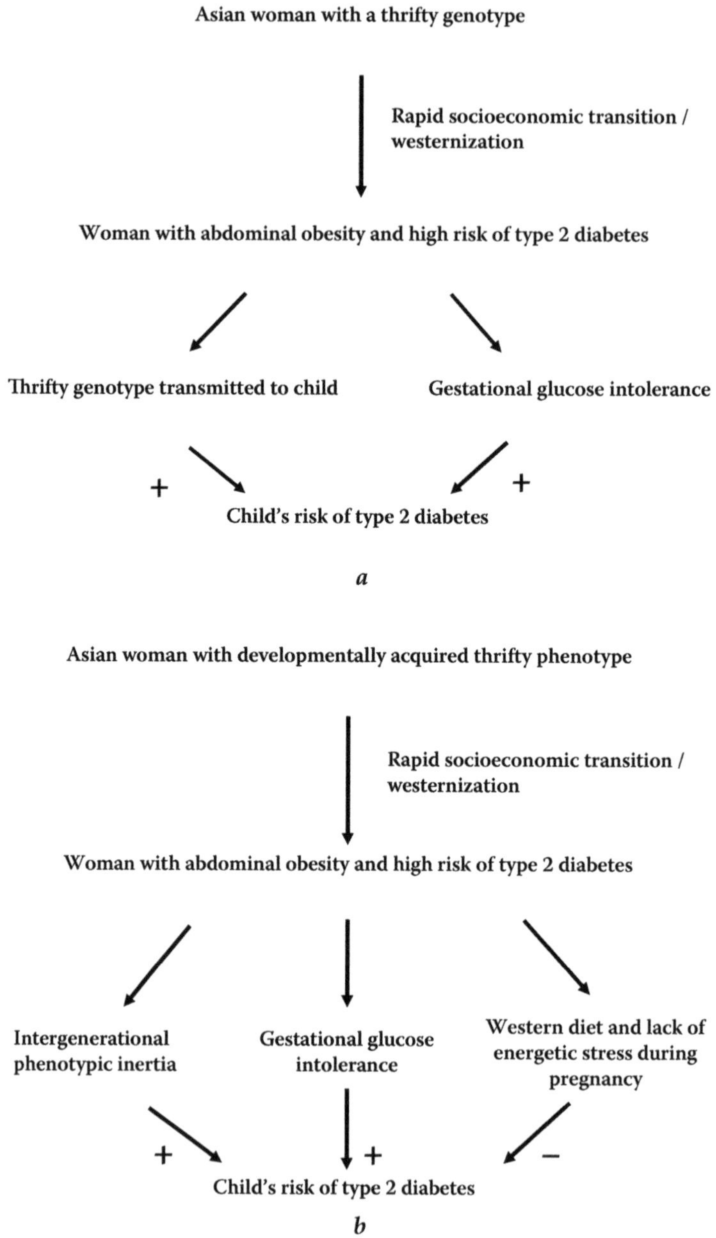

Asian woman with a thrifty genotype

Rapid socioeconomic transition / westernization

Woman with abdominal obesity and high risk of type 2 diabetes

Thrifty genotype transmitted to child   Gestational glucose intolerance

+   Child's risk of type 2 diabetes   +

a

Asian woman with developmentally acquired thrifty phenotype

Rapid socioeconomic transition / westernization

Woman with abdominal obesity and high risk of type 2 diabetes

Intergenerational phenotypic inertia   Gestational glucose intolerance   Western diet and lack of energetic stress during pregnancy

+   Child's risk of type 2 diabetes   +   −

b

Figure 4.2 Possible mechanisms for intergenerational transmission of type 2 diabetes risk in Asian populations: a) applying the thrifty genotype hypothesis; b) applying current thinking about the developmental origins of type 2 diabetes.

Inflammation and Infection

Inflammation, part of the suite of responses that constitute the "innate immunity," acts as the body's first line of defense against infection or injury.[80] For example, in the case of a wound, a key role of the inflammatory response is to neutralise bacterial pathogens and remove them.[81] Chronic inflammation is now known to be closely associated with insulin resistance and type 2 diabetes risk.[82,83] Several studies among healthy South Asian residents in the UK have shown higher concentrations of acute phase proteins associated with the inflammatory response in this group compared to their white counterparts, suggesting greater levels of chronic subclinical inflammation.[84,85] Similarly, elevated levels of markers for inflammation have been described among young healthy adolescents in India.[86]

It seems likely that evolutionarily novel elements of the Western or Westernising environment induce an inappropriate inflammatory response. For example, cigarette smoking is also strongly associated with low-grade systemic inflammation,[87] and Black[88] has argued that chronic psychological stress may be an important factor. However, most other authors have focused on the role of excess adipose tissue.[89,90] It is now well established that obesity is associated with systemic inflammation and that adipose tissue, particularly the visceral adipose tissue that contributes importantly to abdominal obesity, produces and releases a variety of pro-inflammatory and anti-inflammatory factors.[91] The pro-inflammatory molecules produced by adipose tissue are thought to be actively involved in the development of insulin resistance.[91] In turn, insulin resistance appears to contribute to inflammation via effects on the liver, which plays an important role in inflammatory responses,[92] so that there may be positive feedback between insulin resistance and inflammation. Differences in levels of inflammatory markers between people of South Asian and European origin in the West are partly accounted for by greater central obesity among South Asians.[84,85]

It is also possible that exposure to infection may increase type 2 diabetes risk via inflammatory effects. As expected, positive correlations have been found between pathogen burden and increased levels of pro-inflammatory biomarkers C-reactive protein, with a particular focus on the effects of herpes viruses, cytomegalovirus, hepatitis viruses, and some bacterial infections, including *Chlamydia pneumoniae*, *Helicobacter pylori*, and *Porphyomonas gingivalis*.[93,94] Infection with several of these pathogens, which are all persistent within the body, is well established as a risk factor for atherosclerosis and coronary heart disease.[95,96] There is also now evidence that these infections are associated with insulin resistance, and that inflammation may mediate this association.[97] Chronic infection is therefore a likely risk factor for type 2 diabetes. More specifically, infection with hepatitis C has a well-established association with type 2 diabetes, including in Asian populations.[98–100] This appears to be partly because of the direct actions of the virus on the liver and the pancreas, but there is also evidence that inflammation may be responsible for an insulin resistance state in those with the virus.[101,102]

Thus, obesity (particularly abdominal obesity) cigarette smoking, and infection are all likely to contribute to type 2 diabetes risk in Asian populations via their association with inflammation. We have already seen that Asian populations have disproportionately high levels of abdominal obesity, given low, but increasing, levels of BMI. Rates of smoking are also very high among men, and increasing among women.[103] Infectious diseases are generally more prevalent in poorer than in more affluent parts of the world. For example, around 50% of adults may be infected with *H. pylori* in richer countries, whereas the figure is more likely to be 90% in poorer countries,[104] and similar differences in prevalence are estimated for *C. pneumoniae*.[105] Hepatitis C is also more common in Asia than it is in the West.[106] Thus, the hypothesis that a higher burden of infection and associated inflammation in Asian populations may contribute to high rates of type 2 diabetes, when Western lifestyle risk factors are adopted, is plausible. This suggestion has been made specifically in relation to people of South Asian origin living in the UK, where levels of serum gamma globulin, a nonspecific measure of immune activation, suggested a higher exposure to infection than in people of European origin.[107] Migrants and those born in Britain were not distinguished in this analysis, but it is likely that exposure to infection will be different in these two groups.

It has also been argued that certain populations, with greater ancestral exposure to pathogens, may have a genetic propensity to mount a strong inflammatory response, which might increase the risk of type 2 diabetes on exposure to classic risk factors for type 2 diabetes.[92] This idea has excited considerable interest, as have similar suggestions that ancestral exposure to helminths, more common in tropical areas, may have exerted selective pressure on populations from these regions, resulting in more vulnerability to the type of inflammation characteristic of allergy.[108] These ideas remain, however, speculative. Furthermore, it is possible, in line with the thrifty genotype/thrifty phenotype contrast, that predispositions to abdominal adiposity and inflammation are a consequence of developmental effects determining the distribution or activity of adipose tissue and inflammation, possibly in an adaptive fashion.

Understanding of relationships among inflammation, adipose tissue, and insulin resistance is developing fast. It seems likely that we will soon be able to evaluate some of the ideas expressed above, and that an evolutionary perspective on the role of inflammation and its interaction with adipose tissue will be important in improving our understanding of the type 2 diabetes epidemic in Asia and other parts of the world. Since processes regulating inflammation and thriftiness are closely intertwined, it is also likely that we will need to consider both together, rather than separately.[92]

Practical Implications

Given the strength of the evidence showing developmental influences on disease risk, policies aimed at slowing the rate of increase in type 2 diabetes in Asia should be formulated to take account of these effects. Clearly, it is important to target the developing foetus and young child in order to reduce his or her future susceptibility to disease. The importance of good antenatal care and promoting a healthy maternal diet in order to encourage appropriate foetal development in populations with a high risk of type 2 diabetes has previously been emphasised.[109,110] In general, these strategies should be aimed at reducing the risk of either impaired development or development adapted to a predicted poor environment, as indicated by low birth weight. However, in women at risk of gestational diabetes, more specific interventions, with known benefits to the type 2 diabetes risk of her offspring, can be implemented,[109] although the kind of antenatal intervention needed to have this effect may be beyond the scope of many Asian health care systems.

The potential of such programmes should not, however, be overemphasised, as longer-term influences on foetal development will limit the impact of short-term interventions during pregnancy.[110] As we have seen, the development of the foetus is also affected by whatever has previously determined the mother's size, and at its simplest, smaller women have smaller babies. Thus, the nutrition and health of women need to be targeted not just during pregnancy, but also much earlier in the life course.[65] This difficult task has added complexity in the context of populations in transition, where undernutrition and overnutrition coexist in the population, and even in the same household.[111,112] In promoting growth it is important not to promote obesity at the same time. Dietary quality, for example the consumption of sufficient protein, fruits, and vegetables, may be especially important, to promote growth, health, and development in children and to help prevent obesity in adults.[111] There have been efforts in this direction in China.[113]

In his exploration of the impact of ideas about disease causation on diabetes prevention programmes in Native American communities, Benyshek also notes the importance of sensitive transmission of biomedical understandings. For example, he cites a study conducted in communities in California and southern Arizona, which found that "inheritance" or "genetics" was often cited as a reason for high levels of type 2 diabetes, and links this with research identifying views in some of the same communities that diabetes is inevitable and uncontrollable.[114,115] Given the dominance of the thrifty genotype hypothesis in the academic community for several decades, these findings are not surprising. They illustrate, however, the need for academics and policy makers to be aware of the impact that messages about causation can have. For example, policies based on thrifty phenotype effects must be careful not to shift responsibility to pregnant women.[109]

Conclusion

Increasing rates of type 2 diabetes in Asia are a cause of considerable concern, especially in urban India. It is possible that an improved understanding of the susceptibility of Asian populations to type 2 diabetes will help us tackle this epidemic. From an evolutionary perspective, the two dominant explanations for this susceptibility are the thrifty genotype and thrifty phenotype hypotheses. Among Asians, the thrifty genotype hypothesis has been applied particularly to people of South Asian origin. However, progress to identify genes that can account for between-population differences in susceptibility has been slow, bringing the standard thrifty genotype hypothesis, as modified from Neel's original work, into question. The position in relation to the thrifty phenotype hypothesis is very different. In this case, there is good evidence that the mechanisms exist, and theories about why they exist have been elaborated subsequently, with considerable debate about the validity of the adaptationist arguments proposed. Thus, we await further work to perhaps identify thrifty genes in Asians or, more likely, nonthrifty genes in Europeans, and we continue to debate whether developmental effects are adaptive. Further understanding of the ultimate, evolutionary mechanisms underlying develomental effects will increase our understanding of how they work. In the meantime, preventive strategies should focus less on putative thrifty genotype effects and more on better-established developmental effects.

An evolutionary perspective also has the potential to help us understand the role of inflammation in type 2 diabetes, and the relationships among adiposity, inflammation, and insulin resistance. It may be that some populations have more inflammatory tendencies than others because of the selective pressures imposed by the disease ecology experienced by their ancestors, or developmental experiences may affect the inflammatory profile of individuals in later life.

The full explanation is likely to involve a combination of factors affecting metabolism over the life span, and will need to explain not only rapidly rising rates of the disease across Asia, but also variability in prevalence within Asia.

References

1. International Diabetes Federation. 2007. *Diabetes atlas.* http://www.eatlas.idf.org/About_e_Atlas/ (accessed September 21, 2007).
2. World Health Organization. 1999. *Definition, diagnosis and classification of diabetes mellitus and its complications: Report of a WHO consultation. Part 1. Diagnosis and classification of diabetes mellitus.* Geneva: World Health Organisation.
3. Klein, R., and Klein, B. E. K. 1997. Diabetic eye disease. *Lancet* 350:197.
4. Atkins, R. C. 2005. The epidemiology of chronic kidney disease. *Kidney Int.* 67:S14.
5. The Global Lower Extremity Amputation Study Group. 2000. Epidemiology of lower extremity amputation in centres in Europe, North America and East Asia. *Br. J. Surg.* 87:328.

6. Stamler, J., et al. 1993. Diabetes, other risk factors, and 12-yr cardiovascular mortality for men screened in the Multiple Risk Factor Intervention Trial. *Diabetes Care* 16:434.
7. Roper, N. A., et al. 2002. Cause-specific mortality in a population with diabetes—South Tees Diabetes mortality study. *Diabetes Care* 25:43.
8. Roper, N. A., et al. 2001. Excess mortality in a population with diabetes and the impact of material deprivation: Longitudinal, population based study. *BMJ* 322:1389.
9. Wild, S., et al. 2004. Global prevalence of diabetes: Estimates for the year 2000 and projections for 2030. *Diabetes Care* 27:1047.
10. Chuang, L.-M., et al. 2002. The status of diabetes control in Asia—A cross-sectional survey of 24, 317 patients with diabetes mellitus in 1998. *Diabet. Med.* 19:978.
11. Must, A., et al. 1999. The disease burden associated with overweight and obesity. *JAMA* 282:1523.
12. Ramachandran, A., et al. 2001. High prevalence of diabetes and impaired glucose tolerance in India: National Urban Diabetes Survey. *Diabetologia* 44:1094.
13. Wild, S. H., and Byrne, C. D. 2006. Risk factors for diabetes and coronary heart disease. *BMJ* 333:1009.
14. Yoon, K.-H., et al. 2006. Epidemic obesity and type 2 diabetes in Asia. *Lancet* 368:1681.
15. Gregg, E. W., et al. 2004. Trends in the prevalence and ratio of diagnosed to undiagnosed diabetes according to obesity levels in the U.S. *Diabetes Care* 27:2806.
16. King, H., Aubert, R. E., and Herman, W. H. 1998. Global burden of diabetes, 1995–2025. *Diabetes Care* 21:1414.
17. Asia Pacific Cohort Studies Collaboration. 2007. The burden of overweight and obesity in the Asia-Pacific region. *Obes. Rev.* 8:191.
18. Popkin, B. M. 1999. Urbanization, lifestyle changes and the nutrition transition. *World Dev.* 17:1905.
19. Popkin, B. M. 2002. Part II: What is unique about the experience in lower- and middle-income less-industrialised countries compared with the very-high-income industrialised countries? The shift in stages of the nutrition transition in the developing world differs from past experiences. *Public Health Nutr.* 5:205.
20. Pingali, P. 2007. Westernization of Asian diets and the transformation of food systems: Implications for research and policy. *Food Policy* 32:281.
21. Lieberman, L. S. 2003. Dietary, evolutionary, and modernizing influences on the prevalence of type 2 diabetes. *Ann. Rev. Nutr.* 23:345.
22. Wang, Y. Z., et al. 2007. Is China facing an obesity epidemic and the consequences? The trends in obesity and chronic disease in China. *Int. J. Obes.* 31:177.
23. Choi, Y. J., et al. 2006. Rapidly increasing diabetes-related mortality with socio-environmental changes in South Korea during the last two decades. *Diabetes Res. Clin. Pract.* 74:295.
24. Wu, Y. 2006. Overweight and obesity in China. *BMJ* 333:362.
25. Yajnik, C., et al. 2002. Adiposity and hyperinsulinemia in Indians are present at birth. *J. Clin. Endocrinol. Metab.* 87:5575.
26. WHO Expert Consultation. 2004. Appropriate body-mass index for Asian populations and its implications for policy and intervention strategies. *Lancet* 363:157.
27. Yajnik, C. S., et al. 2003. Neonatal anthropometry: The thin-fat Indian baby. The Pune Maternal Nutrition Study. *Int. J. Obes. Relat. Metab. Disord.* 27:173.

28. Neel, J. V. 1962. Diabetes mellitus: A "thrifty" genotype rendered detrimental by "progress"? *Am. J. Hum. Genet.* 14:353.
29. Neel, J. V. 1982. The thrifty genotype revisited. In *The genetics of diabetes mellitus,* ed. J. Köbberling and R. Tattersall, 137. Amsterdam: Academic Press.
30. Neel, J. V., Weder, A. B., and Julius, S. 1998. Type II diabetes, essential hypertension, and obesity as "syndromes of impaired genetic homeostasis": The "thrifty genotype" hypothesis enters the 21st century. *Perspect. Biol. Med.* 42:44.
31. Stannard, S. R., and Johnson, N. A. 2003. Insulin resistance and elevated triglyceride in muscle: More important for survival than "thrifty" genes? *J. Physiol.* 554:595.
32. Zimmet, P., et al. 1977. The high prevalence of diabetes mellitus on a central Pacific island. *Diabetologia* 13:111.
33. Baker, P. T. 1984. Migration, genetics, and the degenerative diseases of South Pacific Islanders. In *Migration and mobility: Biosocial aspects of human movement,* ed. A. J. Boyce, 209. London: Taylor & Francis.
34. McKeigue, P. M., Shah, B., and Marmot, M. G. 1991. Relation of central obesity and insulin resistance with high diabetes prevalence and cardiovascular risk in South Asians. *Lancet* 337:382.
35. Williams, B. 1995. Westernised Asians and cardiovascular disease: Nature or nurture? *Lancet* 345:401.
36. Wells, J. C. K. 2007. Commentary; Why are South Asians susceptible to central obesity? The El Niño hypothesis. *Int. J. Epidemiol.* 36:226.
37. Kagawa, Y., et al. 2002. Single nucleotide polymorphisms of thrifty genes for energy metabolism: Evolutionary origins and prospects for intervention to prevent obesity-related diseases. *Biochem. Biophys. Res. Commun.* 295:207.
38. Barroso, I. 2005. The genetics of type 2 diabetes. *Diabetic Med.* 22:517.
39. Abate, N., et al. 2003. Genetic polymorphism PC-1 K121Q and ethnic susceptibility to insulin resistance. *J. Clin. Endocrinol. Metab.* 88:5927.
40. Carulli, L., et al. 2005. Diabetes, genetics and ethnicity. *Aliment. Pharmacol. Ther.* 22 (Suppl. 2):16.
41. Allen, J. S., and Cheer, S. M. 1996. The non-thrifty genotype. *Curr. Anthropol.* 37:831.
42. Wells, J. C. K. 2007. The thrifty phenotype as an adaptive maternal effect. *Biol. Rev.* 82:143.
43. Young, J. H., et al. 2005. Differential susceptibility to hypertension is due to selection during the out-of-Africa expansion. *PLoS Genet.* 1:730.
44. Paradies, Y. C., Montoya, M. J., and Fullerton, S. M. 2007. Racialised genetics and the study of complex diseases: The thrifty genotype revisited. *Perspect. Biol. Med.* 50:203.
45. Barker, D. J. P., and Osmond, C. 1986. Infant mortality, childhood nutrition and ischaemic heart disease in England and Wales. *Lancet* I:1077.
46. Barker, D. J. P. 1994. *Mothers, babies, and disease in later life.* London: BMJ Publishing Group.
47. Simmons, R. 2005. Developmental origins of adult metabolic disease: Concepts and controversies. *Trends Endocrinol. Metab.* 16:390.
48. Newsome, C. A., et al. 2003. Is birth weight related to later glucose and insulin metabolism? A systematic review. *Diabetic Med.* 20:339.
49. Harder, T., et al. 2007. Birth weight and subsequent risk of type 2 diabetes: A meta-analysis. *Am. J. Epidemiol.* 165:849.
50. Singhal, A., and Lucas, A. 2004. Early origins of cardiovascular disease: Is there a unifying hypothesis? *Lancet* 363:1642.
51. Oken, E., and Gillman, M. W. 2003. Fetal origins of obesity. *Obes. Res.* 11:496.

52. Jacquet, D., et al. 2000. Insulin resistance early in adulthood in subjects born with intrauterine growth retardation. *J. Clin. Endocrinol. Metab.* 85:1401.
53. Rich-Edwards, J. W., et al. 1999. Birthweight and the risk for type 2 diabetes mellitus in adult women. *Ann. Int. Med.* 130:278.
54. Bhargava, S. K., et al. 2004. Relation of serial changes in childhood body-mass index to impaired glucose tolerance in young adulthood. *N. Engl. J. Med.* 350:865.
55. Moore, S. E., et al. 2001. Glucose, insulin and lipid metabolism in rural Gambians exposed to early malnutrition. *Diabetic Med.* 18:645.
56. Gluckman, P. D., et al. 2005. Environmental influences during development and their later consequences for health and disease: Implications for the interpretation of empirical studies. *Proc. R. Soc. London B* 272:671.
57. Weir, G., et al. 2001. β-cell adaptation and decompensation during the progression to diabetes. *Diabetes* 50 (Suppl. 1):S154.
58. Hales, C. N., and Barker, D. J. P. 1992. Type 2 (non-insulin-dependent) diabetes mellitus: The thrifty phenotype hypothesis. *Diabetologia* 35:595.
59. Hales, C. N., and Barker, D. J. P. 2001. The thrifty phenotype hypothesis. *Br. Med. Bull.* 60:5.
60. Kuzawa, C. W. 1998. Adipose tissue in human infancy and childhood: An evolutionary perspective. *Yrbk. Phys. Anthropol.* 41:177.
61. Gluckman, P., and Hanson, M. 2006. *Mismatch: Why our world no longer fits our bodies.* Oxford: Oxford University Press.
62. Wells, J. C. K. 2003. The thrifty phenotype hypothesis: Thrifty offspring or thrifty mother? *J. Theor. Biol.* 221:143.
63. Gluckman, P., and Hanson, M. 2005. *The fetal matrix: Evolution, development and disease.* Cambridge: Cambridge University Press.
64. Hattersley, A. T., and Tooke, J. E. 1999. The fetal insulin hypothesis: An alternative explanation of the association of low birthweight with diabetes and vascular disease. *Lancet* 353:1789.
65. Fall, C. H. D. 2001. Non-industrialised countries and affluence. *Br. Med. Bull.* 60:33.
66. Pollard, T. M. 1997. Environmental change and cardiovascular disease: A new complexity. *Yrbk. Phys. Anthropol.* 40:1.
67. Kramer, M. S. 1987. Determinants of low birth weight: Methodological assessment and meta-analysis. *Bull. WHO* 65:663.
68. Yajnik, C. 2004. Early life origins of insulin resistance and type 2 diabetes in India and other Asian countries. *J. Nutr.* 134:205.
69. Yajnik, C. 2000. Interactions of perturbations in intrauterine growth and growth during childhood on the risk of adult-onset disease. *Proc. Nutr. Soc.* 59:257.
70. Kuzawa, C. W. 2005. Fetal origins of developmental plasticity: Are fetal cues reliable predictors of future nutritional environments? *Am. J. Hum Biol.* 17:5.
71. Yuhua, G. 2000. Family relations: The generation gap at the table. In *Feeding China's little emperors*, ed. J. Jing, 94. Stanford, CA: Stanford University Press.
72. Jing, J. 2000. Food, children, and social change in contemporary China. In *Feeding China's little emperors*, ed. J. Jing, 1. Stanford, CA: Stanford University Press.
73. Retherford, R. D., et al. 2005. How far has fertility in China really declined? *Pop. Dev. Rev.* 31:57.
74. Drake, A. J., and Walker, B. R. 2004. The intergenerational effects of fetal programming: Non-genomic mechanisms for the inheritance of low birth weight and cardiovascular risk. *J. Endocrinol.* 180:1.
75. Pettitt, D. J., et al. 1988. Congenital susceptibility to NIDDM: Role of intrauterine environment. *Diabetes* 37:622.

76. Dabelea, D., et al. 2000. Intrauterine exposure to diabetes conveys risks for type 2 diabetes and obesity: A study of discordant sibships. *Diabetes* 49:2208.
77. Krishnaveni, G. V., et al. 2005. Anthropometry, glucose tolerance, and insulin concentrations in Indian children. *Diabetes Care* 28:2919.
78. Hill, J. C., et al. 2005. Glucose tolerance in pregnancy in South India: Relationships to neonatal anthropometry. *Acta Obstet. Gynecol. Scand.* 84:159.
79. Gallou-Kabani, C., and Junien, C. 2005. Nutritional epigenomics of metabolic syndrome. *Diabetes* 54:1899.
80. Schwarzenberg, S. J., and Sinaiko, A. R. 2006. Obesity and inflammation in children. *Paed. Resp. Rev.* 7:239.
81. Parenteau, N. L., and Hardin-Young, J. 2007. The biological mechanisms behind injury and inflammation: How they can affect treatment strategy, product performance, and healing. *Wounds* 19:87–96.
82. Pickup, J. C. 2004. Inflammation and activated innate immunity in the pathogenesis of type 2 diabetes. *Diabetes Care* 27:813.
83. Wu, W., et al. 2003. The role of C-reactive protein in type 2 diabetes and its macrovascular complications in Chinese. *Diabetes Res. Clin. Pract.* 61:143.
84. Chambers, J. C., et al. 2001. C-reactive protein, insulin resistance, central obesity, and coronary heart disease risk in Indian Asians from the United Kingdom compared with European whites. *Circulation* 104:145.
85. Somani, R., et al. 2006. Complement C3 and C-reactive protein are elevated in South Asians independent of a family history of stroke. *Stroke* 37:2001.
86. Vikram, N. K., et al. 2003. Correlations of C-reactive protein levels with anthropometric profile, percentage of body fat and lipids in healthy adolescents and young adults in urban North India. *Atherosclerosis* 168:305.
87. Yanbaeva, D. G., et al. 2007. Systemic effects of smoking. *Chest* 131:1557.
88. Black, P. H. 2006. The inflammatory consequences of psychologic stress: Relationship to insulin resistance, obesity, atherosclerosis and diabetes mellitus, type II. *Med. Hypotheses* 67:879.
89. De Lorenzo, A., et al. 2007. Normal-weight obese syndrome: Early inflammation? *Am. J. Clin. Nutr.* 85:40.
90. Tilg, H., and Moschen, A. R. 2006. Adipocytokines: Mediators linking adipose tissue, inflammation and immunity. *Nat. Rev. Immunol.* 6:772.
91. Fantuzzi, G. 2005. Adipose tissue, adipokines, and inflammation. *J. Allergy Clin. Immunol.* 115:911.
92. Fernandez-Real, J. M., and Ricart, W. 1999. Insulin resistance and inflammation in an evolutionary perspective: The contribution of cytokine genotype/phenotype to thriftiness. *Diabetologia* 42:1367.
93. Moutsopoulos, N. M., and Madianos, P. N. 2006. Low-grade inflammation in chronic infectious diseases: Paradigm of periodontal infections. *Ann. N.Y. Acad. Sci.* 1088:251.
94. Zhu, J. H., et al. 2000. Effects of total pathogen burden on coronary artery disease risk and C-reactive protein levels. *Am. J. Cardiol.* 85:140.
95. Espinola-Klein, C., et al. 2002. Impact of infectious burden on extent and long-term prognosis of atherosclerosis. *Circulation* 105:15.
96. Smieja, M., et al. 2003. Multiple infections and subsequent cardiovascular events in the Heart Outcome Prevention Evaluation (HOPE) study. *Circulation* 107:251.
97. Fernández-Real, J. M., et al. 2006. Burden of infection and insulin resistance in healthy middle-aged men. *Diabetes Care* 29:1058.
98. Arao, M., et al. 2003. Prevelence of diabetes mellitus in Japanese patients infected chronically with hepatitis C virus. *J. Gastroenterol.* 38:355.

99. Mason, A. L., et al. 1999. Association of diabetes mellitus and chronic heptatis C virus infection. *Hepatology* 29:328.

100. Ryu, J. K., et al. 2001. Association of chronic hepatitis C virus infection and diabetes mellitus in Korean patients. *Kor. J. Int. Med.* 16:28.

101. Shaheen, M., et al. 2007. Hepatitis C, metabolic syndrome, and inflammatory markers: Results from the Third National Health and Nutrition Examination Survey [NHANES III]. *Diabetes Res. Clin. Pract.* 75:320.

102. Maeno, T., et al. 2003. Mechanisms of increased insulin resistance in non-cirrhotic patients with chronic hepatitis C virus infection. *J. Gastroentorol. Hepatol.* 18:1358.

103. Asia Pacific Cohort Studies Collaboration. 2005. Smoking, quitting, and the risk of cardiovascular disease among women and men in the Asia-Pacific region. *Int. J. Epidemiol.* 34:1036.

104. Webb, P. M., et al. 1994. Relation between infection with *Helicobacter pylori* and living conditions in childhood: Evidence for person to person transmission in early life. *BMJ* 308:750.

105. Cook, P. J., and Honeybourne, D. 1995. Clinical aspects of *Chlamydia pneumoniae* infection. *Presse Med.* 24:278.

106. Kim, W. R. 2002. Global epidemiology and burden of hepatitis C. *Microbes Infect.* 4:1219.

107. Fischbacher, C. M., et al. 2003. IgG is higher in South Asians than Europeans: Does infection contribute to ethnic variation in cardiovascular disease? *Arterioscler. Thromb. Vasc. Biol.* 23:703.

108. Le Souëf, P. N., Goldblatt, J., and Lynch, N. R. 2000. Evolutionary adaptation of inflammatory immune responses in human beings. *Lancet* 356:242.

109. Benyshek, D. C. 2005. Type 2 diabetes and fetal origins: The promise of prevention programs focusing on prenatal health in high prevalence Native American communities. *Hum. Org.* 64:192.

110. James, W. P. T. 2002. Will feeding mothers prevent the Asian metabolic syndrome epidemic? *Asia Pac. J. Clin. Nutr.* 11:S516.

111. Garrett, J., and Ruel, M. T. 2005. The coexistence of child undernutrition and maternal overweight: Prevalence, hypotheses, and programme and policy implications. *Matern. Child Nutr.* 1:185.

112. Doak, C. M., et al. 2002. Overweight and underweight coexist within households in Brazil, China and Russia. *J. Nutr.* 130:2965.

113. Zhai, F., et al. 2002. What is China doing in policy-making to push back the negative aspects of the nutrition transition? *Public Health Nutr.* 5(1A):269.

114. Weiner, D. 1999. Ethnogenetics: Interpreting ideas about diabetes and inheritance. *Am. Ind. Cult. Res. J.* 23:155.

115. Kozak, D. L. 1996. Surrendering to diabetes: An embodied response to perceptions of diabetes and death in the Gila River Indian community. *Omega J. Death Dying* 35:347.

5

Seasonality, Climatic Unpredictability, Food Deprivation, and Polycystic Ovary Syndrome

Laurence Shaw
Reproductive Medicine and Endoscopic Surgery,
London Bridge Fertility Gynaecology and Genetics Centre, London

Sarah Elton
Functional Morphology and Evolution Unit,
Hull York Medical School, University of Hull

Contents

Introduction

Stein and Leventhal's finding of polycystic ovaries, which they eponymously described as part of a circumscribed clinical syndrome,[1] has evolved to become a nebulous entity with polycystic ovaries as a prominent component. Biochemical aspects, both endocrine and nonendocrine, have become part of a wider polycystic ovary syndrome (PCOS), recognised as a very common cause of anovulatory infertility.[2-4] While the original Stein-Leventhal syndrome described multiple ovarian cysts, obesity, hirsutism, menstrual abnormalities, and amenorrhoea, symptoms of PCOS also include acne and androgenic alopecia. Obesity, when it occurs, tends to include centrally distributed adipose tissue. These features, singly or together, are not evident in all affected women.[5] Despite this, most show a degree of insulin resistance.[6] PCOS is thus a complex endocrine disorder, for which the diagnosis is not always straightforward.[6,7] However, many practitioners now agree that it can be defined on the basis of at least two of the following features in combination: presence of polycystic ovaries on ultrasound examination, oligo- or anovulation, and clinical or biochemical evidence of androgen excess.[5]

In PCOS, interference with the pituitary ovarian axis results in over-recruitment of follicles to a size of about 5 mm, with no subsequent selection and maturation of a single leading follicle. The resultant cohort of many 5 mm follicles align themselves under the surface of the ovary, but the oocytes never mature and erupt. Ovulation does not occur; luteinizing hormone (LH) remains elevated but never surges. Polycystic ovaries overproduce androgens, although some of the excess androgens observed in women with polycystic ovaries could be adrenal in origin.[6] Elevated androgen levels in women may lead to hirsutism and menstrual irregularities.[8] High levels of androgens also appear to be related to increased risk of cardiovascular disease, high insulin levels, and type 2 diabetes.[9,10] In addition, hyperinsulinaemia can act to further stimulate ovarian androgen production, resulting in an interaction effect between hyperandrogenism and hyperinsulinaemia,[8] although there is no evidence of a similar effect on adrenal androgens.[11] Obesity, observed in a large proportion of women with polycystic ovaries, exacerbates these endocrine disturbances. Weight loss can lead to a reduction in insulin resistance and an improvement in endocrine status that often results in resumption of ovulation.[6] In clinical practise, women with PCOS seeking fertility treatment are frequently advised that weight loss is the first avenue that should be explored.[12,13]

Prevalence of Polycystic Ovaries and PCOS and Phenotypic Differences in the Expression of Associated Traits

The presence of polycystic ovaries does not automatically imply the presence of the associated syndrome. Based on studies conducted in the UK and Australasia, polycystic ovaries themselves (but not necessaily PCOS) are found in over a fifth of females (see review by Hart et al.[5]) and do not necessarily result in diminished fertility.[14] In some populations the prevalence of polycystic ovaries has been observed to be much higher than one in five, with the highest recorded to date, over half, being in randomly selected South Asians living in the UK.[15] Some nonrandomly selected samples also show interpopulation differences. Polycystic ovaries were observed in 45% of Yoruba women from western Nigeria attending a UK fertility clinic, in contrast to 11% in the infertile European women studied.[16] However, other studies examining the prevalence of polycystic ovaries in women seeking fertility treatment recorded no differences between ethnic groups.[17]

Based on UK and Australasian studies, around one in twelve females have polycystic ovary syndrome.[5] Although differential diagnosis of PCOS hampers rigorous interpopulation comparisons, a body of evidence is accumulating that suggests variations in the prevalence of the syndrome exist across different cultural or geographic populations[4,18–20] (see also table 5.1). In the UK, there appears to be a higher prevalence of PCOS in British South Asian women than in their European counterparts.[4] This is also the case in New Zealand, where Indian women with PCOS are overrepresented compared to the general population.[19] An increased prevalence in Caribbean Hispanic and Mexican American women,[21,22] as well as in Aboriginal Australians,[20] has also been suggested. In contrast, one study has found that Chinese women with PCOS living in New Zealand were underrepresented compared to the population as a whole.[19]

Population-based phenotypic differences in the expression of traits associated with PCOS have also been observed. One study of women with PCOS from different populations (Japanese, Italian, and U.S.) indicated that although adrenal androgen excess and insulin resistance appeared to be at similar levels in all three populations, levels of obesity and hirsutism showed considerable interpopulation variation.[23] This is probably related to proximate environmental influences as well as the biological variation observed in the general populations from which these women came,[23] with people from Japan, for example, being on the whole less hirsute than those from the Mediterranean. In some cross-population studies, however, significant endocrine differences have been observed. Mexican American women with PCOS are more insulin resistant and show higher incidences of insulin resistance than their European counterparts,[24] as well as having lower circulating levels of dehydroepiandrosterone sulphate (DHEAS), an indicator of adrenal androgen production.[11] European women from Iceland have higher androstenedione, lower testosterone, less acne, and are less hirsute than European

Table 5.1 Prevalence of Polycystic Ovaries (PCO) and Polycystic Ovary Syndrome (PCOS) in Different Populations

	Population	% Prevalence	Reference
PCO	General population (Western)	21–23%	Sources cited in 5
	British South Asian	52%	15
PCOS	General population (Western)	8%	5
	European New Zealanders	In proportion to general population	19
	Maori New Zealanders	In proportion to general population	19
	Pacific Island New Zealanders	In proportion to general population	19
	Chinese New Zealanders	Underrepresented	19
	Indian New Zealanders	Overrepresented	19
	Aboriginal Australians	26% (preliminary data)	20
	Mexican Americans	13% (self-reported)	22
	Greeks (Lesbos)	7%	18

Note: Polycystic ovaries can be present without the associated syndrome. Current consensus on the diagnosis of polycystic ovary syndrome relies on the presence of at least two of the following in combination: identification of polycystic ovaries on ultrasound, oligo- or anovulation, and biochemical evidence of androgen excess.[5] However, since the syndrome is nebulous, diagnoses have been made on the basis of other criteria, so the data reported here are not necessarily internally consistent or directly comparable.

women with PCOS from Boston.[25] British South Asian women have higher fasting insulin concentrations, lower insulin sensitivity, and exhibit more severe symptoms of PCOS than do UK European women.[4] Middle Eastern women with PCOS also have higher insulin levels than Western European women, independent of age and BMI.[26] In New Zealand, Maori and Pacific Island women with PCOS are most likely to be overweight and have the highest rates of insulin resistance.[19]

Genetic and Developmental Predispositions to PCOS

PCOS is very likely to have a genetic basis, as it appears to run in families[27,28] and there is greater concordance of PCOS-related traits in monozygotic than in dizygotic twins,[29] although some of the evidence from twin studies is equivocal.[30] However, interpopulation differences in prevalence and phenotypic expression of PCOS also lend support to the notion that PCOS has a genetic component. Nonetheless, identifying specific genes is proving to be far from straightforward.[28] The associations among the calpain-10 gene, insulin resistance, and type 2 diabetes led to research into whether the gene also has a link to PCOS,[31,32] but the results to date have been inconclusive. One relatively small study identified an association between the calpain-10 112/121-haplotype combination and an increased risk (based on odds ratios) of PCOS in African Americans and Europeans.[31] A second, slightly larger study found no robust association between the calpain-10 gene and PCOS.[32] Much larger studies might be required to detect clear links, if they exist, between the two.[32] However, the Pro[12]Ala polymorphism in the PPARγ gene might have a role in modifying the insulin resistance of European women with PCOS, as those with the Pro/Ala genotype appear to have increased insulin sensitivity compared to those with the Pro/Pro genotype.[33] However, this finding does not extend to African American women.[33] Polymorphisms of the insulin receptor substrate-1 (IRS-1) gene appear to be associated with the phenotypic expression of PCOS in some populations,[34] including Europeans,[35] but not Taiwanese women.[36] However, there is strong evidence for a PCOS susceptibility locus that maps close to the D19S884 dinucleotide repeat marker, linked to the insulin receptor gene on chromosome 19p13.2.[37] On current evidence, therefore, it is likely that a small number of genes involved in regulating steroid biosynthesis and glucose homeostasis contribute to PCOS,[27] although a great deal of work still remains to be done in identifying candidate genes.[28]

There is also likely to be a strong developmental component to PCOS. Females with low birth weight apparently have greater chance of precocious pubarche, a risk factor for PCOS, and also the development of hyperinsulinaemia and hyperandrogenism,[38] although a subsequent study on a larger cohort did not support this finding.[39] Comparison of South Asian and European women born and resident in the UK with South Asian women who live in the UK but were born in Pakistan suggests that the relatively poor early environment experienced by first-generation Pakistani immigrants to the UK is an important contributory factor in the relatively high levels of free androgens observed in these women.[10] Low birth weight has also been highlighted as an important precursor of dyslipidaemia and insulin resistance later in life (e.g., references 40–42). One hypothesis is that poor fetal and infant growth can promote the development of PCOS in individuals with a genetic susceptibility, especially when exposed to a nutritional environment of excess later in life. It is also possible that exposure to excess androgens during development plays a central role in the development of PCOS.[28] Prenatally androgenised animal models (sheep and rhesus macaques) display many of the traits associated with PCOS in humans.[43,44] In humans, exposure to excess androgens, possibly as a result of the action of genes that regulate the relevant pathways, might occur during the short-lived activation of the hypothalamic-pituitary-gonadal axis in infancy as well as at puberty.[28] This in turn might lead to endocrine disturbance, including insulin resistance and elevated LH, and abdominal adiposity, the effects of which can be mediated or magnified by diet.[28]

Evolutionary Perspective on PCOS: Seasonality and Unpredictable Environments

The strong role that genetic factors apparently play in PCOS plus its high general prevalence worldwide could indicate an evolutionary basis to the syndrome. In other words, at some point in our evolutionary history, PCOS might have been selectively advantageous, resulting in the high levels observed today. A number of authors have argued along these lines, particularly in terms of the thrifty genotype.[7,45–47] It is hypothesised that women with PCOS who have a high chance of being obese and anovulatory during times of normal or excess food availability, and who begin to ovulate on weight loss, have a selective advantage by being able to reproduce during periods of food shortage, when other women become anovulatory.[7,46] When faced with a modern Westernised lifestyle characterised by abundant food and limited physical activity, however, the thrifty genes that might have been advantageous under certain environmental conditions contribute to PCOS and become disadvantageous.[45]

A similar argument has been played out for another condition of insulin resistance, type 2 diabetes. It has been suggested that modern populations, specifically the Pima and some Pacific Island groups, with high levels of insulin resistance were at some point in the past subjected to environments with extremely poor food availability, possibly with feast-famine cycles.[48,49] As with PCOS, genes that today predispose individuals to insulin resistance could have therefore conferred a selective advantage for survival, but in times of relative abundance, they cease to do this and instead result in serious health problems.[48] Long periods of food shortage during the colonisations of the Americas and Pacific have been suggested as the primary selective agent for such a thrifty genotype in some populations.[48,49] However, this explanation cannot be easily extended to another population with very high levels of insulin resistance, South Asians, who are unlikely to have experienced colonisation-induced famine on the scale hypothesised for Native Americans and Pacific Islanders. Instead, it is possible that general undernutrition could have resulted in selection for individuals with thrifty genes across many human populations. Comprehensive data for interpopulation differences are not available, but on current evidence it seems probable that Western Europeans, who are frequently used as a reference in cross-population studies of insulin resistance and associated disorders, have unusually low levels of insulin resistance compared to most other human groups.[50,51] This implies that they are better adapted than many human groups to the Westernised diets and lifestyles now being adopted worldwide, although it is not clear whether this adaptation occurred recently and relatively rapidly[51] or much earlier.[50]

If an evolutionary basis to conditions of insulin resistance is assumed, it is possible that adaptation to seasonality, rather than to more extended, migration-induced feast-famine periods, accounts for the high levels of PCOS evident in many populations. Much of human evolution occurred in seasonal environments, and seasonal fluctuations have the potential to be important selective pressures.[52] Seasonality has a profound effect on human biology and health, and has been correlated with changes in the incidences of a range of conditions, including infectious disease, peptic ulcer, asthma, stroke, and congenital anomalies.[53] Seasonal variations in fertility, birth, and reproductive performance have also been observed.[54,55] The term *seasonality* covers a range of phenomena, including photoperiod, climate, and resource availability, and it is possible that some or all of these factors contribute to differential fertility and reproduction.[54,55] However, there appears to be a particularly close link between negative energy balance and lowered birth rates.[55] In these circumstances, birth rates might be suppressed through physiological responses to undernutrition that result in anovulation.[55,56] In women with PCOS, however, the weight loss that accompanies a reduced energy intake, even if it is relatively modest, may well facilitate rather than impair ovulation, allowing them to conceive at times when other women are less likely to. This could constitute a *food deprivation hypothesis*.

The transition between anovulation and ovulation might occur quite quickly in times of seasonal food shortage, as caloric reduction can lead to relatively rapid weight loss. In obese native Hawaiians, replacement of a Westernised diet by a traditional, relatively low-calorie (1,600 kcal), low-fat diet led to an average weight loss of 7.8 kg over 3 weeks, equating to a BMI reduction of 2.6 kg/m². [57] In obese subjects under severely restricted caloric regimens (between 400 and 800 kcal/day), weight losses of between 1.4 and 2.5 kg per week have been observed in both males and females, leading in some cases to a weight loss of nearly 15 kg over 6 weeks.[58] Clinical studies of the impacts of weight loss on fertility have also demonstrated that pregnancy can occur reasonably soon after starting controlled diet and exercise programmes. In a study of obese infertile women on such a programme (around 80% of whom had PCOS), the mean weight loss was 10.2 kg, the vast majority resumed spontaneous ovulation within 6 months, and nearly 30% of the total sample also achieved a medically unassisted pregnancy in that time.[59] Although this study was undertaken over a 6-month period, other research has shown that the metabolic profile of women with PCOS can improve within 4 to 8 weeks of starting a diet and exercise programme,[60] and ovulation can occur on the loss of approximately 5.5 to 6.5 kg or 2 to 5% of previous body mass (as reviewed by Moran and Norman[61]). In some populations, weight loss in women during the "hungry" season can be as much as 5 kg, although modal values appear to be closer to 2 to 3 kg (reviewed in Ulijaszek[55] and Ferro-Luzzi and Branca[62]). Interestingly, greater weight loss has been seen in women with higher BMIs than in those who are more lean.[62] Women with PCOS living in traditional subsistence societies are unlikely to reach the levels of obesity observed in developed or rapidly developing populations, but they might well have raised BMIs compared to other women in the same community. Further work is required to assess the prevalence and presentation of PCOS in such populations, as well as the phenotype of women with the condition. However, extrapolation from observations on general responses to seasonality indicate that it would be very possible for anovulatory women with PCOS living in environments with seasonal fluctuation in resource availability to lose sufficient weight to ovulate, then conceive.

Conception during the hungry season could have advantages and disadvantages for mother and child. In rural Gambia, infants conceived in the wet season, at the time when fertility is lowest, appeared to have lower mortality rates than those conceived at other times.[55] If this holds for other populations, and if women with PCOS are most likely to conceive when the fertility of other women is reduced, there is potentially an immediate fitness benefit associated with PCOS. By conceiving at the time of greatest food shortage, it is probable that at the most energetically expensive time in pregnancy, the third trimester,[63] there will be a reasonable supply of food, even though the most abundant period occurred in early pregnancy. In Gambia, infants of mothers whose third trimester coincided with the wet season had lower birth weights, a risk factor for infant mortality,[55] again highlighting the potential advantages of conceiving during the wet season. However, this birth timing may be disadvantageous in terms of lactation and subsequent growth into adulthood, with evidence from Gambia tentatively indicating that females born in the dry season (between April and August) grow into lighter and smaller adults.[64]

The food deprivation hypothesis presented here relies on seasonal food shortage and consequent weight loss. However, it is likely that in many societies, food reserves are carefully managed in order to prevent severe and dramatic food shortages. Such behaviour may significantly reduce the period of time in which energy intake is well below energy expenditure, and in turn decrease any selective advantage of conception during this time. Clearly, in-depth, systematic studies are required to adequately test whether there are links between seasonality and PCOS. One alternative that nonetheless draws on food deprivation is that PCOS is an adaptation to climatic unpredictability. In areas of the world with high reported levels of PCOS, including the Indian subcontinent, parts of the Americas, and Australia, indigenous populations are faced with frequent climatic unpredictability, such as drought, failure of monsoons, hurricanes, cyclones, and El Niño effects. All of these can lead to severe food shortages, which in turn might impact, through the mechanisms discussed briefly above, on fertility in women with and without PCOS. Indeed, famine can have a major impact upon the fertility of a population,[65] and those caused in the past by climatic events are likely to have been frequent and severe enough to act as selective pressures on humans, with one possible result being the emergence of high levels of PCOS. In addition to maintaining the population during these periods of extreme food shortage, babies born to mothers with a tendency toward insulin resistance might have a greater ability to survive famine periods. Women with gestational diabetes tend to have babies who are of above-average weight at birth.[66] In one population with high levels of insulin resistance, South Asians, infants have a tendency to have low body weights but have a high proportion of adipose tissue.[67] Infant mortality is a common and significant result of famine,[68,69] so if babies are able to "buffer" themselves through either a larger body mass or greater adiposity laid down during gestation, they may have a better chance of survival. Again, the role of unpredictable climate and resulting food deprivation as a selective pressure for PCOS requires further, systematic testing, but on current evidence, it cannot be discounted that PCOS is adaptive in environments with fluctuating or unpredictable resources.

This notwithstanding, conception during periods of hunger seems paradoxical. It may be important to distinguish here between famine and chronic malnutrition, as the physiological mechanisms that regulate fertility in each of these situations differ. Famine may result in starvation amenorrhoea, in which ovulation ceases entirely as a result of either extremely low energy intake or turnover.[70] Under conditions of chronic malnutrition, however, conception is usually still possible, but fertility is managed in other ways, such as through an increase in the length of time before postpartum resumption of ovarian function,[70] which influences birth spacing. Thus, one question is whether having a PCOS phenotype is advantageous not in environments characterised by chronic malnutrition but in conditions that precipitate starvation anovulation. In times of constant plenty, follicular recruitment in women with PCOS is excessive. The follicle-stimulating hormone (FSH) drive cannot mature such a high number of follicles and they arrest at about 5 cm diameter. The total surface area of granulosa provides oestrogen feedback to the pituitary, which produces LH, resulting in the characteristic hormone profile of PCOS. Weight loss in women with obese polycystic anovulation often results in ovulation, so in famine situations these phenotypes may be the first to recruit follicles in response to improving nutrition. Thus, it is not starvation that stimulates ovulation in this alternative *refeeding hypothesis*, but the return of food.

A third possibility is a *transgenerational privation hypothesis*. If a population suffers persistent, severe, yet subfatal, privation (chronic malnutrition), it would be beneficial to have an enhanced androgenic and hence anabolic state that increases the efficient use of food for protein synthesis or fat storage. Clearly, females would have greater vulnerability than males, so those women with higher endogenous androgens might be better off. Such advantages would act directly, via nutrition (as described above), and also indirectly, as their female offspring would tend toward a preferred anabolic state. This is supported by the observation that prenatally androgenised laboratory macaques develop PCOS-like traits.[44] Moreover, these offspring would possess the characteristic of preferred early ovulation, outlined in the refeeding hypothesis. However, in environments with limited resources, it must also be considered that emotional stress might interact not only with reproduction but also with the creation of androgenic states in adult women and their female offspring. Food deprivation is inherently emotionally stressful, and the effects of stress might reinforce infertility or subfertility.[71] Stress has also been shown to influence metabolic abnormalities in adults with metabolic syndrome.[72] Importantly, stressful situations, including those caused by undernutrition, can lead to increased cortisol stimulation in neonates, encouraging rapid increases in body weight that may eventually result in pre- and postpubertal hyperandrogenism and hyperinsulinaemia and an increased probability of PCOS.[72] These observations draw attention to the proximate factors that could contribute, either on their own or in combination with genetic adaptations, to the prevalence of PCOS in certain populations.

A limitation of evolutionary medicine is often the lack of testable hypotheses. Thus, it would be particularly interesting to compare the incidence of PCOS in populations who have experienced recurrent famine with those that have not, or groups that have marked seasonal privation with those that have very little variation. Seasonal energy stress appears to have the greatest impact in India and parts of sub-Saharan Africa,[62] so it might be expected that if PCOS is an adaptation to seasonality, the prevalence would be highest in populations from these areas. South Asians certainly show very high levels of polycystic ovaries[15] and PCOS,[4] although as the subcontinent has considerable geographic differences in the intensity of seasonal stresses, more detailed examination of the prevalence of PCOS in different regions would be advantageous, especially if viewed in the context of past population movements. Data on indigenous African populations are scarce. One of the few, albeit nonrandom, studies of polycystic ovaries in African women found that the prevalence of polycystic ovaries was high in infertile Yoruba women,[16] although no endocrine data were reported. The Yoruba traditionally inhabit east Benin and western Nigeria, a region that Ferro-Luzzi and Branca[62] (p. 161) label as having low to moderate seasonality. If these women are also found to have the associated metabolic syndrome, this observation does not necessarily support the general hypothesis that polycystic ovaries are an adaptive response to extreme seasonal resource fluctuation. Outside Africa, high levels of PCOS are reported in aboriginal Australians[20] and Mexican Americans.[22] Neither of these groups is usually thought to be under extreme seasonal resource stress, although it has been pointed out that traditional views of Aboriginal Australian subsistence might overestimate the resources available at certain times of the year and underestimate the effort expended in procuring them.[73]

It is possible that the emergence of PCOS predated current population distributions or even modern humans themselves. It has been argued that due to the demands of growing large-brained fetuses, female primates require a suite of reproductive adaptations that predispose to ovulatory dysfunction similar to that seen in PCOS.[74] Thus, the origins of the condition (whether or not it has subsequently been advantageous in seasonal environments) might be relatively ancient. Similarly, because of steadily decreasing global temperatures and increasing aridity, seasonality is likely to have been a significant selective pressure as far back as the Late Miocene and Early Pliocene.[52] It is impossible to test whether PCOS existed in early hominins and if it could have conferred a benefit in the changing environments of the Late Miocene, Pliocene, and Pleistocene. However, whichever hypothesis is considered, primate models might shed some light on the evolutionary basis of the condition. Although prenatally androgenised laboratory macaques develop PCOS-like traits,[44] to date there is no evidence for a similar syndrome in wild-living monkeys. Establishing whether PCOS occurs spontaneously in nonhuman primates, and whether it is found in species such as baboons and macaques, and indeed other mammals, that evolved and currently live in seasonal environments, would provide an important comparative perspective on the condition in modern humans and potentially their ancestors.

Why Is PCOS Not at Even Higher Levels in Modern Populations?

Regardless of the circumstances under which high levels of PCOS evolved, it is plausible that PCOS gives reproductive benefits in environments with fluctuating resources. Women with PCOS, who generally have high androgen levels, also appear to have greater bone mineral density[75] and muscle mass[45] than non-PCOS women. This might provide additional advantages in terms of greater capacity for physical work and endurance,[45] which may be especially desirable in regions where subsistence requires considerable energy expenditure. Given these possible benefits, an obvious question is why PCOS is not found at even higher levels. One controversial explanation is that since direct evidence of fertility and potential fecundity in human females is concealed, morphological features associated with reproduction, such as fat patterning, might be very important in mate choice and perceived attractiveness,[76] and therefore reproductive success. Many women with PCOS, even when they are nonobese, exhibit fat patterning that is most commonly seen in males, and it has been argued that as a result of this, they might be less attractive to potential mates.[76] However, this argument is weakened considerably by the simple observation that there is likely to be a strong genetic element to PCOS and that women with the condition do reproduce. More probable is that although PCOS confers an advantage at certain times, it is disadvantageous at other times, when food is more readily available. The total reproductive window is therefore reduced for women with PCOS. Alternatively, it is possible that as food supplies become increasingly buffered against seasonal or climatic fluctuations, the polymorphism has moved from being balanced to transient, with the deleterious effects of PCOS outweighing the advantages, leading to selection against individuals with a genetic predisposition to PCOS. Under this model, it is possible, for example, that the prevalence of PCOS in European countries was at higher levels in the past, as has been hypothesised for type 2 diabetes.[51]

Critical Analysis of the Evolutionary Perspective on PCOS

The notion that PCOS is a manifestation of the selective pressures in our evolutionary past is an attractive one. However, despite having over 40 years to test Neel's original thrifty genotype hypothesis,[48] formulated to account for type 2 diabetes rather than PCOS, the search for candidate thrifty genes has been problematic and no consensus has yet been reached about the validity of the concept[77] (see also chapter 4). Particular criticism has come from proponents of the thrifty phenotype hypothesis, who argue for a strong developmental, rather than genetic, basis to insulin resistance.[42] However, the thrifty phenotype is not without major limitations, not least the lack of solid evidence found to date, the uncritical use of birth weight as an indicator of fetal growth restriction, and the need for more sophisticated studies to elucidate patterns and mechanisms.[78] Barker[79] (p. 250) argued that, under the thrifty genotype hypothesis, high levels of type 2 diabetes would be evident in South Asian populations until a more primitive way of life was adopted once more, reasoning that could presumably be extended to PCOS. However, recent consensus work on insulin resistance and associated conditions, including PCOS, indicates that there is an underlying genetic component that is "primed" by proximate influences, some of the most important of which may well occur during fetal or early life.[28] This suggests that elements of thrifty genotype and thrifty phenotype are not mutually exclusive. In addition, by removing the selective pressure of uncertain food supplies, under an evolutionary model the PCOS polymorphism may be expected to move over time into a transient state, as the genes that predispose to conditions of insulin resistance are selected against.

This notwithstanding, one major theoretical problem with the thrifty genotype hypothesis as it is often presented for PCOS is the need to invoke group selection to explain one of the major advantages of the condition. The idea that women with PCOS are able to "rescue" the population in times of energy shortage through losing weight, ovulating, and subsequently conceiving, stresses the benefits to the group but not to the individual. Although group selection might well be more important in human evolution than is currently acknowledged,[80] it remains an unpopular concept in evolutionary biology, partly due to the popularity of "selfish gene" models, and also because it is rarely or never observed in real populations.[81] However, the hypothesis does not stand or fall on the validity of group selection. There might be fitness benefits to the individual in conceiving at a time when other women do not, as infant mortality might be lower, at least in some populations.[55] Interestingly, and adding weight to this, Bangladeshi babies who survived the 1974–1975 famine had significantly lower mortality in the months following the famine than would normally be expected for those age cohorts, suggesting that selection had taken place.[68] Thus, although the advantages to the group might be tangible and important, advantages to the individual are also possible under the PCOS thrifty genotype hypothesis.

In the context of evolutionary medicine, there is still much work to be done to test the thrifty genotype hypothesis, especially for PCOS. The three hypotheses presented in this chapter require more data than are currently available in order to support or refute them. Given sufficient resources, it should be possible to examine links between seasonality and PCOS, as populations that live under extreme seasonal pressures still exist, as do those that have few seasonal pressures. One potential region for future study is the highly seasonal Gambia, where decades of research into maternal health have already been undertaken. One further challenge within this framework would be assessing whether it is the refeeding element that is crucial for restoration of fertility. It may be harder to test whether climatic unpredictability has influenced PCOS prevalence, as it not only requires knowledge of current climatic activity and its interaction with food supply, but also needs accurate reconstruction of past climates and subsistence practises. The same is true of the transgeneration privation hypothesis. This issue goes to the heart of the limitation of evolutionary medicine: that much selection has occurred in periods for which there are little or no accurate data, and under population regimes that can, at best, only be approximated.

Do Evolutionary Hypotheses Relating to PCOS Aid Clinical Practise?

The wealth of research into PCOS that has been undertaken in the past decade has shed much light on the causes and treatment of PCOS, particularly in the context of infertility. The recognition that polycystic ovaries and the related syndrome are highly prevalent has led a number of authors to suggest an evolutionary basis for PCOS,[7,45–47] and although the hypotheses require much more rigorous testing, the underlying idea is plausible, especially in the context of fluctuations in resource availability caused by either seasonal stresses or climatic unpredictability. However, diseases and medical conditions can have proximate and ultimate causes, and focusing on the evolutionary background to a condition may not necessarily lead to the most effective treatment and management of it.

Given that PCOS is argued to be adaptive in environments in which there are periods of extreme and sometimes rapidly occurring food shortage, and that weight loss appears to trigger ovulation in many women with the syndrome, "crash" dieting could be seen by some as a useful and relatively easy way of restoring fertility. Removed from the context in which PCOS evolved, however, this strategy is likely to produce a harmful cycle of yo-yo weight loss and gain, which may result in increased BMI and possibly harmful metabolic complications.[82] Instead, endocrinologists managing conditions of insulin resistance recommend long-term lifestyle modification that is sustainable and achievable.[60] Overemphasising the possible evolutionary basis of PCOS might conversely lead women with the syndrome and their health care providers to adopt a fatalistic attitude toward its management. Concern has been raised about genetic determinism in clinical treatment for the related condition, type 2 diabetes, especially with respect to marginalised and deprived Native American groups.[77,83] There is the potential for this to occur in the context of PCOS, for example, in South Asian women living in Europe and North America. South Asian women exhibit a higher prevalence of PCOS than is found in the general Western population,[4] but like other immigrant populations, they may have difficulty in accessing necessary health care.[84] Such barriers, whether they be socioeconomic, linguistic, or cultural, could be reinforced by stressing the possible evolutionary basis to the syndrome and creating the illusion that it is not worth seeking treatment because PCOS is inevitable in South Asian women.

Nonetheless, a recognition that individuals who have a genetic predisposition to the disease can have their risk reduced through prevention of maternal malnutrition might be a useful public health strategy. This type of approach recognises the genetic/evolutionary backdrop to the condition while seeking to control the proximate environmental influences that affect its development and severity. However, even the adoption of this approach is far from straightforward. Maternal size acts as a constraint on fetal size, and nutritional intervention in pregnancy may in fact lead to an increased rather than decreased risk of insulin resistance in the offspring, due to the so-called thin-fat baby phenomenon.[78] This might be especially marked in South Asian infants, who have a tendency to have less muscle mass and greater fat mass—the thin-fat baby, a body composition that predisposes to insulin resistance.[67,85] Thus, improvement in intergenerational nutrition might well be necessary to seriously address insulin resistance,[78] a management option that might also be useful in tackling PCOS.

Such are the possibly negative impacts of evolutionary hypotheses in PCOS. On the positive side, comprehension of the origins of diseases can be highly beneficial to patients. Contrary to the crash diet risk is the belief by PCOS sufferers that they are "fat because they have PCOS," and that if the disease is treated, then they will be thin. It has already been mentioned that the first-line treatment for PCOS is weight loss. If the patient attends with the feeling that they will get a cure and thereby become thin, she will react poorly to the advice she is given. With a sympathetic explanation that in another time her efficient use of food, a scarce commodity, would have made her the first to conceive, there may be less stigma and a more constructive acceptance of her need to reach an optimal BMI. Crash dieting is not recommended; it is not suggested here, after all, only that famine improves fertility in women with PCOS. Another arm to the ideas presented in this chapter is that reintroduction of food following a famine-induced anovulation initiates follicular recruitment. So, careful explanation of this to the patient could emphasise exercise as the more helpful part of lifestyle change over and above diet.

A positive and long-term benefit to evolutionary understanding is as a guide to future research, both medical and anthropological. If there is a strong belief that this is a genetically driven disease, then continued search for the genes concerned might lead to a more fundamental treatment than exists at the moment. If not actual gene therapy, it may be that the source protein might hold the key to future treatment developments. However, what is or are the genes? Where are they located? Do the responsible genes code for insulin resistance, lipid metabolism, follicular recruitment sensitivity, or do they work at the level of the switch for rendering follicles atretic (i.e., as part of the maturation and degeneration process)? Understanding the evolution of the disease helps in the search for the answers to these questions. From the answers will come future treatments.

Conclusions

PCOS is a nebulous and complex endocrine condition. This is reflected by the fact that its definition has changed frequently during the second half of the twentieth century. Currently, possession of two characteristics is required from a list of recognised features in order to make the diagnosis. Its familial nature, the level of twin concordance, and variation in different populations lend weight to a genetic basis to this condition. However, it is accepted that gene expression is subject to environmental influence, including during fetal or early life. The syndrome links reproductive performance with nutrition in the form of glucose and lipid metabolism, factors at the heart of survival. If genetics are at the core of such a common condition, the question of whether there is an evolutionary advantage to the possession of the gene or genes is inevitably raised.

Three hypotheses have been proposed in this chapter. The first concerns the observation that PCOS sufferers tend to ovulate when weight is lost. Indeed, that forms the primary treatment of the condition. Perhaps there might be an advantage for members of a population to conceive during a food shortage. The observation in Gambia of lower mortality among infants conceived in the wet season, a period of food deprivation, lends weight to this. This food deprivation hypothesis sees a positive enhancement in ovulation when food is short and, as such, might suggest that the primary effect of the gene is related to glucose metabolism. The second hypothesis suggests that individuals with a PCOS-like condition living in a traditional environment would not be as well fed, even in times of plenty, as they would in an industrialised environment. In this scenario, the advantage occurs at the end of a period of food deprivation as the food source increases. These women are the first to ovulate in response to this refeeding. An anabolic propensity for food storage is an appropriate adjuvant preparation for early pregnancy. This refeeding hypothesis implies that the gene might be one that drives ovarian follicular recruitment sensitivity to FSH. The third hypothesis suggests that the origin of the gene may predate modern humans, possibly being related to the onset of seasonal environments in the Late Miocene, or the climatic fluctuations that characterised Pleistocene environments. A beneficial strategy favouring insulin resistance might have improved survival at that time. This transgenerational privation hypothesis again suggests a gene inducing insulin resistance.

Food deprivation lies at the heart of these hypotheses, as insulin resistance buffers the individual against food shortage. Although the advantages of insulin resistance are described elsewhere in relation to type 2 diabetes mellitus, it is likely to be at its most profound if associated with follicular recruitment. Significant pressure on food resources would have been needed to favour these genes. In the case of the transgenerational hypothesis, the strategy might have been part of a wider suite of adaptations necessary to deal with long-term environmental change. In the case of the food deprivation and refeeding hypotheses, fluctuations in resource availability leading to food deprivation probably would be severe but frequent. Climatic unpredictability is one source of this, but the most regular occurrence of food deprivation is, of course, seasonality. Variability in the frequency of PCOS from population to population suggests that inhabiting seasonal environments might have contributed to selection for the PCOS genotype. Studies comparing PCOS in indigenous peoples from highly seasonal and less seasonal areas might throw some light on this. As the diagnosis is so ill-defined, negative correlations might be less relevant than positive ones.

One further question is whether the adaptation is primarily for nutritional survival with a reproductive add-on, or whether the opposite is true. Discovering where the genes are located and for which proteins they code would help to elucidate the evolution of PCOS and also aid patient education. Even if alterations in intergenerational maternal nutrition help to reduce the overall prevalence of the syndrome or severity of the symptoms, greater knowledge of the genetic and evolutionary basis to PCOS would guide future treatment strategies. Only within such an approach does there lie a cure, as opposed to the symptomatic treatment currently available.

Acknowledgements

We are very grateful to the two reviewers of this paper for their insights and suggestions.

References

1. Stein, I. F., and Leventhal, M. L. 1935. Amenorrhea associated with bilateral polycystic ovaries. *Am. J. Obstet. Gynecol.* 29:181–91.
2. Franks, S. 1995. Medical progress article: Polycystic ovary syndrome. *New Engl. J. Med.* 333:853–61.
3. Balen, A. 1999. Pathogenesis of polycystic ovary syndrome—The enigma unravels? *Lancet* 354:966–67.

4. Wijeyaratne, C. N., Balen, A. H., Barth, J. H., and Belchetz, P. E. 2002. Clinical manifestations and insulin resistance (IR) in polycystic syndrome (PCOS) among South Asians and Caucasians: Is there a difference? *Clin. Endocrinol.* 57:343–50.

5. Hart, R., Hickey, M., and Franks, S. 2004. Definitions, prevalence and symptoms of polycystic ovaries and polycystic ovary syndrome. *Best Pract. Res. Clin. Obstet. Gynecol.* 18:671–83.

6. De Leo, V., la Marca, A., and Petraglia, F. 2003. Insulin-lowering agents in the management of polycystic ovary syndrome. *Endocrinol. Rev.* 24:633–67.

7. Balen, A., and Michelmore, K. 2002. What is polycystic ovary syndrome? Are national views important? *Hum. Reprod.* 17:2219–27.

8. Livingstone, C., and Collison, M. 2002. Sex steroids and insulin resistance. *Clin. Sci.* 102:151–66.

9. Lui, Y., et al. 2001. Relative androgen excess and increased cardiovascular risk after menopause: A hypothesized relation. *Am. J. Epidemiol.* 154:489–94.

10. Pollard, T., Unwin, N. C., Fischbacher, C. M., and Chamley, J. K. 2006. Sex hormone-binding globulin and androgen levels in immigrant and Bristish-born premenopausal British Pakistani women: Evidence of early life influences. *Am. J. Hum. Biol.* 18:741–47.

11. Kauffman, R. P., Baker, V. M., DiMarino, P., and Castracane, V. D. 2006. Hyperinsulinemia and circulating dehydroepiandrosterone sulfate in white and Mexican American women with polycystic ovary syndrome. *Fertil. Steril.* 85:1010–16.

12. RCOG. 2005. Patient information sheet. http://www.rcog.org.uk/resources/public/pdf/pcos_patient_info_0106.pdf (accessed November 23, 2006).

13. ASRM. 2005. Patient information sheet. http://www.asrm.org/Patients/FactSheets/PCOS.pdf (accessed November 23, 2006).

14. Clayton, R. N., et al. 1992. How common are polycystic ovaries in normal women and what is their significance for the fertility of the population? *Clin. Endocrinol.* 37:127–34.

15. Rodin, D. A., Bano, G., Bland, J. M., Taylor, K., and Nussey, S. S. 1998. Polycystic ovaries and associated metabolic abnormalities in Indian subcontinent Asian women. *Clin. Endocrinol.* 49:91–99.

16. Wada, I. et al. 1994. High ovarian response in Yoruba African women during ovulation indiction for assisted conception. *Hum. Reprod.* 9:1077–80.

17. Kousta, E., et al. 1999. The prevalence of polycystic ovaries in women with infertility. *Hum. Reprod.* 14:2720–23.

18. Diamanti-Kandarakis, E., et al. 1999. A survey of the polycystic ovary syndrome in the Greek Island of Lesbos: Hormonal and metabolic profile. *J. Clin. Endocrinol. Metab.* 84:4006–11.

19. Williamson, K., Gunn, A. J., Johnson, N., and Milsom, S. R. 2001. The impact of ethnicity on the presentation of polycystic ovarian syndrome. *Aust. N.Z. J. Obstet. Gynecol.* 41:202–6.

20. Davis, S. R., et al. 2002. Preliminary indication of a high prevalence of polycystic ovary syndrome in indigenous Australian women. *Gynecol. Endocrinol.* 16:443–46.

21. Solomon, C. G. 1999. The epidemiology of polycystic ovary syndrome. Prevalence and associated disease risks. *Endocrinol. Metab. Clin. North Am.* 28:247–63.

22. Goodarzi, M. O., et al. 2005. Polycystic ovary syndrome in Mexican-Americans: Prevalence and association with the severity of insulin resistance. *Fertil. Steril.* 84:766–69.
23. Carmina, E., et al. 2002. Does ethnicity influence the prevalence of adrenal hyperandrogenism and insulin resistance in polycystic ovary syndrome? *Am. J. Obstet. Gynecol.* 167:1807–12.
24. Kauffman, R. P., et al. 2002. Polycystic ovarian syndrome and insulin resistance in white and Mexican American women: A comparison of two distinct populations. *Am. J. Obstet. Gynecol.* 187:1362–69.
25. Welt, C. K., et al. 2006. Defining constant versus variable phenotypic features of women with polycystic ovary syndrome using different ethnic groups and populations. *J. Clin. Endocrinol. Metab.* 91:4361–68.
26. Al-Fozan, H., Al-Futaisi, A., Morris, D., and Tulandi, T. 2005. Insulin responses to the oral glucose tolerance test in women of different ethnicity with polycystic ovary syndrome. *J. Obstet. Gynecol. Can.* 27:33–37.
27. Franks, S., et al. 1997. The genetic basis of polycystic ovary syndrome. *Hum. Reprod.* 12:2641–48.
28. Franks, S. 2006. Genetic and environmental origins of obesity relevant to reproduction. *Reprod. Biomed. Online* 12:526–31.
29. Vink, J. M., Sadrzadeh, S., Lambalk, C. B., and Boomsma, D. I. 2006. Heritability of polycystic ovary syndrome (PCOS) in a Dutch twin-family study. *J. Clin. Endocrinol. Metab.* 91:2100–4.
30. Jahanfar, S., et al. 1995. A twin study of polycystic ovary syndrome. *Fertil. Steril.* 63:478–86.
31. Ehrmann, D. A., et al. 2002. Relationship of calpain-10 genotype to phenotypic features of polycystic ovary syndrome. *J. Clin. Endocrinol. Metab.* 87:1669–73.
32. Haddad, L., et al. 2002. Variation within the type 2 diabetes susceptibility gene calpain-10 and polycystic ovary syndrome. *J. Clin. Endocrinol. Metab.* 87:2606–10.
33. Hara, M., et al. 2002. Insulin resistance is attenuated in women with polycystic ovary syndrome with the Pro^{12}Ala polymorphism in the PPARγ gene. *J. Clin. Endocrinol. Metab.* 87:772–75.
34. Sir-Petermann, T., et al. 2001. G972R polymorphism of IRS-1 in women with polycystic ovary syndrome. *Diabetologia* 44:1200–1.
35. El Mkadem, S. A., et al. 2001. Role of allelic variants Gly972Arg of IRS-1 and Gly1057Asp of IRS-2 in moderate-to-severe insulin resistance of women with polycystic ovary syndrome. *Diabetes* 50:2164–68.
36. Lin, T.-C., et al. 2006. Abnormal glucose tolerance and insulin resistance in polycystic ovary syndrome amongst the Taiwanese population—Not correlated with insulin receptor substrate-1 Gly972Arg/Ala513Pro polymorphism. *BMC Med. Genet.* 7:36.
37. Urbanek, M., et al. 2005. Candidate gene region for polycystic ovary syndrome on chromosome 19p13.2. *J. Clin. Endocrinol. Metab.* 90:6623–29.
38. Ibanez, L., Potau, N., Francois, I., and de Zegher, F. 1998. Precocious pubarche, hyperinsulinism and ovarian hyperandrogenism in girls: Relation to reduced fetal growth. *J. Clin. Endocrinol. Metab.* 83:3558–62.
39. Laitinen, J., et al. 2003. Body size from birth to adulthood as a predictor of self-reported polycystic ovary syndrome symptoms. *Int. J. Obstet.* 27:710–15.
40. Hales, C. N., et al. 1991. Fetal and infant growth and impaired glucose tolerance at age 64. *BMJ* 303:1019–22.

41. Barker, D. J. P., et al. 1993. Type 2 (non-insulin dependent) diabetes mellitus, hypertension and hyperlipidaemia (syndrome X): Relation to reduced fetal growth. *Diabetologia* 36:62–67.
42. Hales, C. N., and Barker, D. J. P. 1992. Type 2 (non-insulin-dependent) diabetes mellitus: The thrifty phenotype hypothesis. *Diabetologia* 35:595–601.
43. Birch, R., Robinson, J. E., Hardy, K., and Franks, S. 2001. Morphological differences in preantral follicle distribution between normal and androgenised ovine ovaries. *Endocrinol. Abst.* 2:P74.
44. Abbott, D. H., Foong, S. C., Barnett, D. K., and Dumesic, D. A. 2004. Nonhuman primates contribute unique understanding to anovulatory infertility in women. *Inst. Lab. Anim. Resour. J.* 45:116–31.
45. Holte, J. 1998. Polycystic ovary syndrome and insulin resistance: Thrifty genes struggling with over-feeding and a sedentary lifestyle? *J. Endocrinol. Invest.* 21:589–601.
46. Gleicher, N., and Barad, D. 2006. An evolutionary concept of polycystic ovarian disease: Does evolution favour reproductive success over survival? *Reprod. Biomed. Online* 12:587–89.
47. Wood, L. E. P. 2006. Obesity, waist-hip ratio and hunter-gatherers. *BJOG* 113:1110–16.
48. Neel, J. V. 1962. Diabetes mellitus: A "thrifty" genotype rendered detrimental by "progress"? *Am. J. Hum. Genet.* 14:353–62.
49. Bindon, J. R., and Baker, P. T. 1997. Bergmann's rule and the thrifty genotype. *Am. J. Phys. Anthropol.* 104:201–10.
50. Baschetti, R. 2006. Diabetes susceptibility. *CMAJ* 174:1597–98.
51. Gerstein, H. C., and Waltman, L. 2006. Why don't pigs get diabetes? Explanations for variations in diabetes susceptibility in human populations living in a diabetogenic environment. *CMAJ* 174:25–26.
52. Foley, R. A. 1993. The influence of seasonality on hominid evolution. In *Seasonality and human ecology*, ed. S. J. Ulijaszek and S. S. Strickland, 17–37. Cambridge: Cambridge University Press.
53. Johnston, F. E. 1993. Seasonality and human biology. In *Seasonality and human ecology*, ed. S. J. Ulijaszek and S. S. Strickland, 5–16. Cambridge: Cambridge University Press.
54. Rosetta, L. 1993. Seasonality and fertility. In *Seasonality and human ecology*, ed. S. J. Ulijaszek and S. S. Strickland, 65–75. Cambridge: Cambridge University Press.
55. Ulijaszek, S. J. 1993. Seasonality of reproductive performance in rural Gambia. In *Seasonality and human ecology*, ed. S. J. Ulijaszek and S. S. Strickland, 76–88. Cambridge: Cambridge University Press.
56. Rosetta, L. 1990. Biological aspects of fertility among Third World populations. In *Fertility and resources*, ed. J. Landers and V. Reynolds, 18–34. Cambridge: Cambridge University Press.
57. Shintani, T., Hughes, C. K., Beckham, S., and O'Connor, H. K. 1991. Obesity and cardiovascular risk intervention through the *ad libitum* feeding of traditional Hawaiian diet *Am. J. Clin. Nutr.* 53:1647S–51S.
58. Saris, W. H. M. 2001. Very-low-calorie diets and sustained weight loss. *Obstet. Res.* 9:295S–301S.
59. Clark, A. M., et al. 1998. Weight loss in obese infertile women results in improvement in reproductive outcome for all forms of fertility treatment. *Hum. Reprod.* 13:1502–5.

60. Norman, R. J., Davies, M. J., Lord, J., and Moran, L. J. 2002. The role of lifestyle modification in polycystic ovary syndrome. *Trends Endocrinol. Metab.* 13:251–57.

61. Moran, L. J., and Norman, R. J. 2002. The obese patient with infertility: A practical approach to diagnosis and treatment. *Nutr. Clin. Care* 5:290–97.

62. Ferro-Luzzi, A., and Branca, F. 1993. Nutritional seasonality: The dimensions of the problem. In *Seasonality and human ecology*, ed. S. J. Ulijaszek and S. S. Strickland, 149–65. Cambridge: Cambridge University Press.

63. Butte, N. F., et al. 2004. Energy requirements during pregnancy based on total energy expenditure and energy deposition. *Am. J. Clin. Nutr.* 79:1078–87.

64. Cole, T. J. 1993. Seasonal effects on physical growth and development. In *Seasonality and human ecology*, ed. S. J. Ulijaszek and S. S. Strickland, 89–106. Cambridge: Cambridge University Press.

65. Dyson, T. 1991. On the demography of South Asian famines: Part I. *Population Studies* 45:5–25.

66. Gillman, M. W., et al. 2003. Maternal gestational diabetes, birth weight and adolescent obesity. *Pediatrics* 111:e221–26.

67. Yajnik, C. S., et al. 2002. Neonatal anthropometry: The thin-fat Indian baby. The Pune Maternal Nutrition Study. *Int. J. Obstet.* 27:173–80.

68. Razzaque, A., Alam, N., Wai, L., and Foster, A. 1990. Sustained effects of the 1974–5 famine on infant and child mortality in a rural area of Bangladesh. *Population Studies* 44:145–154.

69. Scott, S., Duncan, S. R., and Duncan, C. J. 1995. Infant mortality and famine—A study in historical epidemiology in northern England. *J. Epidemiol. Commun. Health* 49:245–52.

70. Ellison, P. T. 2003. Energetics and reproductive effort. *Am. J. Hum. Biol.* 15:342–51.

71. Schenker, J. G., Meirow, D., Schenker, E. 1992. Stress and human reproduction. *Eur. J. Obstet. Gynecol. Reprod. Biol.* 45:1–8.

72. Diamanti-Kandarakis, E., and Economou, F. 2006. Stress in women: Metabolic syndrome and polycystic ovary syndrome. *Ann. N.Y. Acad. Sci.* 1083:54–62.

73. Ulijaszek, S. J. 2001. Potential seasonal ecological challenge of heat strain among Australian Aboriginal people practicing traditional subsistence methods: A computer simulation. *Am. J. Phys. Anthropol.* 116:236–45.

74. Barnett, D. K., and Abbott, D. H. 2003. Reproductive adaptations to a large-brained fetus open a vulnerability to anovulation similar to polycystic ovary syndrome. *Am. J. Hum. Biol.* 15:296–319.

75. Yuksel, O., et al. 2001. Relationship between bone mineral density and insulin resistance in polycystic ovary syndrome. *J. Bone Miner. Metab.* 19:257–62.

76. Kirchengast, S. 2005. Evolutionary and medical aspects of body composition characteristics in subfertile and infertile women. *Acta Med. Lituanica* 12:22–27.

77. Fox, D. 1999. The famished road. *New Scientist* 2212:38–43.

78. Adair, L. S., and Prentice, A. M. 2004. A critical evaluation of the fetal origins hypothesis and its implications for developing countries. *J. Nutr.* 134:191–93.

79. Barker, D. J. P. 1998. The fetal origins of coronary heart disease and stroke: Evolutionary implications. In *Evolution in health and disease*, ed. S. C. Stearns, 246–50. Oxford: Oxford University Press.

80. Wilson, D. S., and Sober, E. 1994. Reintroducing group selection to the human behavioural sciences. *Behav. Brain Sci.* 17:585–608.

81. Zahavi, A. 2005. Is group selection necessary to explain social adaptations in microorganisms? *Heredity* 94:143–44.

82. Kajioka, T., Tsuzuku, S., Shimokata, H., and Sato, Y. 2002. Effects of intentional weight cycling on non-obese young women. *Metabolism* 51:149–54.

83. Benyshek, D. 2005. Type 2 diabetes and fetal origins: The promise of prevention programs focusing on prenatal health in high prevalence Native American communities. *Hum. Organiz.* 64:192–200.

84. Choudhry, U. K., et al. 2002. Health promotion and participatory action research with South Asian women. *J. Nurs. Scholar.* 34:75–81.

85. Yajnik, C. S., et al. 2002. Adipocity and hyperinsulinaemia in Indians are present at birth. *J. Clin. Endocrinol. Metab.* 87:5575–80.

6

Evolution and Endocrinology
The Regulation of
Pregnancy Outcomes

Virginia J. Vitzthum
Anthropology Department and the Kinsey Institute,
Indiana University and Institute for Primary and
Preventative Health Care, Binghamton University

Contents

Introduction

Some time this past year, the earth's human population rose past 6.5 billion living persons. Roughly 130 million babies were born in 2006, which translates into about 350,000 per day. Impressive as these figures are, they represent but a fraction of an invisible industry. For every human birth, as many as four conceptions never reached fruition—nearly 1.4 million each and every day, and more than half a billion lost human conceptions each

year. Having breathed a sigh of relief on behalf of the planet that such potential was not realised, one is left to wonder at the extraordinary inefficiency of
this biological economy. Why hasn't natural selection during 6 million years
of human evolution honed a sleeker assembly line?

Few physicians, having faced the disappointment of patients experiencing
recurrent pregnancy losses, would design such a production system. Understandably, most feel compelled to intervene on a patient's behalf to circumvent nature's apparent wastefulness. But therein lies the greatest difficulty.
Which treatments might achieve the desired goals? At what cost? And with
what unintended consequences? These questions are, of course, pertinent to
almost any medical treatment. Physicians are accustomed to weighing the
risks and benefits, and then ameliorating, typically with extraordinary skill,
nature's manufacturing errors.

Evolutionary anthropologists take a different view, beginning with the
premise that biological economies do make sense, albeit in a currency that
may have nothing to do with human goals or suffering. The currency, of
course, is reproductive fitness—that is, the proportion of one's genes, relative
to the genes of others, that are represented in subsequent generations. Hence,
to extend the economic metaphor a bit further, the rise and fall of one's account
during one's lifetime is immaterial as long as the final balance is higher than
those of one's conspecifics. In this context, occasional "wastefulness" may
be a strategy to maximize final wealth.* Such life history strategies (that is,
the timing of those activities essential to survival and reproduction, and the
allocation of resources to each) are usually unconscious. Over time natural
selection has favoured those physiological mechanisms by which the most
successful life history strategies are implemented, and these mechanisms
respond accordingly given the resources available throughout an organism's
lifetime.

These arguments from evolutionary theory are not, perhaps, as far
removed from the difficult decisions faced by physicians and their patients
as they may appear. I begin this exploration of the relevance of the evolutionary paradigm to clinical practice with a brief review of what is known about
pregnancy loss from a medical vantage point, including the temporal patterning of loss over the course of gestation and the principal proximate (i.e.,
immediate) causes of loss. I then consider the possible ultimate (i.e., evolutionary) causes of pregnancy loss, which explain variation in survival of any
one conception in terms of an organism's total fitness advantage (i.e., lifetime
reproductive success [LRS]). I begin with a synopsis of the shared theoretical
foundation of evolutionary models and then direct particular attention to
what has been recently learned of the physiological mechanisms by which
pregnancy outcomes may be regulated.

---

* Theoretically, the final balance can be assessed as the relative proportion of one's genes
  in the immediately subsequent generation or the proportion observed in any subsequent generation.

Efforts to determine these and other life history mechanisms in a variety of organisms have coalesced into the emerging cross-disciplinary field of evolutionary endocrinology. Among those concerned with human biology, there are two especially compelling reasons for discerning potential life history mechanisms in as much detail as possible. First, the kinds of observations and experiments that allow for the testing of evolutionary models in other organisms are only rarely possible in humans. Hence, elucidating the physiological mechanisms by which life history strategies could be implemented is one means to test hypotheses generated by such models, and hence ultimately to test the models themselves. But, understandably, this goal is likely to be of only peripheral interest to medical practitioners.

Rather, it is the possibility of improved health care that is of greater relevance in the present context and that is considered in the closing section of this paper. An evolutionary view of human health does not preclude medical intervention. It is a nonsequitur to argue as some have—often by invoking a more or less eloquent version of the maxim "Don't mess with Mother Nature"—that one should not trifle with the evolved features of human biology and behaviour. In fact, it is because biological economies make sense that insights from evolutionary science can add to the medical arsenal.[1-6] Appreciating the investment trade-offs faced by an organism and the costs of reproduction incurred by women could contribute to achieving realistic reproductive goals, a clearer understanding of women's physiology during pregnancy, and improvements in women's and children's emotional and physical health.

Pregnancy Nomenclature

Because there is a great deal of inconsistency in the terminology referring to pregnancy and the various possible outcomes of a conception, it is worth taking the time to establish a common language before proceeding further. The World Health Organisation's definition of fetal loss is "death before the complete expulsion or extraction from its mother of a product of conception, irrespective of the duration of pregnancy; the death is indicated by the fact that after such separation, the fetus does not breathe or show any other evidence of life, such as beating of the heart, pulsation of the umbilical cord, or definite movement of voluntary muscles."[7]

Though the WHO definition, in principle, has been widely adopted since its introduction in 1950, *de facto* definitions of pregnancy loss are often qualified with regard to gestational age in both epidemiological studies and clinical settings. For example, the U.S. National Center for Health Statistics[8] restricts the calculation of fetal death rates to those concepti that have persisted for >20 weeks; late fetal deaths are those occurring after 28 weeks gestation.

Recently, the Special Interest Group for Early Pregnancy (SIGEP) of the European Society of Human Reproduction and Embryology (ESHRE) proposed[9] that a fetus be defined ultrasonically by heart activity or a crown-rump length of >10 mm; hence, fetal loss is only applicable to those concepti that have first met these criteria. At the same time, they also proposed that a pregnancy lost prior to 6 weeks gestation that had been identified only by a rise in hCG (human chorionic gonadotrophin, which rises with implantation and falls with pregnancy loss) be called a biochemical loss, that those lost at >12 weeks gestation where fetal measurement was followed by a loss of fetal heart activity be called late pregnancy loss, and that the group falling between these two categories be called early pregnancy loss. Though these definitions may be useful in clinical settings in industrialised countries, most human conceptions take place far removed from the necessary diagnostic tools.

In the extensive literature on pregnancy outcomes, late pregnancy loss has referred to those occurring at >12 weeks, >20 weeks, or >28 weeks. Early pregnancy loss may refer to those occurring before any of these cutoffs, or be restricted to those losses before 6 weeks or to those losses before awareness of pregnancy (occult or subclinical pregnancies). Terminations before 6 weeks have also been classed as very early pregnancy loss. Hence a conceptus lost at 21 weeks may be designated, depending on the scheme, as either a late pregnancy loss or an early fetal loss.

Because of the ample opportunity to generate even further confusion, in this chapter I will consider any conception not ending in a live birth to be a *pregnancy loss* regardless of the duration of pregnancy or the stage of the development or the method of detection (if detected). *Miscarriage* will not be used, and *abortion* will refer only to induced abortion. In the present context, all discussion of pregnancy loss will be restricted to terminations not brought about by induced abortion, although it should be noted that the WHO definition does not make this distinction. I will avoid the qualifiers *early* and *late* in favour of specifying durations in days or weeks.

WHO defines the period of gestation as beginning with the first day of the last menstrual period (LMP) and ending with the day of birth or day of termination of pregnancy.[7] By this definition, typical gestation in humans is 40 completed weeks (i.e., 280–286 days). In fact, gestation cannot have begun before ovulation, which cannot have taken place, on average, until about 14 days after LMP (Figure 6.1). Nonetheless, gestational age (GA) as defined by WHO is in very wide use, and it seems unwise to start redefining it now. Rather, as necessary in the present discussion, I will specify the postconception duration (PCD). The advantage of this term is that it better reflects the actual time span during which reproductive events are taking place. Hence, the PCD at implantation varies from 6 to 12 days (mean = 9 days[10]) in successful pregnancies, and the typical PCD to a live birth is 38 completed weeks (i.e., 266–272 days). In the absence of additional information, PCD can be estimated as the time since LMP minus 14 days. The use of hormonal indicators (for example, over-the-counter kits that measure the rise in lutenizing hormone in urine that signals ovulation; Figure 6.1) allow much more precise estimates of PCD.

Estimates of the Probability of Pregnancy Loss

In 1975 the *Lancet* published a letter[11] with "back of the envelope" calculations of how many births might be expected, given reasonable assumptions regarding coital frequency and such, among married women aged 20–29 in England and Wales during 1971. Comparing this estimate to the actual number born, the authors concluded that "Nature" had terminated at least 78% of all conceptions in this population. Although there was scant empirical evidence at the time to support such a high rate of pregnancy loss, more recent analyses are consistent with this estimate.

   In 1990, Boklage[12] combined the published results of several observational studies of pregnancy loss in industrialised populations and developed a parametric model from which he estimated a total pregnancy loss of about

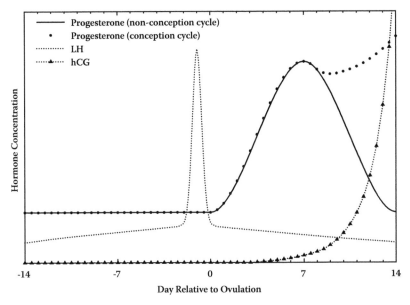

Figure 6.1   Hormonal profiles during conception and non-conception cycles (ovulation = 0 days PCD, LMP = –14 days, mid-luteal = 7 days PCD, implantation = 9 days PCD). The peri-ovulatory period (centered at day 0) is characterized by a sharp high surge in lutenizing hormone (LH). Before ovulation, progesterone is typically low and flat, rising after ovulation to a mid-luteal peak and then falling to basal levels in the absence of a conception. If fertilization occurs, rising hCG, produced by the conceptus, will rescue the corpus luteum and progesterone will continue to rise.

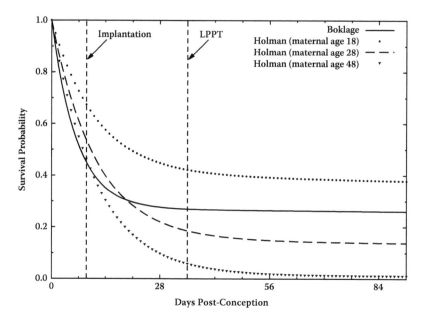

Figure 6.2 Survival probabilities for pregnancy from gestation through term estimated from Boklage's[12] model and Holman's[13,14] model for three age groups: 18-year-olds, 28-year-olds, and 48-year-olds. At a given PCD, the risk of pregnancy loss to that point in time = 1-survival probability. LPPT = Luteoplacental progesterone-transition.

76% (Figure 6.2). Holman[13,14] monitored nearly 500 noncontracepting married Bangladeshi women and collected urine samples, later assayed for hCG, from 1,561 menstrual cycles. Of 329 followed pregnancies, 151 went to term, 84 ended before the woman was aware of the pregnancy, 10 ended after she had become aware she was pregnant, 3 ended by induced abortion, and 81 were ongoing at the study's end. From these data, he estimated that the probability of pregnancy loss in 20-year-old women is about 0.55, in 30-year-old women about 0.84, and in 40-year-old women about 0.96 (Figure 6.2). Assuming a population with an equal distribution of conceptions across these ages, the total pregnancy loss would be about 80%.

Despite the general agreement of these two studies, it should be remembered that their estimates rely on some assumptions that may prove not to hold as more data become available. At this time, the detection of natural conceptions in the large-scale studies that are necessary for estimating rates of pregnancy loss relies on measuring the rise of hCG that occurs with implantation. Hence, preimplantation losses remain unrecognised. It is entirely possible that preimplantation loss rates are significantly lower than currently thought. It is very unlikely, however, that the average rate of pregnancy loss is higher than 80% because the monthly probability of producing a conception that is clinically recognised (i.e., apparent fecundability) has been estimated by several studies to be about 0.20 to 0.30.[15] Assuming true fecundability (i.e., the monthly probability of conceiving) is 1.0, then the upper limit on loss from conception to clinical recognition would be about 0.70 to 0.80.

The Distribution of Losses over the Course of Gestation

Preimplantation Loss Rates

However large the probability of pregnancy loss may be, it is widely agreed that the overwhelming majority of these are terminated within the first few weeks after conception. Many of these may occur even before implantation has begun, but because preimplantation natural conceptions are virtually undetectable in epidemiological studies, pregnancy loss in this period must be extrapolated from models. As of 9 days PCD (the average time to implantation in successful pregnancies[10]), Boklage's model[12] estimates that about 55% of all conceptions have already been lost (Figure 6.2). Holman's model[13,14] predicts 28-year-old Bangladeshi women lose about 46% of their conceptions before implantation (Figure 6.2).

Postimplantation Loss Rates

As noted earlier, there have been various schemes for classifying postimplantation pregnancy losses. Until the advent of highly sensitive tests for urinary hCG, neither a woman nor her clinician typically recognised the occurrence of a conception before 4 weeks PCD (6 weeks LMP), hence pregnancy losses during this period were referred to as subclinical, occult, early, or very early. However, the arbitrariness of a 6-week cutoff for distinguishing early from late pregnancy loss has become more apparent with further study, particularly in light of the increasing clinical capacity to identify conceptions within a few days of the start of implantation. Rather, biological markers may prove more useful for defining gestational periods, particularly as the etiology of pregnancy loss is likely to vary over the course of gestation as a consequence of biological changes. The initiation of implantation is one such useful biological marker, recognizable from a rise in hCG.

Another such marker is the shift in the dominant source of progesterone. Progesterone is critical to uterine development, implantation, and pregnancy maintenance.[16,17] During the luteal phase (i.e., following ovulation), the corpus luteum, formed from the ruptured follicle, is the principal source of progesterone until about 5–6 weeks after conception, at which time placental production begins to predominate[18,19] (Figure 6.1). During this initial period, but not afterwards, either surgically removing the corpus luteum or biochemically blocking progesterone with mifepristone leads to pregnancy failure.[20,21] Thus, the transition from maternal to placental production of progesterone marks a significant shift in the control of pregnancy maintenance.

While it may be difficult to actually observe this shift, distinguishing pregnancy losses bracketed by implantation and the luteo-placental-progesterone transition (LPPT) at about 5 weeks PCD is both biologically justifiable and consistent with evolutionary arguments (discussed later). Two recent studies adopted this criterion in their analyses of pregnancy loss.[22–26]*

Relying on sensitive tests for detecting the rise and fall of urinary hCG, several large prospective studies[10,27–30] in women attempting natural conception have estimated the rate of pregnancy loss from shortly after implantation has begun through either 5 or 6 weeks LMP (about 21–28 days PCD). The exact breadth of this window depends on the assay, the sampling frequency, and the criteria for recognizing a conception and loss. A more sensitive assay will detect more conceptions closer to the time of implantation, and hence report a higher rate of pregnancy loss. More stringent criteria may miss some conceptions and losses, and lead to reporting a lower rate. Nonetheless, despite these and other differences in study design, these studies of industrialised populations[10,27–30] have reported similar pregnancy loss rates for this gestational period, ranging from about 25% to about 30% of implanted concepti. Because of the various biases specific to the different study designs, it is not possible to say if the true values are actually similar among all these samples, or if they are widely disparate. Boklage's model[12] estimated the pregnancy loss for the period from 9 through 35 days PCD to be about 40% (Figure 6.2).

Anthropologists studying nonindustrialised populations have also reported high levels of pregnancy loss. A study of rural Bolivian women[24–26] observed that 31% of implanted concepti (detected by a rise in hCG) were lost prior to the LPPT (35 days PCD). Based on hazard models (which are expected to generate higher and more accurate rates than calculation of the fraction of all observed pregnancies that are subsequently lost), Holman[13,14] estimated the rate of loss to be about 65% during the period from implantation through 35 days PCD in 28-year-old Bangladeshi women (Figure 6.2). Among settled Turkana agriculturalists in Africa, Leslie and colleagues[31] found the rate of loss to be about 70% of those implanted pregnancies that had been hormonally detected at some time before 12 weeks LMP (about 70 days PCD).

Post-LPPT Loss Rate

There are many studies of pregnancy loss rates in "clinically recognised" conceptions, that is, pregnancies that have lasted long enough to be recognised as having occurred by the mother or her clinician (i.e., usually >6

---

* Because gestation is a continuous process, the boundaries defining these periods are necessarily fuzzy. Nonetheless, as with the classification of gestation into trimesters, the periods are conceptually and analytically useful.

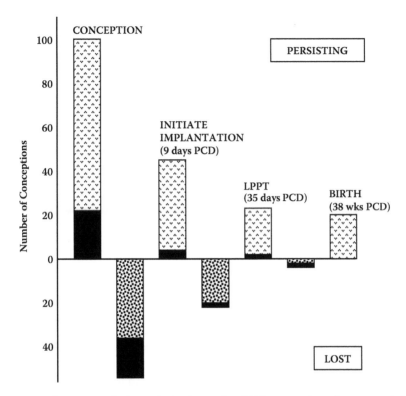

Figure 6.3 Outcomes of a hypothetical sample of 100 conceptions. Losses are below the horizontal line; conceptions persisting to each stage are above the line. That segment of either lost or persisting conceptions that are likely to be genetic defects are blackened.

weeks LMP). Pregnancy loss rates from about the time of the LPPT through term are typically reported to be about 10–15%.[15]

In sum, based on the collective evidence, if one started with 100 conceptions, one could expect 20 live births. About 55 of the conceptions would not have implanted; of the remaining 45, about 22 would have been terminated prior to LPPT; of the remaining 23, about 3 would be lost post-LPPT (Figure 6.3).

Principal Proximate Causes of Pregnancy Loss

Despite the high frequency with which they occur, the causes of most pregnancy losses are uncertain,[32,33] principally because the vast majority of all conceptions are unrecognised by either the woman or her clinician. Conceptions ending before implantation are not detectable in epidemiological studies, and an unknown proportion of the those ending postimplantation (post-I) but prior to LPPT also go undetected even with the most sensitive assays.

Most of the etiological evidence concerns that fraction of losses (about 15%) occurring in clinically recognised pregnancies (i.e., those terminating post-LPPT or >5 weeks PCD). Roughly half of these losses (i.e., about 8% of clinically recognised pregnancies) are associated with genetic abnormalities.[34] A review of the earliest cytogenetic studies led Bishop[35] to propose that chromosomal abnormalities are responsible for most pregnancy losses, particularly those terminating within a few weeks after conception, as unviable concepti are weeded out.* Although Bishop's hypothesis is both plausible and widely accepted, extrapolating a dramatically increasing proportion of abnormal concepti backwards from LPPT to conception remains conjectural.†

Maternal age is a well-established risk factor for post-LPPT loss[15] and appears to be a risk factor for pregnancy loss during earlier stages of gestation. In Bangladeshi women, the estimated total risk (from conception to term) for 40-year-olds was nearly double that for 20-year-olds (0.96 versus 0.55).[13,14] The relative age-associated difference in risk persisted, more or less, from conception throughout pregnancy (Figure 6.2). A prospective study of post-I/pre-LPPT loss in Bolivian women estimated risk during this period at 7% for 20-year-olds, rising to 75% in 40-year-olds.[26] In contrast to these two studies, maternal age was not found to increase the risk of loss between implantation and 6 weeks LMP in a study of U.S. women (North Carolina Early Pregnancy Study [NCEPS][36]), perhaps because of the limited age range of the participants.

The age-associated risk for pregnancy loss has long been attributed to the increasing risk of chromosomal abnormalities with maternal age, evidenced by a higher proportion of such defects among the pregnancy losses of older women (85% of all losses at >40 years maternal age versus 42% of losses at <20 maternal age).[15] This risk for genetic defects has been thought to be the consequence of the aging of a fixed supply of oocytes that is established during the woman's own fetal development. Consistent with this model, fertility rates are more highly correlated with the age of ovum donors than with the age of the recipient uterus.[37-40] On the other hand, there is new and compelling evidence for a population of differentiating and regressing follicles in adult women.[41-44] Perhaps the age-associated increase in post-LPPT loss and the increase in the proportion of chromosomal abnormalities among these losses are not simply or solely the inevitable consequences of old eggs. Clearly more work is needed to resolve the inconsistencies among these findings. As summed up by Pellestor and colleagues,[45] "there is no clear evidence of any simple explanation of the relationship between maternal aging and the occurrence of aneuploidy [chromosomal defects]."

Maternal use of cigarettes or other drugs are known risk factors for post-LPPT loss, and there is some evidence that high levels of caffeine or alcohol consumption may be additional risk factors.[32,34] A recent Danish study[46] reported

---

* See Holman and Wood[14] and Wood[15] for detailed discussions of this hypothesis.
† See additional discussion on this point in the section "The Role of Maternal Evaluation of Resources in Pregnancy Loss."

that very high levels of alcohol consumption around the time of conception were associated with an increased risk of loss among all detected conceptions. Sample sizes in this study precluded distinguishing any statistically significant increase in risk specific to losses occurring prior to 6 weeks LMP. Likewise, the NCEPS did not find an association between any of these behavioural risk factors and risk of pregnancy loss occurring prior to 6 weeks LMP.[36]

Several endocrine, autoimmune, and thrombotic abnormalities are associated with recurrent pregnancy loss, but these conditions are uncommon.[32] Though maternal infection is a known risk factor, it also does not appear to account for a substantial proportion of pregnancy losses, at least in industrialised countries. But where infectious disease is endemic, the story is likely to be different. Both malaria and HIV, for example, are known risk factors for pregnancy loss, and changing economic practices can expose populations to disease. The very high rate of pregnancy loss among newly settled Turkana agriculturalists may be attributable to malaria.[31] In industrialised populations, ethnicity and socioeconomic class are risk factors for post-LPPT pregnancy loss, but it is unclear exactly which specific characteristics underlie the associations.[15]

Life History Theory (LHT)

Life history theory is evolutionary economics. It concerns the fundamental challenge faced by all life forms: What is the optimal allocation of finite resources within a finite lifetime? The different possible schedules of investments in growth, reproduction, and somatic maintenance are referred to as life history strategies. These schedules can differ with respect to the age and size at initiating reproduction; the number, quality, and timing of offspring; and the age at death, among other life history variables. The relative optimality of a life history strategy is measured in the currency of reproductive success (fitness). Natural selection favours the strategy that results in an organism leaving relatively more copies of its genes in subsequent generations than would have been left by following some other strategy. With enough time in a given environment, natural selection will result in most members of a population following the optimal strategy, to varying degrees. Changes in environmental conditions can prompt selection for a different optimal life history strategy. Life history variables are often quite flexible in their expression, depending upon the specific environmental conditions encountered by the organism.*

An organism inevitably faces trade-offs in the allocation of the resources available to it. Energy devoted to somatic maintenance may prolong life but would then not be available for reproduction. On the other hand, a new investment in reproduction can place both one's own survival and future reproductive opportunities at risk. Hence, foregoing a current reproductive

---

* To learn more about LHT, see references 47–49.

opportunity, or terminating one before too much has been invested, can be the best lifelong strategy.

The formalisation of these trade-offs began with Fisher,[50] who defined reproductive value (RV) as the mean future reproductive success (i.e., how many offspring the organism can expect to produce) for those of a given age and sex in that population. Williams[51] divided RV into that which is immediately at stake (the current reproduction, CR) and the residual reproductive value (RV = CR + RRV). He defined *a* to be the proportional gain to CR, and *c* to be the proportional cost to RRV, when a positive allocation decision is made by the organism (e.g., forage, ovulate, defend, mate). In the case of a negative allocation decision (e.g., do not forage, do not ovulate), he defined *b* to be the proportional decrement to CR. Hence,

$$RV \text{ given positive decision: } RVp = (1 + a)CR + (1 - c)RRV$$
$$RV \text{ given negative decision: } RVn = (1 - b)CR + RRV$$

If *RVp* (the gain from investing in current reproduction plus the reduced future residual reproduction due to the current investment) exceeds *RVn* (the reduced current reproduction by not investing plus the future residual reproduction), then selection will favour the positive allocation decision and it will become the normative decision. Alternatively, the negative allocation decision would become normative if *RVn* > *RVp*.

Perhaps the best-studied example of such a trade-off in humans is the investment in breast feeding one's youngest offspring.[52] Depending upon the frequency, duration, and intensity of breast feeding, and on the age of the nursling, lactation suppresses ovulation, thus preventing investment in another pregnancy. In the absence of breast feeding, human females ovulate, on average, about 6 weeks after giving birth. Such a swift shift of investment from a newborn to a potential future progeny would, however, risk the survival of the newborn. On the other hand, too long an investment in breast feeding risks future reproductive potential. The trade-off between these two costs must be balanced. Although it has been speculated that breast feeding could be a long-term "natural contraceptive," life history theory argues against such an expectation. In fact, efforts to extend LAM (the lactational amenorrhea method) much beyond 6 months postpartum have had mixed results.

It is in a female's best interests to allocate resources to as many viable offspring as possible while apportioning just enough to each to ensure its survival to reproductive maturity.* A female mammal has the greatest con-

---

* In many, perhaps most, species, a female's investment in her offspring ends with (or even before) the offspring's reproductive maturation. In some species, female investment may extend beyond reproductive maturation in order to increase the probability that her offspring produce offspring of their own. In these cases, apportionment of resources may reflect very long-term as well as shorter-term strategies. Such an investment strategy has been argued to occur in humans (the grandmother hypothesis[53]).

trol over the decision to invest in a given reproductive opportunity during the time before a conception occurs. Absence of ovulation is, for example, an irreversible rejection of current investment. Once a conception occurs, allocation decisions are no longer entirely up to her. In sexual reproduction, offspring are usually not identical to each other and each carries only half of its mother's genes. Trivers[54] examined the competing interests created by this nonidentity. Natural selection favours any offspring that is able to increase its own chance of survival, even if at some cost to its siblings and contrary to the mother's optimal allocation strategy. In humans, as in other mammals, the conceptus's intimate link to the maternal nutrient supply via the placenta offers a unique opportunity to exploit maternal resources. At the same time, natural selection also favours maternal mechanisms to counteract the offspring's efforts to gain more than its "fair share" (i.e., the optimum allocation from the mother's view).

Williams's elegant formulation of trade-offs and Trivers's seminal insights into the competing interests of mothers and progeny both prompted an impressive legacy of scientific inquiry—each has been cited at least 600 times.

The Role of Maternal Evaluation of Resources in Pregnancy Loss

Life history theory makes it evident that organisms must evaluate resource availability in order to decide when and how much to invest in reproduction. In this context, *resources* should be understood in the broadest sense. Depending upon the species, resources comprise the energy available to support a growing conceptus; the availability of a suitable habitat, mate, and social group; and the physical and psychological status of the mother. Time remaining until the termination of reproductive capacity (which is equal to time till death in most species but is time till menopause in women) may also be thought of as a critical resource. All of these resources contribute to the relative values of RVp and RVn at any given reproductive opportunity. Hence, selection will favour reproductive investment whenever $RVp > RVn$, even if some single resource (e.g., food availability) does not appear to be particularly propitious. Likewise, selection will favour delaying investment in reproduction whenever $RVn > RVp$, even if some relevant resources are favourable.

Wasser and Barash[55] explored the potential advantages of delayed reproduction in female mammals under a broad array of circumstances, including variation in resource availability. Their formulation (the Reproductive Suppression Model) is a special case of Williams's model.[51] They proposed that age (an indicator of time remaining until the end of reproduction) and its associated physical changes are perhaps the most reliable predictors of RV, while nutritional status, disease, and psychosocial stress are less reliable cues. They examined the evidence for social suppression, resulting from "one's interactions with, and the reproductions of, other individuals." Under conditions of strong social competition, they argued that it could be beneficial to delay or terminate reproduction until less socially competitive times. In addition, a female might also increase her own reproductive success by manipulating reproduction in the other members of her social group.

Trevathan[5] has brought to the fore the extraordinary and unique sociality of human parturition. Unlike other mammals, human births in isolation carry a higher risk of mortality than those accomplished with a supportive attendant. Trevathan proposed that the evolution of bipedalism in early humans increased the risks of childbirth and that selection favoured those early hominid females that sought assistance and companionship to reduce the emotional stress and pain, and aid in the passage of the neonate. The apparent importance of such assistance to ensure the survival of both the infant and mother argues that adequate psychosocial resources could be as important as energetic resources in a human female's decision whether to invest in a current opportunity for reproduction.

Drawing on the literature from genetics and developmental biology, Vitzthum[56-58] developed the Flexible Response Model (FRM) of reproductive functioning to explain variation among women in reproductive suppression under seemingly comparable conditions. This work was prompted by an apparent paradox: Why are the ovarian cycles of healthy U.S. women easily disrupted by dieting and exercise while women in nonindustrialised countries have high fertility despite arduous physical labour and poor nutrition? The answer lies in the advantage gained by U.S. women (who usually experience good conditions) of delaying reproduction in the face of temporarily poor conditions: they can expect things to get better soon. Most women in poor countries, on the other hand, cannot expect conditions to improve. Hence, delaying carries no benefit.

Vitzthum argued that an organism judges environmental conditions based on the conditions experienced as it matured. The best conditions encountered before adulthood are likely to be about the best conditions one will experience in a lifetime. It makes no sense to delay reproduction under the prevailing conditions. Of course, if things get even worse, then these women can also be expected to delay their reproductive investments for a time. However, as long as the probability of a successful conception in these poorer conditions is greater than 0, women can be expected to acclimate* over time and to resume reproductive functioning. Organisms able to acclimate will have a reproductive advantage over those that do not. The FRM applies to both pre- and postconception reproductive decisions. Because poor conditions were the norm during their maturation, women accustomed to such may maintain their pregnancies in the face of stressors that would prompt termination in better-off women.

---

* Acclimation is the process of adjustment that returns an organism to homeostasis (i.e., normal functioning) even if the factor that originally disrupted homeostasis persists.

Williams's model[51] laid the groundwork for understanding the trade-offs between current and future reproductive investments. Trevathan's[5] and Wasser and Barash's[55] work emphasised the role of socially dependent resources in mammalian reproductive decisions. Vitzthum's model[56-58] proposed that organisms integrated the information gained from the prevailing conditions experienced during maturation into resource evaluation during adulthood. Each of the models offered insights into the evolutionary logic behind pregnancy loss and the dependency of such loss on maternal evaluation of resources.

The long-standing dogma that most human pregnancy losses, especially those in the earliest weeks of gestation, are due to genetic defects would suggest that variation in maternal resources plays a very minor role in pregnancy loss. Recent evidence, however, argues that genetic abnormalities are not necessarily the principal cause of pre-LPPT losses. Taken collectively, various studies place the rate of chromosomal abnormalities at about 15% and 8% in human oocytes and spermatozoa, respectively.[45,59] Hence, assuming random fertilisation and no disfavouring of abnormal gametes, at conception the maximum proportion of abnormal concepti from these causes is 22%. From fertilisation through LPPT, the risk of pregnancy loss is estimated to be about 73% in Boklage's model[12] and about 81% in 28-year-olds in Holman's model[13,14] (Figure 6.2). Hence, there is a large proportion of pregnancy losses that are not readily attributable to chromosomal defects (Figure 6.3). There may, in fact, be even fewer abnormal gametes than suggested by the available data. The oocytes examined for genetic defects were typically from (often older) women undergoing assisted reproduction. Hence, 15% maybe an overestimate of the true rate in the larger population. On the other hand, there may be genetic defects not detectable with current methods, and hence the true rate may be even higher than 15%. In either case, the present evidence argues that it cannot be assumed that the bulk of pre-LPPT losses are due to genetic abnormalities.

The life history models discussed above logically lead us to predict that there are physiological mechanisms for maternal evaluation of the resources available for investment in reproduction. Without such mechanisms it would be impossible for an organism to implement life history strategies. Evidence for the physiological mechanisms of resource evaluation in women, considered next, comes from recent prospective studies of human conception and pregnancy loss.

Evolutionary Endocrinology: Maternal Evaluation of Resources

> It would be instructive to know...by what physiological mechanisms a just apportionment is made between

> the nutriment devoted to the gonads and that devoted
> to the rest of the parental organism.[50]

The endocrine system plays a central role in orchestrating an organism's response to external signals and stimuli.[60–64] Hormonal changes can, for example, trigger or suppress gene transcription and modify metabolic rates. In women, evidence from laboratory, clinical, and field studies collectively indicate that reductions in energy intake or increases in energy expenditure (i.e., changes in maternal resources) are associated with decreased levels of ovarian progesterone and estradiol,[65] and that low levels of these steroids are associated with a lower probability of conception.[66]

A plausible working hypothesis is that modulation of reproductive effort by a given woman involves varying these hormones. From an evolutionary perspective, lowering ovarian steroid levels in response to a reduction of maternal resources is a potentially adaptive strategy to delay reproduction until these resources are either replenished or until the cost to RV of waiting is greater than the cost of not trying to reproduce. The endocrinological mechanisms that appear to be responsible for preconception reproductive decisions in response to maternal resources would also appear to be logical candidate mechanisms for modulating postconception reproductive investment.

Two prospective studies of conception and pregnancy loss have tested the hypothesis that pregnancy loss will be associated with relatively lower levels of ovarian steroids. The North Carolina Early Pregnancy Study (NCEPS) recruited 221 women attempting to become pregnant naturally.[10,36,66,67] They self-collected daily urine samples subsequently assayed for progesterone and estrogen metabolites, and hCG. Twenty women had a successful term pregnancy subsequent to experiencing a pregnancy loss. Contrary to the hypothesis that ovarian steroid variation will be associated with pregnancy outcome, the hormonal profiles of the successful and lost conceptions did not differ from ovulation through early implantation.[67]

Project REPA (Reproduction and Ecology in Provincía Aroma) is a multidisciplinary longitudinal study of reproductive functioning and health among rural Aymara families indigenous to the Bolivian *altiplano*.[24–26] Of 316 adult female participants, 191 menstruating women in stable sexual partnerships were visited every other day to record menstrual status and collect a saliva sample for subsequent assay for progesterone. Tests for urinary hCG detected 65 conceptions, of which 23 were observed to term, 27 were lost before the third trimester, and 1 was medically aborted; 14 women withdrew from the study while still pregnant, principally due to waning participant interest. Similar to the findings in the NCEPS, progesterone levels through the luteal phase and during early implantation (i.e., through day 12) in those conceptions lost prior to the LPPT (i.e., <35 days PCD) were comparable to progesterone levels in pregnancies that persisted longer (Figure 6.4).[25]

In sum, despite numerous observations that ovarian steroids vary with maternal energetic status, and may be acting as a mechanism to regulate

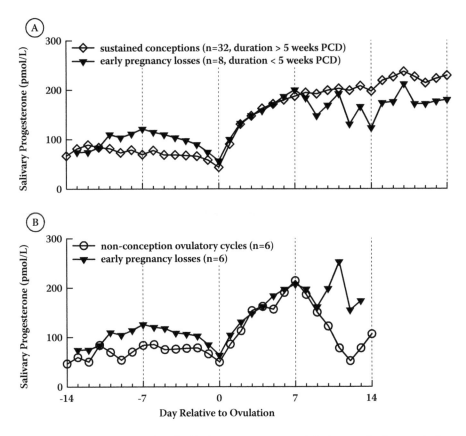

Figure 6.4 Progesterone levels (daily geometric mean) before and after ovulation (day 0) in naturally occurring conception and non-conception cycles. Pre-ovulatory (adrenal) progesterone was significantly higher in the sample of conceptions lost prior to LPPT compared with those pregnancies persisting for at least 5 weeks after conception (Panel A) and compared to non-conception ovulatory cycles (Panel B). Post-ovulatory (luteal) progesterone levels were comparable in all three samples. From [25].

reproductive investment prior to conception, both studies of pregnancy loss failed to support the hypothesis that variation in ovarian steroids acts to modulate reproductive investment during the period from conception through early implantation.

Another putative mechanism for maternal evaluation of resources involves cross-talk between the hypothalamic-pituitary-ovarian (HPO) and hypothalamic-pituitary-adrenal (HPA) axes. During most of the follicular phase (i.e., before ovulation), the adrenal cortex is the main source of progesterone,[68–70] typically produced at levels much lower than those of the progesterone produced by the ovaries following ovulation (Figure 6.1).

Although adrenal progesterone had been thought to play no role in reproduction, there is now evidence to refute that assumption. For example, ovariectomised monkeys and postmenopausal women, when experimentally subjected to stress in the presence of replacement estradiol, exhibited a rise in glucocorticoids and adrenal progesterone that triggered surges in LH.[71,72] Such excessive or poorly timed LH surges were hypothesised to impair follicular development. A number of factors, including emotional stress, physical activity, and food restriction, can stimulate the HPA. Hence, a plausible hypothesis is that elevations in cortisol or adrenal progesterone act as signals of maternal resources and may be part of a mechanism to moderate maternal investment in reproduction.

Evidence that this mechanism is acting in pregnancy loss comes from two studies. In a recent prospective investigation of the relationship between stress and pregnancy outcome in rural Guatemalan women, 24 participants self-collected thrice-weekly urine samples later assayed for cortisol, gonadotrophins, hCG, and ovarian steroids.[22,23] Of 92 menstrual cycles, 22 conceptions were observed, of which 9 went to term and 13 were lost within 47 days after ovulation (about 47 days PCD). In ovulatory cycles, rises in cortisol were accompanied by relatively higher follicular-phase progesterone and LH, which the authors suggested could impair fecundity.[22] Moreover, those conception cycles with high levels of cortisol had a 2.7 times (95% confidence interval = 1.2–6.2) greater risk of post-I/pre-LPPT loss than those with normal cortisol levels. While a third of the normal cortisol conceptions were lost, 90% of those conceptions characterised by elevated cortisol were terminated prior to the LPPT.[23]

Project REPA in Bolivia observed that progesterone levels during the follicular phase were significantly higher in those pregnancies that terminated prior to the LPPT compared to those pregnancies that persisted beyond this transition (Figure 6.4).[25] Cycles that ended in a pregnancy loss were also compared to prior ovulatory cycles from the same women. Luteal-phase progesterone was comparable in the paired samples, but follicular-phase progesterone was significantly higher in the cycle that ended in loss. Hence, elevated follicular-phase progesterone did not typically characterize these women's cycles. Rather, the elevation in follicular-phase progesterone was specific to the pregnancy loss cycle, as would be expected if the elevated progesterone is a signal of context-specific maternal resources.[25]

The temporal pattern of pregnancy loss in the Bolivian study also supports the hypothesis that the availability of maternal resources plays a role in the maternal decision to terminate investment in reproduction. The planting and harvesting seasons had a 3.7 times greater risk of loss than the other seasons, and agropastoralists were 9 times more likely to experience post-I/pre-LPPT loss than those engaged in some other livelihood.[26] These increased risks for pregnancy loss were attributed to arduous physical labour, although the authors noted that inadequate food reserves and greater psychological or immunological stress could also be contributory factors.

Seasonality in pregnancy loss is not restricted to populations engaged in demanding physical labour. Observed losses in the NCEPS also displayed a marked seasonality, with a peak (at least four times greater than the trough) occurring some time from September through December in three consecutive years.[73] The reason for the seasonal variation was uncertain. But unless there are as yet unrecognised reasons for seasonality in chromosomal abnormalities, these temporal variations argue that there are factors other than unrecognised genetic errors responsible for a significant proportion of post-I/pre-LPPT pregnancy losses, at least in some populations. From an evolutionary viewpoint, these factors are most likely to be varying availability of the resources necessary for reproduction.

The Role of Offspring Quality in Pregnancy Loss

Even with abundant resources, offspring must merit maternal investment. The proportion of pregnancy losses with chromosomal abnormalities is highest early in pregnancy, dropping sharply by the end of the first trimester.[15] A defective conceptus that is unlikely to mature and contribute genes to subsequent generations should be rejected as quickly as possible if doing so frees the parent to invest in offspring of higher quality.[74-78] Although many genetic defects preclude normal development, and hence can be expected to terminate of their own accord, selection would favour any maternal mechanism, presumably hormonal, that detects and eliminates poor-quality offspring early in development.

Despite the evolutionary logic of these arguments, the age-associated increasing risk of having a live birth with genetic defects would appear to challenge the concept of a maternal screening mechanism. Proposals that the putative screening mechanism weakens, for whatever reason, with maternal aging have suffered from a lack of unequivocal empirical support.[79-86] In particular, the strong evidence for age-associated increases in genetic defects in gametes has been taken to be a sufficient explanation, without invoking an age-sensitive screening mechanism, for the increasing risk of these defects in live births. However, comparing older to younger women, the increase in chromosomal abnormalities in live births is perhaps twenty-five-fold, whereas the available evidence suggests that genetic abnormalities in oocytes are only about three- to ninefold greater.[45,87-89] Clearly, an age-related increase in poor-quality oocytes is not a sufficient explanation for the increased risk of genetic defects at birth among the offspring of older women.[89]

Drawing upon the work of Haig,[77] Kloss and Nesse,[86] and others, Forbes[89] has argued that a hormonal screening mechanism should not be expected to weed out all defective concepti. To do so, all low-quality offspring would have to be characterised by a hormonal level that is lower than that of all better-quality offspring. Rather, he proposed, it is more likely that most low-quality offspring are below some threshold that is still within the range of normal offspring (Figure 6.5). Maternal rejection of those embryos below

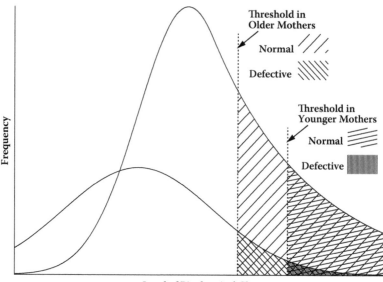

Figure 6.5 Threshold model for maternal evaluation of offspring quality. Normal embryos are characterized by a range of the hormonal cue (e.g., hCG); defective embryos are characterized by a lower, and perhaps broader, range of the cue. In younger women, the threshold is set high; though normal embryos will be rejected, the probability of allowing a defective conceptus to continue is very low. In older women, the threshold is set lower; though the risk of carrying a defective embryo to term is several-fold higher than in a younger woman, the risk of rejecting a normal embryo is much reduced. (Modified and reprinted from *Trends in Ecology and Evolution*, 12, L.S. Forbes, The evolutionary biology of spontaneous abortion in humans, pp. 446, Copyright 1997, with permission from Elsevier.)

the threshold would carry the unavoidable risk of eliminating desirable offspring. But the cost of such an error appears to be quite low. Women typically ovulate in the cycle following a pregnancy loss that has occurred prior to the LPPT. In the NCEPS, clinically recognised pregnancies occurred in the subsequent cycle in 35% of those women who had experienced a pregnancy loss within 6 weeks LMP.[90]

Forbes[89] noted that rejecting a viable conceptus, however, would be expected to be more costly as a woman ages and the opportunities for reproduction dwindle. Hence, any screening mechanisms would be expected to relax with increasing maternal age in order to reduce the probability of terminating a viable offspring (Figure 6.5). The cost, of course, is an increased risk of allowing a poor-quality embryo to continue. As Forbes discussed, it seems unlikely that bearing an offspring carrying a serious genetic defect is adaptive. Even if born alive, these offspring are at elevated risk for neglect and infanticide. However, the chance of bearing such an offspring is still only one in twenty at the oldest maternal ages;[87,88] the odds remain strongly in favour for a normal offspring.

Evolutionary Endocrinology: Evaluating Offspring Quality

Haig[76-78] proposed that adequate production of hCG by the conceptus is a principal cue by which human mothers evaluate offspring quality. In the absence of conception, the corpus luteum regresses; progesterone, having peaked in the mid-luteal phase, subsequently falls to basal levels; and menstrual bleeding signals the end of the cycle (Figure 6.1). If conception occurs, embryos must secrete enough hCG to bind to the LH receptors of the corpus luteum to prevent its regression. This "rescue" of the corpus luteum ensures the continued production of ovarian progesterone, essential to the maintenance of a pregnancy. The production of adequate levels of hCG, which may begin as early as the four-cell stage of embryogenesis,[91] is a measure of the embryo's capacity to carry out transcription, translation, and glycosylation (that is, the steps in protein production), the most minimal requirements of viability.[77] Mothers can be expected to reject early any embryo not meeting this minimal criterion.

An unknown proportion of concepti fail this first test, and menstruation renews the cycle. For those embryos who have managed rescue of the corpus luteum and initiated implantation, the threat of maternal termination still looms large (Figure 6.3). At least 30% of these, probably more, will be terminated before the LPPT despite the conceptus's attempt to gain influence over progesterone secretion by rapidly raising the level of hCG (Figure 6.1). Once the placenta has become the dominant source of progesterone, shifting a principal locus of pregnancy control from the mother to the offspring, hCG stops rising. Peaking at about 8 weeks PCD, hCG falls to about a quarter of its maximum level, subsequently fluctuating little during the remainder of a normal pregnancy.[15]

Data from the NCEPS[10,92] demonstrate the dependency of pregnancy continuation on both a sufficient rise in hCG and the timing of that increase. Implantation was assumed to have begun with a rise in urinary hCG to >0.015 ng/ml. All of the conceptions that persisted longer than 6 weeks LMP (about 28 days PCD) surpassed this hCG threshold at some point during 6–12 days after ovulation (mean = 9.2 days); 84% of these surpassed the hCG threshold on days 8, 9, or 10 after ovulation. Detected pregnancies subsequently lost within 6 weeks LMP surpassed the hCG threshold some time during days 6–18 (mean = 10.5 days). The later the hCG threshold was surpassed, the greater the risk of pregnancy loss. Of 102 pregnancies implanting by day 9, 13% were lost prior to 6 weeks LMP; of the 11 pregnancies implanting on day 12 or later, 9 (82%) were lost within 6 weeks LMP.

The timing of the progesterone rise that followed the hCG rise was also predictive of pregnancy termination.[92] Only 5% of those implanted conceptions that had a progesterone rise within 2 days of implantation (45% of the total) were lost prior to 6 weeks LMP. Of those with a progesterone rise later than 2 days after implantation, 18% were lost; of those with no associated rise in progesterone, 22% were lost. All of the implanted conceptions that showed a decline in progesterone were lost within 6 weeks LMP (28 days PCD). Because preimplantation progesterone levels were comparable in conception cycles that ended in loss and in those that were successful, the failure to show an appropriately timed progesterone rise was attributed by the authors to the strength of the hCG signal rather than to some failure of the corpus luteum to respond to the hCG signal. The findings of a study of artificially inseminated women who otherwise conceived naturally[93–95] generally agreed with those of the NCEPS. In conceptions ending in pregnancy loss, hCG production was delayed and lower than in conceptions that proceeded normally.

In sum, the first quality test for a human offspring comprises conceptus production of hCG in sufficiently high levels within a brief window of opportunity to produce a marked early rise in progesterone. Intriguingly, trisomy 21 (Down syndrome) is characterised by a markedly elevated level of hCG, an attribute that is now the basis for prenatal screening for this aneuploidy.[88] This hormonal signature may help to explain the high rate of trisomy 21 in live births,[89] although it is not known if hCG is unusually elevated in these concepti during implantation. Since nearly all trisomy 21 conceptions arise from new mutations,[88] this possible circumventing of the maternal quality mechanism would most likely be a chance association rather the result of natural selection. However, the advantages of hCG as a gateway mechanism are apparently not outweighed by the fitness disadvantage of producing an infertile offspring that "cheats" on the quality test. The fact that most trisomy 21 conceptions are, despite elevated hCG, terminated before birth suggests that there is more than one mechanism for evaluating the quality of human concepti.

Of What Use?

> Tender loving care and health advice are the only interventions [for recurrent miscarriage] that do not require [more study]. All other proposed therapies… are of no proven benefit or are associated with more harm than good.[96]

Earlier I suggested that an appreciation of the investment trade-offs women face, and of the costs of reproduction that they incur, could contribute to achieving realistic reproductive goals, a clearer understanding of women's physiology during pregnancy, and improvements in women's emotional and physical health.

In stark contrast to an illness model, an evolutionary analysis of pregnancy loss brings to the fore the potentially substantial benefits to a woman and to her other offspring (either already born or future) that may be realised as the result of the natural termination of a given pregnancy. Hence, pregnancy loss is not necessarily a pathology but rather may be an appropriate outcome arising from physiological mechanisms that evaluate both maternal resources and offspring quality. The woman is not necessarily broke, and there may be no reason to fix her.

Although recurrent pregnancy loss (three consecutive losses) is suggestive of a malfunctioning of the reproductive process, only a small proportion of recurrent loss is clearly attributable to maternal pathology. The remainder may result from some unrecognised impairment or the odds of having a series of independent losses simply by chance. Given a risk of 0.2 for the loss of a detected pregnancy, the risk of three consecutive losses is 0.8%, accounting for many of the 1% of women who have had recurrent losses.* In fact, 75% of women who have experienced recurrent loss subsequently have a successful pregnancy without having had any medical interventions other than "psychological support and tender loving care."[32,96,97]

The evidence that maternal resources may be as important as offspring quality also suggests that improving the maternal environment can reduce the probability of pregnancy loss. Physical activity, nutritional adequacy, and psychosocial factors all probably play a role in pregnancy outcome, but the specifics of these relationships are not yet known. Wasser and Barash[55] emphasised the likely value of supportive psychosocial therapy for infertility clients, a proposal supported by some subsequent studies.[96,97] Trevathan[5] has persuasively argued for recognizing and accommodating the evolved psychosocial demands of human pregnancy. Nepomnaschy and colleagues[22,23] have demonstrated that the stress of daily life is associated with elevations in cortisol levels that increase the risk of loss. Vitzthum and colleagues[25,26] have observed an association between arduous labour, elevated adrenal progesterone, and increased risk of loss. Data from Bolivia and North Carolina demonstrate that season is a risk factor,[26,73] but additional study is needed to determine the reasons for such seasonality. Whatever the specific etiology, perhaps restricting one's attempts at conception to a single season because of scheduling constraints (e.g., summer vacation) may inadvertently increase the risk of loss in some women. Changing these schedules may prove clinically beneficial in some cases.

The evidence that hCG is part of a screening mechanism might suggest ways to circumvent this quality control so as to reduce the risk of pregnancy loss, but such an intervention could come at the cost of increasing genetic abnormalities in progeny. Rather, if possible, it may prove more valuable to strengthen the apparent relaxing of the screening mechanism that is associated with maternal aging and thereby reduce the risk of genetic abnormalities in live births. It should also be helpful to determine which other hormonal mechanisms act as checks on conceptus quality. This might lead to better prenatal screening for genetic defects, such as is now available for trisomy 21.

---

* The risk of loss in clinically recognised pregnancies is often stated to be about 15%, but this figure predates technologies, available at local pharmacies, able to detect pregnancies not long after implantation.

Clearly there is much work yet to be done on the behavioural and environmental covariates, and associated hormonal correlates, of pregnancy outcome. Although the existence of maternal mechanisms for evaluating resource availability and detecting offspring quality is consistent with life history theory, we still need solid empirical evidence of these. Maternal age is a well-recognised factor in pregnancy outcomes, yet the reasons behind this relationship are far from certain. While an evolutionary approach to health is unlikely to lead to quick or miracle cures, it can provide new insights that enhance the medical arsenal and lead to its more judicious application. In the case of reproductive goals, an evolutionary lens can potentially reduce feelings of guilt or inadequacy, which themselves may actually interfere with a successful pregnancy outcome, and can provide demonstrably logical and evidence-based arguments for patience. In our quick-fix world, the wisdom of 60 million years of mammalian evolution may prove to be, at least in some cases, the right medicine.

Acknowledgements

This chapter is dedicated to Ian, his parents' (and my) reward for patiently listening to evolutionary arguments. My abiding gratitude goes to Dr. Hilde Spielvogel and Lic. Esperanza Caceres (Instituto Boliviano de Biologia de Altura, La Paz), my principal collaborators in Project REPA and in many other studies during 20 years of friendship. With respect and admiration, I thank the many Bolivian women and their families who made our work possible. I am much obliged to the organizers of the symposium for their invitation and to my fellow participants and two reviewers for their thoughtful commentary. To Dr. Jonathan Thornburg (Max Planck Institut für Gravitationsphysik) goes my heartfelt appreciation for never allowing an assumption to go unchallenged nor a sentence to be easily misparsable.

References

1. Williams, G. C., and Nesse, R. M. 1991. The dawn of Darwinian medicine. *Q. Rev. Biol.* 66:1.
2. Nesse, R. M., and Williams, G. C. 1994. *Why we get sick: The new science of Darwinian medicine.* New York: Times Books.
3. Trevathan, W. R., Smith, E. O., and McKenna, J. J., eds. 1999. *Evolutionary medicine.* Oxford: Oxford University Press.
4. Stearns, S. C., ed. 1998. *Evolution in health and disease.* Oxford: Oxford University Press.
5. Trevathan, W. R. 1999. Evolutionary obstetrics. In *Evolutionary medicine*, ed. W. R. Trevathan, E. O. Smith, and J. J. McKenna. Oxford: Oxford University Press.
6. Vitzthum, V. J., and Ringheim, K. 2005. Hormonal contraception and physiology: A research-based theory of discontinuation due to side effects. *Studies Fam. Planning* 36:13.

7. World Health Organisation. 2001. Definitions and indicators in family planning, maternal and child health, and reproductive health. www.euro.who.int/document/e68459.pdf. Accessed 1 December 2006.

8. National Center for Health Statistics, Centers for Disease Control and Prevention. 2006. www.cdc.gov/nchs/datawh/nchsdefs/fetaldeath.htm. Accessed 1 December 2006.

9. Farquharson, R. G., Jauniaux, E., and Exalto, N. 2005. On behalf of ESHRE Special Interest Group for Early Pregnancy (SIGEP), updated and revised nomenclature for description of early pregnancy events. *Hum. Reprod.* 20:3008.

10. Wilcox, A. J., Baird, D. D., and Weinberg, C. R. 1999. Time of implantation of the conceptus and loss of pregnancy. *N. Engl. J. Med.* 340:1796.

11. Roberts, C. J., and Lowe, C. R. 1975. Where have all the conceptions gone? *Lancet* 305:498.

12. Boklage, C. E. 2000. Survival probabilities of human conceptions from fertilization to term. *Int. J. Fertil.* 35:75.

13. Holman, D. J. 1996. Total fecundability and pregnancy loss in rural Bangladesh. PhD dissertation, Pennsylvania State University.

14. Holman, D. J., and Wood, J. W. 2001. Pregnancy loss and fecundability in women. In *Reproductive ecology and human evolution*, ed. P. T. Ellison, chap. 1. Hawthorne, NY: Aldine de Gruyter.

15. Wood, J. W. 1994. *Dynamics of human reproduction: Biology, biometry, demography*, chap. 6–7. New York: Aldine de Gruyter.

16. Norwitz, E. R., Schust, D. J., and Fisher, S. J. 2001. Implantation and the survival of early pregnancy. *N. Engl. J. Med.* 345:1400.

17. Sunder, S., and Lenton, E. A. 2000. Endocrinology of the peri-implantation period. *Bailliere's Clin. Obstet. Gynaecol.* 14:789.

18. Mendizabal, A. F., et al. 1984. Hormonal monitoring of early pregnancy by a direct radioimmunoassay of steroid glucuronides in first morning urine. *Fertil. Steril.* 42:737.

19. Lenton, E. A. 1988. Pituitary and ovarian hormones in implantation and early pregnancy. In *Implantation*, ed. M. Chapman, G. Grudzinskas, and T. Chard, 17. London: Springer-Verlag.

20. Csapo, A. I., et al. 1972. The significance of the human corpus luteum in pregnancy maintenance. *Am. J. Obstet. Gynecol.* 112:1061.

21. Baulieu, E. E. 1988. A novel approach to human fertility control: Contragestion by the anti-progesterone RU486. *Eur. J. Obstet. Gynecol. Reprod. Biol.* 28:125.

22. Nepomnaschy, P. A., et al. 2004. Stress and female reproductive functioning: A study of daily variations in cortisol, gonadotrophins, and gonadal steroids in a rural Mayan population. *Am. J. Hum. Biol.* 16:523.

23. Nepomnaschy, P. A., et al. 2006. Cortisol levels and very early pregnancy loss in humans. *Proc. Natl. Acad. Sci. U.S.A.* 103:3938.

24. Vitzthum, V. J., Spielvogel, H., and Thornburg, J. 2006. Interpopulational differences in progesterone levels during conception and implantation in humans. *Proc. Natl. Acad. Sci. U.S.A.* 101:1443.

25. Vitzthum, V. J., Spielvogel, H., and Thornburg, J. 2006. A prospective study of early pregnancy loss in humans. *Fertil. Steril.* 86:373.

26. Vitzthum, V. J., Spielvogel, H., and Thornburg, J. N.d. Risk factors for early pregnancy loss in a rural Bolivian population: Agropastoralism and maternal age. Under review.

27. Ellish, N. J., et al. 1996. A prospective study of early pregnancy loss. *Hum. Reprod.* 11:406.

28. Zinaman, M. J., et al. 1996. Estimates of human fertility and pregnancy loss. *Fertil. Steril.* 65:503.
29. Wang, X., et al. 2003. Conception, early pregnancy loss, and time to clinical pregnancy: A population based prospective study. *Fertil. Steril.* 79:577.
30. Van Montfrans, J. M., et al. 2004. Basal FSH concentrations as a marker of ovarian aging are not related to pregnancy outcome in a general population of women over 30 years. *Hum. Reprod.* 19:430.
31. Leslie, P. W., Campbell, K. L., and Little, M. A. 1993. Pregnancy loss in nomadic and settled women in Turkana, Kenya: A prospective study. *Hum. Biol.* 65:237.
32. Regan, L., and Rai, R. 2000. Epidemiology and the medical causes of miscarriage. *Bailliere's Clin. Obstet. Gynaecol.* 14:839.
33. Macklon, N. S., Geraedts, J. P. M., and Fauser, B. C. J. M. 2002. Conception to ongoing pregnancy: The "black box" of early pregnancy loss. *Hum. Reprod. Update* 8:333.
34. Goddijn, M., and Leschot, N. J. 2000. Genetic aspects of miscarriage. *Bailliere's Clin. Obstet. Gynaecol.* 14:855.
35. Bishop, M. W. H. 1964. Paternal contribution to embryonic death. *J. Reprod. Fertil.* 7:383.
36. Wilcox, A. J., Weinberg, C. R., and Baird, D. D. 1990. Risk factors for early pregnancy loss. *Epidemiology* 1:382.
37. Volarcik, K., et al. 1998. The meiotic competence of in-vitro matured human oocytes is influenced by donor age: Evidence that folliculogenesis is compromised in the reproductively aged ovary. *Hum. Reprod.* 13:154.
38. Abdalla, H. I., et al. 1997. Age of uterus does not affect pregnancy or implantation rates: A study of egg donation in women of different ages sharing oocytes from the same donor. *Hum. Reprod.* 12:827.
39. Legro, R. S., et al. 1995. Recipient's age does not adversely affect pregnancy outcome after oocyte donation. *Am. J. Obstet. Gynecol.* 172:96.
40. Yaron, Y., et al. 1995. In vitro fertilization and oocyte donation in women 45 years and older. *Fertil. Steril.* 63:71.
41. Johnson, J., et al. 2004. Germline stem cells and follicular renewal in the postnatal mammalian ovary. *Nature* 428:145.
42. Bazer, F. W. 2004. Strong science challenges conventional wisdom: New perspectives on ovarian biology. *Reprod. Biol. Endocrinol.* 2:28.
43. Bukovsky, A., et al. 2004. Origin of germ cells and formation of new primary follicles in adult human ovaries. *Reprod. Biol. Endocrinol.* 2:20.
44. Bukovsky, A., et al. 2005. Oogenesis in cultures derived from adult human ovaries. *Reprod. Biol. Endocrinol.* 3:17.
45. Pellestor, F., Anahory, T., and Hamamah, S. 2005. Effect of maternal age on the frequency of cytogenetic abnormalities in human oocytes. *Cytogenet. Genome Res.* 111:206.
46. Henriksen, T. B., et al. 2004. Alcohol consumption at the time of conception and spontaneous abortion. *Am. J. Epidemiol.* 160:661.
47. Charnov, E. 1993. *Life history invariants.* Oxford: Oxford University Press.
48. Stearns, S. C. 1992. *The evolution of life histories.* Oxford: Oxford University.
49. Hill, K., and Hurtado, A. M. 1996. *Ache life history: The ecology and demography of a foraging people.* New York: Aldine de Gruyter.
50. Fisher, R. A. 1930. *The genetical theory of natural selection.* New York: Dover.
51. Williams, G. C. 1966. Natural selection, the cost of reproduction, and a refinement of Lack's principle. *Am. Nat.* 100:687.

52. Vitzthum, V. J. 1994. The comparative study of breastfeeding structure and its relation to human reproductive ecology. *Yrbk. Phys. Anthropol.* 37:307.

53. Hawkes, K. 2003. Grandmothers and the evolution of human longevity. *Am. J. Hum. Biol.* 15: 380.

54. Trivers, R. L. 1974. Parent-offspring conflict. *Am. Zool.* 14:249.

55. Wasser, S. K., and Barash, D. P. 1983. Reproductive suppression among female mammals: Implications for biomedicine and sexual selection theory. *Q. Rev. Biol.* 58:513.

56. Vitzthum, V. J. 1990. *An adaptational model of ovarian function.* Research Report 90-200. Population Studies Center, University of Michigan, Ann Arbor.

57. Vitzthum, V. J. 1997. Flexibility and paradox: The nature of adaptation in human reproduction. In *The evolving female: A life history perspective*, ed. M. E. Morbeck, A. Galloway, and A. Zihlman, 242–58. Princeton, NJ: Princeton University Press.

58. Vitzthum, V. J. 2001. Why not so great is still good enough: Flexible responsiveness in human reproductive functioning. In *Reproductive ecology and human evolution*, ed. P. T. Ellison, chap. 8. Hawthorne, NY: Aldine de Gruyter.

59. Templado, C., Bosch, M., and Benet, J. 2005. Frequency and distribution of chromosome abnormalities in human spermatozoa. *Cytogenet. Genome Res.* 111:199.

60. Davidowitz, G., and Nijhout, H. F. 2004. The physiological basis of reaction norms: The interaction among growth rate, the duration of growth and body size. *Integrat. Comp. Biol.* 44:443.

61. Dufty, A. M., Clobert, J., and Moller, A. P. 2002. Hormones, developmental plasticity and adaptation. *Trends Ecol. Evol.* 17:190.

62. Finch, C. E., and Rose, M. R. 1995. Hormones and the physiological architecture of life history evolution. *Q. Rev. Biol.* 70:1.

63. Wade, G. N., Schneider, J. E., and Li, H.-Y. 1996. Control of fertility by metabolic cues. *Am. J. Physiol (Endocrinol. Metabol.)* 270:E1.

64. Worthman, C. M. 1995. Hormones, sex, and gender. *Ann. Rev. Anthropol.* 24:593.

65. Vitzthum, V. J. 2008. Evolutionary models of women's reproductive functioning. In press. Vol. 37.

66. Baird, D. D., et al. 1999. Preimplantation urinary hormone profiles and the probability of conception in healthy women. *Fertil. Steril.* 71:40.

67. Baird, D. D., et al. 1991. Hormonal profiles of natural conception cycles ending in early, unrecognized pregnancy loss. *J. Clin. Endocrinol. Metab.* 72:793.

68. Judd, S., et al. 1992. The source of pulsatile secretion of progesterone during the human follicular phase. *J. Clin. Endocrinol. Metab.* 74:299.

69. Eldar-Geva, T., et al. 1998. The origin of serum progesterone during the follicular phase of menotropin-stimulated cycles. *Hum. Reprod.* 13:9.

70. De Geyter, C., et al. 2002. Progesterone serum levels during the follicular phase of the menstrual cycle originate from crosstalk between the ovaries and the adrenal cortex. *Hum. Reprod.* 17:933.

71. Ferin, M. 1999. Stress and the reproductive cycle. *J. Clin. Endocrinol. Metabol.* 84:1768.

72. Puder, J. J., et al. 2000. Stimulatory effects of stress on gonadotropin secretion in estrogen-treated women. *J. Clin. Endocrinol. Metabol.* 85:2184.

73. Weinberg, C., et al. 1994. Is there a seasonal pattern in risk of early pregnancy loss? *Epidemiology* 5:484.

74. Temme, D. H. 1986. Seed size variability: A consequence of variable genetic quality among offspring? *Evolution* 40:414.

75. Kozlowski, J., and Stearns, S.C. 1989. Hypotheses for the production of excess zygotes: Models of bet-hedging and selective abortion. *Evolution* 43:1369.

76. Haig, D. 1990. Brood reduction and optimal parental investment when offspring differ in quality. *Am. Nat.* 136:550.
77. Haig, D. 1993. Genetic conflicts in human pregnancy. *Q. Rev. Biol.* 68:495.
78. Haig, D. 1998. Genetic conflicts of pregnancy and childhood. In *Evolution in health and disease*, ed. S. C. Stearns. Oxford: Oxford University Press, 77–90.
79. Erickson, J. D. 1978. Down syndrome, paternal age, maternal age and birth order. *Ann. Hum. Genet.* 41:289.
80. Ayme, S., and Lippman-Hand, A. 1982. Maternal-age effect in aneuploidy: Does altered embryonic selection play a role? *Am. J. Hum. Genet.* 34:558.
81. Carothers, A. D. 1983. Evidence that altered embryonic selection contributes to maternal-age effect in aneuploidy: A spurious conclusion attributable to pooling of heterogeneous data? *Am. J. Hum. Genet.* 35:1057.
82. Warburton, D., Stein, Z., and Kline, J. 1983. In utero selection against fetuses with trisomy. *Am. J. Hum. Genet.* 35:1059.
83. Ayme, S., and Lippman-Hand, A. 1983. Maternal-age and altered embryonic selection: A reply to Carothers and to Warburton, Stein, and Kline. *Am. J. Hum. Genet.* 35:1064.
84. Hook, E. B. 1983. Down syndrome rates and relaxed selection at older maternal ages. *Am. J. Hum. Genet.* 35:1307.
85. Stein, Z., Stein, W., and Susser, M. 1986. Attrition of trisomies as a maternal screening device. An explanation of the association of trisomy 21 with maternal age. *Lancet* 1:944.
86. Kloss, R. J., and Nesse, R. M. 1992. Trisomy: Chromosome competition or maternal strategy? *Ethol. Sociobiol.* 13:283.
87. Hook, E. B. 1981. Rates of chromosome abnormalities at different maternal ages. *Obstet. Gynecol.* 58:282.
88. Olsen, C. L., Cross, P. K., and Gensburg, L. J. 2003. Down syndrome: Interaction between culture, demography, and biology in determining the prevalence of a genetic trait. *Hum. Biol.* 75:503.
89. Forbes, L. S. 1997. The evolutionary biology of spontaneous abortion in humans. *Trends Ecol. Evol.* 12:446.
90. Wilcox, A. J. et al. 1988. Incidence of early loss of pregnancy. *N. Engl. J. Med.* 319:189.
91. Hansis, C. et al. 2002. Assessment of β-HCG, β-LH mRNA and ploidy in individual human blastomeres. *Reprod. Biomed. Online* 5:156.
92. Baird, D. D., et al. 2003. Rescue of the corpus luteum in human pregnancy. *Biol. Reprod.* 68:44.
93. Lasley, B., et al. 1985. Urinary hormone levels at the time of ovulation and implantation. *Fertil. Steril.* 43:861.
94. Stewart, D., et al. 1993. The relationship between hCG and relaxin secretion in normal pregnancies vs. peri-implantation spontaneous abortions. *Clin. Endocrinol.* (Oxford) 38:379.
95. Stewart, D., et al. 1993. Enhanced ovarian syeroid secretion before implantation in early human pregnancy. *J. Clin. Endocrinol. Metab.* 76:1470.
96. Jauniaux, E., et al. 2006. Evidence-based guidelines for the investigation and medical treatment of recurrent miscarriage. *Hum. Reprod.* 21:2216.
97. Clifford, K., Rai, R., and Regan, L. 1997. Future pregnancy outcome in unexplained recurrent first trimester miscarriage. *Hum. Reprod.* 12:387.

7

Evolutionary Paediatrics
A Case Study in Applying Darwinian Medicine

Helen Ball
Parent Infant Sleep Lab and Medical Anthropology Research
Group, Department of Anthropology, Durham University

Contents

Introduction

One of the principal goals of evolutionary medicine is to identify and examine consequences to the health of modern humans that are derived from incompatibilities between the lifestyles and environments in which humans currently live, and the conditions under which human biology previously evolved.[1,2] A related approach specific to infant and child health, known as ethno-paediatrics (the comparative study of parents and infants across cultures) explores the way different caregiving styles affect the health, well-being, and survival of infants and children.[3] The confluence of evolutionary medicine and ethno-paediatrics has provided fertile ground for the emergence of what we have termed evolutionary paediatrics—an approach to infant and child health that draws upon cross-species, cross-cultural, historical, and palaeoanthropological evidence to inform critical examination of Western postindustrial and biomedical models of infant care. Over the course of the last decade biological anthropologists and paediatric clinicians with an interest in evolution have been instrumental in the development of this area of study, which (as I will later demonstrate) is now finding a foothold within aspects of mainstream infant and child health care. One of the key features of the growing success of evolutionary paediatrics has been the effort of its proponents to use the evidence generated by their studies to ameliorate the iatrogenic effects of mismatches between evolved mother-infant biology and Western postindustrial/biomedical infant care practices.[4–6] Successes have been achieved by challenging both parents and practitioners to reexamine key assumptions about infant care and development in light of evolutionary perspectives on maternal-infant behaviour and physiology. Where necessary, this has necessitated the proponents of evolutionary medicine directly involving themselves in clinical trials that test the validity of interventions informed by an evolutionary perspective.[7]

Encouraging Infant Independence and the Separation of Mothers and Babies

The consequence of what physical anthropologist Sherwood Washburn christened the "human evolutionary obstetric dilemma"[4,8–10]—how to accommodate the passage of the head of an increasingly large-brained infant through a bipedally adapted pelvis—was for human infants to be born sooner than would be expected for a primate of our brain size.[11] Martin[11] estimates that human gestation should be at least twice its current duration were it not for the constraint imposed by the maternal pelvis on infant brain size at birth. This truncation of human gestation means our infants are neurologically underdeveloped at birth in comparison with other hominoids (being born with 25% of their adult brain size, compared with >50% in other apes) and require a prolonged period of caregiver dependency while brain growth is prioritised.[3,11,12] Given their state at birth, human infants are said to be secondarily altricial, and their vulnerability, if abandoned during this period, suggests there would have been powerful selective pressure in our evolutionary past to ensure mothers and infants did not become separated.[3,12] Truly altricial infants (i.e., those born hairless and helpless with closed eyes and ears) are cached in nests while their mothers forage, but are rarely left alone, remaining in physical contact with nest mates who provide warmth and physiological regulation. Among such species infants are adapted to remain silent, and to inhibit defecation in their mother's absence, thereby avoiding advertisement of their location to predators. At the other end of the mammalian spectrum, precocial infants, born singly, with well-developed hearing and vision and the ability to walk soon after birth, are able to flee from danger and maintain close proximity with their mothers.[3] Even the infants of great apes, although unable to walk and climb in the immediate neonatal period, are able to cling to their mother's fur, and thereby avoid the dangers of separation, within a few weeks of birth.[12] Human infants, however, show none of these characteristics. As precocial infants with secondarily altricial traits, human neonates are unable to hide, flee, or cling to their mothers in the face of danger, and are dependent upon their caregiver to ensure they are kept in close proximity. The distinctive human infant cry (unique among primates) is hypothesised to have evolved as a separation distress call to elicit the attention of adults,[13] but this instinctive response to separation and fear also increases vulnerability to predation. In our hominin past, those infants who were separated from their mothers would simply not have survived,[14] and mother-infant separation is unlikely to have been a parenting strategy pursued by our ancestors.

Although the risks of death or injury in the face of predators are greatly diminished (but not entirely lost; see BBC News[15]) in the Western postindustrial environment, separation of an infant from its mother still carries harmful consequences. Due to their neurological immaturity, newborn humans are poorly equipped to regulate their own temperature or breathing patterns for the first few months of life, and experience greater physiological stability and energy conservation when in physical contact with a caregiver than when alone.[16-18] It is unsurprising, therefore, that separation is physiologically stressful for infants—as it also is for many mothers.[19] Culturally, however, infant independence has been valued as a parental goal in many Euro-American countries,[20-24] and separating infants from their mothers for sleep, transport, and periods of quiet wakefulness is a normal, widespread, and seemingly desirable custom. Both infant development and parental competence are often measured by the yardstick of infant independence. But a cross-cultural perspective reveals that postindustrial humans are unusual in encouraging the early independence of their offspring, and in nonindustrial societies infants remain in contact with their mother's body, day and night, for the initial months,[23,25-27] and in contact with the bodies of allocarers (such as fathers and siblings) for several months more.

This view of infants as separate and independent entities from their mothers is historically novel and culturally circumscribed, becoming prevalent only in the twentieth century in North America and parts of Europe where the production of a self-soothing solitary sleeping infant epitomised the socially desirable outcome of successful parenting for at least half a century (early 1930s to early 1980s).[28,29] Prior to this time it is clear from both commentaries on normal family life[30] and the emerging genre of physician-authored advice books to mothers appearing in the 1800s that the mother's body was the primary source of comfort and the normal sleeping place for the dependent infant.[28]

The change in perspectives regarding the nature of infancy throughout the twentieth century in the United States and western Europe has been described and documented in detail by several historians of child rearing.[28,29,31] The transformation of the "role" of infancy culminated in the increasing popularity of so-called scientific infant care, advocated by the infant care experts of the interwar era. These men (e.g., John B. Watson, Frederick Truby King, and Arnold Gessell) espoused the virtues of instilling self-control and self-reliance from an early age by means of scheduled feeding, minimal picking up and cuddling, and an avoidance of "spoiling" (e.g., refusal to respond to an infant's cries).[28,29] A "good" baby was predominantly comatose and undemanding, and early independence from the mother was a developmental goal to be achieved rapidly by infants, particularly at night. Mothers and physicians alike were advised that infants should be sleeping through the night by the age of 3 months, and lack of attainment of such developmental goals signified a wilful, noncompliant infant—or worse still—an inconsistent and overindulgent mother.[28,29] The goal of Western parenting was the training of an infant who learned from an early age to expect and to require minimal parental attention. The emergence and subsequent reinforcement of this parenting strategy did not occur randomly, but was a product of a particular historical era and cultural environment that valued independence, regulation, and control as desirable characteristics. It had little to do with the needs of infants and much to do with the needs of parents and the aspirations of the societies in which they lived.[20]

The Role of Hospital Birth in Separating Mothers and Babies

Popular acceptance of the desirability of infant independence is unlikely to have been accomplished were it not accompanied by a significant development in the experience of childbirth, which had serious repercussions for early infant care and mother-infant relationships. Throughout history and prehistory childbirth was a hazardous and liminal activity.[32] For Victorian women (many of whom experienced particular childbearing difficulties due to pelvic deformities consequential to rickets) the fear of pain and death in childbirth loomed large.[32,33] It was against this backdrop that, in the late nineteenth and early twentieth centuries, the practice of administering chloroform to women in labour was initiated—initially for humane reasons, but subsequently at women's request. Such anaesthesia could only be delivered in a hospital setting, and in the years that followed mothers increasingly chose to deliver their babies in hospitals in order to absent themselves from the fear and pain of childbirth. Chloroformed women were unable to care for their babies for several days while recovering from the effects of the gas, but women saw the use of anaesthetic to be a boon and eagerly requested it. Nurseries were introduced into hospitals as a consequence of mothers' incapacitation and babies had to be cared for by nursing staff.[33]

In the late thirties, the anaesthetic-amnesiac known as twilight sleep was introduced to women in labour; a combination of scopolamine and morphine, it provided a painless conscious birth, but removed the mother both mentally and emotionally from the birth experience, most being subsequently unable to remember the event itself. Twilight sleep societies established in the United States and the National Birthday Trust—headed by prime minister's wife Mrs. Stanley Baldwin—in the UK campaigned for women's rights to a pain-free birth, and its use became widespread.[23,34,35] This heavy dose of narcotics and amnesiacs completely incapacitated labouring women, causing them to lose control and resulting in many being literally strapped to their beds to prevent injury. With the introduction of intravenous barbiturates to this cocktail women were again rendered practically unconscious during labour. Recovery was a long process because of the drugs, and infant care was impossible. Twilight sleep and barbiturates also caused difficulties for the babies, who were born sleepy and unable to respond or suck, and many were force-fed in the first days after birth. Even respiratory movements were suppressed and babies in the nurseries had to be watched carefully to ensure that they did not stop breathing.[36,37]

In the initial decades of the twentieth century hospital deliveries increased geometrically due to the appeal of unconscious childbirth (figure 7.1), despite the fact that hospitals were not safe places for women to deliver. In many medical facilities infections were transmitted from woman to woman by physicians who did not wash their hands between patients,[32] and although the pain of childbirth was ameliorated, the prospect of death increased as maternal mortality from puerperal fever reached epic proportions (7/1,000 births in the United States between 1900 and 1930). In 1880 Pasteur demonstrated that the streptococcal microbes he had identified 20 years previously were the cause of puerperal fever; however, it was not until almost 1940, when aseptic practices and sulphur antibiotics were introduced into

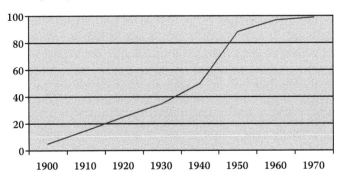

Figure 7.1 Proportion of U.S. infants delivered in hospital from 1900 to 1970.

clinical practice, that hospital birth mortality rates dropped below those of home deliveries. Now the proportion of hospital births increased exponentially to a zenith in 1973, when 99% of all U.S. births took place in a hospital, and under the control of a physician.[37]

In the hospital environment of the mid-twentieth century the separation of mothers and infants following birth had become routine. Campaigns to reduce medicated childbirth, such as those spearheaded by Grantley Dick-Reed and Fernand Lamaze, reduced narcotic use in labour throughout the fifties and sixties.[33,36,37] The rationale for the continued transfer of newborn infants to the hospital nursery was now justified with reference to infection control. Although mothers were no longer completely incapacitated, newborn babies were removed to a safe place for observation, and mothers were encouraged to rest following delivery—viewing their infants through glass partitions and meeting them only at scheduled feeding times.[19] Throughout this period physicians advocated feeding infants via formula milk so that their food intake could be scientifically managed.

The Consequences of Mother-Infant Separation

This untested and unprecedented intervention in human reproductive biology and behaviour—the separation of mothers and infants following birth—subjected Euro-American mothers and infants to experiences that contrast markedly with the close and prolonged postnatal contact of mothers and infants across the anthropoid primates, and across human societies worldwide.[3,25] The lens of hindsight has shown this large-scale disruption of normal mother-infant interaction to be associated with subsequent negative outcomes such as sudden infant death syndrome,[38,39] colic and excessive infant crying,[40] maternal postnatal depression,[41] shaken baby syndrome, and other forms of infant abuse,[42] with consequences potentially as far-reaching as antisocial behaviour and delinquency in adolescence.[43]

A comparative evolutionary perspective reveals that the development of an early physical relationship between mothers and infants, where infants remain in direct contact with their mother's body night and day for periods of weeks or months following birth, is vital for both infant survival and normal development. The results of Harlow's innovative but traumatic research with infant monkeys demonstrated dramatically the importance for infants of the physical comfort provided by contact with their mother's bodies— even when the mother was an inanimate cloth-covered surrogate.[44,45] Subsequent clinical studies regarding the effects on infants of separation from their mothers confirm the importance of close physical contact, not just in terms of psychological development but in terms of basic physiological functioning and the expression of innate survival reflexes.[46–49]

The negative outcomes of early mother-infant separation are not, however, confined to its effects on infants. Arguably, the most harmful outcome of the near-universal uptake of hospital births, medicated deliveries, and mother-newborn separation was the fall in the proportion of mothers initiating breast feeding. In the United States and western Europe breast feeding rates (which had once been almost universal) fell dramatically, reaching a nadir of 20–22% initiation rate in the United States between 1956 and 1972.[50] The massive popular acceptance of artificially formulated milk for infants in this part of the world in the mid-twentieth century signified an unprecedented cultural shift in infant care, the repercussions of which circumnavigated the globe. When, in the 1980s, research began to demonstrate the detrimental health consequences of feeding babies with artificial formula,[51-53] mechanisms were sought to reverse the breast feeding decline, and it soon became apparent that mother-infant separation in the post-birth period undermined both the initiation and establishment of breast feeding.[48]

The Importance of Skin-to-Skin Contact

In the immediate postnatal period, human infants who have been born following an unmedicated labour and who are placed directly onto their mother's abdomen exhibit innate nipple-seeking behaviour,[46] during which they crawl and squirm up their mother's bodies, locate the nipple by head bobbing, and spontaneously latch and suckle without assistance[54] over the course of their first hour of life. They are guided to the nipple by smell.[47,55] Infants who are delivered following a medicated labour involving the opioid analgesics in common use today (e.g., diamorphine or pethidine) make little or no attempt to crawl, and those that try are disorganised, uncoordinated, and unsuccessful in gaining the nipple.[46,56] Unmedicated infants perform an instinctive pattern of hand movements during nipple seeking that is associated with an increase in maternal oxytocin levels, and is similar to those observed in other mammals where massage of the mammary tissue facilitates milk letdown.[57] Mothers and babies who experience unhurried skin-to-skin contact immediately following delivery have a far greater chance of both establishing successful breast feeding and prolonging breast feeding duration.[58,59] The benefits of skin-to-skin contact in the immediate postnatal period are now increasingly recognised and incorporated into perinatal care. Although one recent randomised trial reported no significant difference in breast feeding percentages for mothers and infants who did and did not receive skin-to-skin contact, a recent systematic review of early skin-to-skin contact concluded there was clinical benefit in avoiding early mother-infant separation,[48] yet clinical research has examined limited types of reinforcement of postdelivery mother-baby contact, reflecting the restricted opportunities for contact in the hospital environment.

By the end of the twentieth century, recognition of the importance of breast feeding to infant health, and the role of separation in preventing the effective establishment of breast feeding, led to moves to close newborn nurseries in the hospitals of many European countries (e.g., Sweden, UK), although the United States still lags behind.[60] With the closure of nurseries came a shift to mothers and babies rooming-in, with the baby located at the mother's bedside during the day (but removed to a communal nursery at night) or (more recently) all day and night, with mothers performing all aspects of their infant's care. Research that has examined the effects of rooming-in in comparison to nursery care has concentrated primarily on sleep and breast feeding initiation, with studies demonstrating that separation of infants to neonatal nurseries resulted in less frequent breast feeding[61] and greater like-lihood of breast feeding failure,[39] but no increase in maternal sleep or alert-ness.[62,63] These findings contradicted the traditional argument that recently delivered mothers sleep better if their infants are cared for by others. Infants who spent their nights in nurseries were also found to sleep significantly less and to cry more than those at their mother's bedside.[64] Following a review of these studies, the UNICEF UK Baby Friendly Initiative concluded: "Mothers and babies should stay together at all times unless medically indicated or the mother makes a fully informed choice."[65] Round-the-clock rooming-in is now standard practice in all Baby Friendly accredited hospitals, and in pro-gressive nations such as Sweden,[66] but still suffers resistance from hospital staff and mothers in some circumstances, as discussed by researchers in sev-eral countries (see references 67–69). The evidence concerning the impact of mother-baby separation on breast feeding drives the current emphasis on practices such as skin-to-skin contact following delivery, and rooming-in on the postnatal ward.[70–73]

The view from evolutionary paediatrics suggests that although the situ-ation regarding mother-infant separation following delivery has improved following hospital deliveries with the advent of skin-to-skin contact and rooming-in, current hospital procedures still do not go far enough in facilitat-ing the expression of the mother-infant behavioural interactions that stimu-late normal lactational physiology. The process of lactogenesis (the initiation of lactation) involves two phases: Lactogenesis I—differentiation of alveolar epithelial cells for milk production—occurs during pregnancy and precedes copious milk production. Lactogenesis II—copious production of all milk components, i.e., when the milk comes in—normally occurs 3 days after parturition.[74] Very high levels of prolactin and glucocorticoid are required for milk production to begin in earnest. Following delivery of the placenta, maternal progesterone levels fall, and prolactin levels (which are suppressed throughout pregnancy by progesterone) begin to rise.[74] In the early postnatal period, each time the infant stimulates the nipple via suckling or touch, the mother experiences a rapid increase in prolactin secretion.[58,59] In these early days of breast feeding the amount of prolactin released is directly related to the intensity of nipple stimulation,[75] and breast feeding at night is associated with greater prolactin release than daytime feeding[76,77] due to the circadian nature of prolactin production. Researchers have found that the time of first breast feeding and the frequency of breast feeding on the second postpar-tum day are positively correlated with milk volume on day 5, suggesting that frequent stimulation of prolactin secretion in the period between birth and lactogenesis II increases the efficiency of subsequent milk production.[74,76] These findings lead to the conclusion that frequent nipple stimulation, frequent suckling, and particularly frequent attempted and successful feeds at night, in the period immediately preceding lactogenesis II, will lead to an earlier onset of lactation and more prolific milk supply.

In addition to being critical for breast feeding initiation, high initial prolactin levels are also important for successful long-term lactation. According to the prolactin receptor theory, the maintenance of lactation after lactogenesis II (galactopoeisis) is dependent upon the successful development of prolactin receptors, which occurs in the early postpartum period and also depends upon frequent feeding.[74,78] These prolactin receptors are crucial in maintaining lactation following the switch from endocrine to autocrine control. This suggests that frequent early feeding attempts will not only lead to effective establishment of milk production but enhance its continued maintenance.

Although it is currently assumed that rooming-in provides mothers and babies with the chance to feed frequently on the postnatal wards of modern Western hospitals (and rooming-in is clearly more favourable than nursery care), an evolutionary perspective provoked us to question whether a plastic bassinette at the mother's bedside provided sufficient opportunity for the close physical contact needed to reinforce the benefits afforded by delivery room skin-to-skin contact. From both physiological and evolutionary viewpoints, the clinical model of rooming-in as a mechanism for keeping mothers and babies together, and thereby facilitating frequent feeding attempts, would appear less than ideal. Neither nonhuman primate nor (by extrapolation) hominin mothers would place their infants to sleep in separate containers located an arm's reach or more from their own sleep location. Why, then, should we expect this to be an optimum environment for the development of appropriate mother-infant interaction for modern humans? Evolutionary paediatrics suggests that reinforcement of postdelivery skin-to-skin contact implies closer physical contact than rooming-in can provide. Cross-species and cross-cultural evidence indicates that newborns are in constant physical contact with their mother's body[79]—in fact, cross cultural evidence indicates that it is not simply immediate skin-to-skin contact and early suckling that facilitates development of the mother-infant interaction in most societies, but prolonged postnatal close contact in the days and nights following birth.[80] It is likely, therefore, that despite recent alterations in maternity care practices, a proportion of breast feeding failure remains an iatrogenic consequence of the physical separation of mothers and infants imposed by a rooming-in scenario on the postnatal ward.

Based upon evolutionary assumptions, plus physiological and cross-cultural evidence, I hypothesised that (1) providing mothers and infants with the facilities for unhindered contact on the postnatal ward will facilitate breast feeding initiation more successfully than rooming-in, and (2) the benefit of unhindered contact on the development of the behavioural and physiological relationship between mother and baby will give rise to improved long-term breast feeding outcomes. To test these hypotheses, my collaborators and I designed and conducted a randomised control trial (RCT) using overnight video monitoring to examine two forms of unhindered mother-infant contact (also known as bedding-in) on the postnatal ward, in comparison with 24-hour rooming-in. As we were aware that safety issues would need to be addressed in the context of bedding-in before such a practice could be considered clinically acceptable, we designed the trial to examine both breast feeding and infant safety outcomes. As secondary measures, we examined the impact of the three sleep conditions on maternal and infant sleep in the hospital, and explored whether they also affected sleeping arrangements at home over the infants' first 6 months of life.

Case Study: The Trial

This study was conducted on the postnatal ward of the Royal Victoria Infirmary, Newcastle-upon-Tyne, UK, a tertiary-level teaching hospital delivering five thousand babies annually in the northeast of England. Prior to commencement of the trial, approval was obtained from the clinical governance and research ethics committees of the Newcastle and North Tyneside Acute Hospitals Trust. Based on previously obtained pilot data, a target sample of sixty participants (twenty in each arm of the trial) was calculated as being necessary to provide 80% power to detect a difference of at least 5% in breast feeding frequency between groups. Complete details of the trial protocol following CONSORT guidelines can be found in our clinical report of this trial.[81]

Methods

Pregnant women with a prenatal intention to breast feed (but little* or no previous breast feeding experience) were recruited at 32+ weeks gestation at introductory breast feeding workshops. Prenatal inclusion criteria specified that we could include in the trial only healthy, non-smoking women, pregnant with a single infant, anticipating a normal vaginal delivery at the participating hospital. Volunteers who returned completed consent and enrolment forms and who met the above inclusion criteria were anonymously randomised (following standard RCT procedures) to one of three postnatal sleep conditions: baby in the mother's bed with a cot-side attached (bed condition), baby in side-car crib attached to mother's bed (crib condition), and baby in stand-alone cot adjacent to mother's bed (cot condition). Postnatal ward staff were alerted to a woman's participation in the study by a sticker placed on her medical notes. Following delivery and transfer to the postnatal ward, staff contacted the research team who ascertained the mother's continued willingness and eligibility to participate in the trial.

---

* "Little" breast feeding experience referred to women who had attempted breast feeding with a previous infant but did not breast feed after leaving the hospital.

Postnatal exclusion criteria specified that an ill baby or mother, or a woman experiencing caesarean delivery, or receipt of intravenous or intramuscular opiate analgesics in the preceding 24 hours, was not allowed to take part.

Eligible participants were provided with a single room and the appropriate equipment for their randomly allocated sleep condition, and signed a further consent form to indicate their willingness to be filmed on their first two postpartum nights. A small camcorder was mounted on the top of a 2 m monopod clamped to the frame at the foot of the mother's hospital bed and adjusted to capture an image of the upper half of the bed and the whole of the infant's cot (where appropriate). The camera's nighttime recording facility was activated to film in the dark; it was connected to a long-play video recorder (VCR) and time-code generator housed in a custom-built case located under the foot of the mother's bed. The VCR was operated by the mother using a remote control, and she was requested to start recording whenever she was ready to settle down for sleep. Filming then continued for 8 hours or until the mother stopped the tape. Mothers were requested to keep their infants in the designated location when they were asleep, but we did not specify how or where mothers should feed their infants. After two nights of filming were completed, research staff dismantled the camera equipment and offered mothers the opportunity to review their own tapes prior to giving final consent for their use in the study. Mothers participated in a semistructured debriefing interview regarding their postnatal experience, and medical records were accessed for labour and delivery details. After they were discharged from the postnatal ward, mothers participated in follow-up telephone interviews at 2, 4, 8, and 16 weeks regarding infant feeding and sleeping practices at home. As a gratuity for their participation in the study, mothers each received a gift voucher for £10 of baby supplies, and a tape of clips from their videos.

Videotapes were coded ethologically at Durham University's Parent-Infant Sleep Lab using Noldus® Observer 5 software and employing a behavioural taxonomy developed during a pilot study. The tapes were coded by three trained observers, each of which coded equal proportions of tapes from each condition to minimise any effects of observer bias. Observers undertook regular intra- and interobserver reliability testing, maintaining kappa scores reflecting at least 90% reliability. Statistical analyses were conducted using SPSS, and all analyses employed pair-wise comparisons between conditions. Medians and nonparametric tests are used for non-normally distributed data.

Results

Sample and Analyses
Between March 2003 and December 2004, 144 eligible pregnant women were enrolled to participate in the trial and randomly allocated to a postnatal con-

dition (bed = 48, crib = 50, cot = 46). Following delivery, fifty-five mothers were excluded due to postnatal ineligibility and twenty-five were lost due to unavailability of a single room or camera equipment, or communication failure between postnatal and research staff. Sixty-four mother-infant pairs were filmed; two of the participants, however, became ineligible during the intervention period (one mother unwell, one infant admitted to SCBU), and one set of tapes was unusable. Therefore, final video analyses were conducted using 61 participants (bed = 18, crib = 23, cot = 20). All mothers who participated in filming were interviewed on the ward and subsequently via telephone at 2, 4, 8, and 16 weeks. The characteristics of the mothers and infants in each of the three groups are shown in table 7.1.

We defined compliance with the allocated condition as the infant spending greater than 50% of the observed time in the assigned sleep location. Although we could not ethically require mothers to comply with their assigned condition, we requested that they try it for at least an hour. Of the sixty-one participants who were filmed, five did not meet the criteria for compliance with

Table 7.1 Characteristics of the Three Randomised Groups

	Mother's Bed (n = 18)	Side-Car Crib (n = 23)	Stand-Alone Cot (n = 20)
Mean age mother ± SD (years)	32.8 (± 2.7)	31.4 (± 5.3)	30.9 (± 3.7)
Mean infant age ± SD (hours)	15.4 (± 9.8)	16.6 (± 6.9)	17.6 (± 6.3)
Mean gestation ± SD (days)	283.9 (± 9.6)	283.2 (± 6.9)	280.6 (± 7.9)
Mean birth weight ± SD (kg)	3.3 (± 0.4)	3.4 (± 0.4)	3.5 (± 0.3)
Ethnicity n (%)			
White European	16 (88.9%)	22 (95.7%)	20 (95.0%)
Asian	2 (11.1%)	1 (4.4%)	1 (5.0%)
Labour n (%)			
Spontaneous	16 (88.9%)	17 (77.3%)	16 (80.0%)
Induced	2 (11.1%)	5 (21.7%)	4 (20.0%)
Mean 5 min APGAR ± SD	9.3 (± 0.6)	9.3 (± 0.4)	9.2 (± 0.4)
Mean time since prebirth maternal sleep ± SD (hours)	39.9 (± 10.2)	35.0 (± 18.7)	36.4 (± 17.6)
Mean duration of maternal sleep postdelivery ± SD (hours)	1.4 (± 2.2)	1.8 (± 2.2)	1.7 (± 1.6)
Mean duration of initial contact ± SD (min)	25.9 (± 18.4)	21.1 (± 20.2)	21.8 (± 13.1)

Table 7.2 Analysis Methods for Randomised Control Trials

For clinical evaluation purposes, analysis of a randomised controlled trial (RCT) is preferably conducted as intention to treat (ITT), whereby all participants are analysed in the groups to which they were assigned, regardless of crossover or degree of compliance. ITT analysis allows determination of the potential efficacy of the intervention in a real-world scenario (i.e., where a portion of patients will not follow treatment instructions). In this study it provides the opportunity to assess the effect of implementing an evolutionarily informed strategy for postnatal care in the real world.

A per-protocol (PP) analysis of an RCT considers only those participants who complied with their random allocation (i.e., those who completed the study as per the protocol). This gives an assessment of the effect of the intervention under a best-case scenario—in this study it tells us about the maximum potential effect on breast feeding of following an evolutionarily informed model of postnatal care.

their allocated condition on either night, and a further fourteen did not comply on one night. We used these data to conduct both intention to treat (ITT) and per protocol (PP) analyses (see table 7.2). The results of the ITT analyses for this study are available in our clinical report,[81] and we present PP analyses here to directly test the hypotheses that prolonged nighttime contact between mother and infant in the postnatal period will enhance breast feeding initiation and subsequent duration. The number of nights of data included in this per protocol analysis (i.e., where mothers and infants fully complied with their allocated sleep condition) are shown in table 7.3. Where only one night of data were available for a mother-infant pair, the missing night was entered into the analyses as missing data. Significance was set at $p < 0.05$.

Proximity

The proportion of observation periods that mothers and babies in the three groups spent in each of four proximity categories (physical contact, baby's arm's reach, mother's arm's reach, beyond touch) can be seen in figure 7.2. Together with table 7.4, displaying the results of two-tailed pair-wise t tests between all groups for all proximity categories, this confirms that bed babies spent a significantly greater duration in contact with their mothers, crib babies spent significantly longer within their mother's reach, and cot babies spent a significantly greater proportion of the night unable to touch or be

Table 7.3 Compliance with Randomly Allocated Condition

Condition Allocated	Nights Filmed	Nights Complied
Bed	31	22 (71%)
Crib	38	19 (50%)
Cot	36	25 (69%)

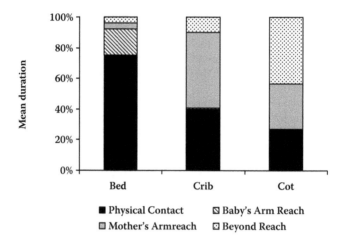

Figure 7.2 Mean percentage duration per night of observed proximity by allocated condition.

Table 7.4 Results of Pair-Wise Comparisons for Proximity Categories
(Two-Tailed T Tests)

	Bed vs. Crib	Bed vs. Cot	Crib vs. Cot
Physical contact	*p = 0.000*	*p = 0.000*	*p = 0.072*
Baby's reach	*p = 0.015*	*p = 0.003*	*p = 0.008*
Mother's reach	*p = 0.000*	*p = 0.000*	*p = 0.018*
Beyond touch	*p = 0.152*	*p = 0.000*	*p = 0.000*

touched by their mothers. This verifies that the allocated conditions provide mothers and infants with different opportunities for physical contact.

Breast Feeding Initiation

Figure 7.3 summarises data on the frequency with which mothers attempted to initiate breast feeding (i.e., offered the nipple), successful breast feeding, and overall breast feeding effort exhibited by mothers and babies who complied with their allocated sleep condition. Breast feeding data were not normally distributed, so medians and nonparametric tests are used. Offering the nipple to the baby is self-explanatory. Successful breast feeding bouts were defined as those where the baby latched on to the nipple and clear sucking and swallowing was observed for at least 5 seconds. Breast feeding effort represents the total number of feeding attempts observed, whether or not they resulted in successful sucking and swallowing. A feeding attempt was scored each time the baby attempted to latch on to the nipple, with or without assistance from the mother. Nipple presentations to which the infant did not respond were not included in the calculation of breast feeding effort;

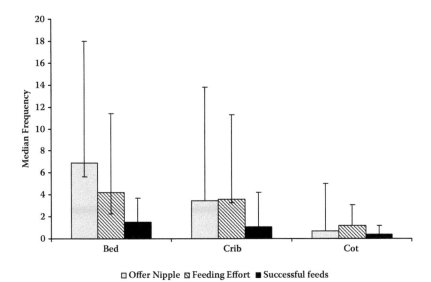

Figure 7.3 Median frequency per hour (and interquartile range) of breast feeding variables for mothers and infants who complied with allocated sleep condition.

however, attempts by the baby to obtain the nipple of his or her own accord (e.g., when the mother was asleep) were included.

There was much interindividual variation in feeding frequency, with some mother-infant dyads practicing repeatedly throughout the night, and others never feeding during the observation period. All infants in the bed condition accomplished some feeding, and figure 7.3 reflects a general pattern of more attempts by the mother to initiate feeds, more effort by the baby to engage in feeding, and consequently more successful feeds for the two groups in closest proximity to one another during the night. Table 7.5 summarises the results of pair-wise comparisons between each of the conditions using two-tailed Mann-Whitney U tests. No significant differences were found in any feeding frequency measures between the bed and the side-car crib condition. The frequency of all feeding measures was significantly greater for those mothers and babies who remained in the bed condition than for those who remained in the cot condition. The differences between crib and cot teetered on the edge of significance for feeding effort and successful feed frequency.

The median proportional duration of time spent engaged in successful feeds and all feeding attempts was calculated for each condition. Figure 7.4 shows that mothers and infants in the bed condition spent approximately five times as long engaged in feeding activities as mothers and babies in the cot condition, with the crib condition being intermediate between the two. Only the differences between bed and cot were significant (successful breast

Table 7.5   Significance Levels for Comparisons between Conditions (Two-Tailed Tests)

	Bed vs. Crib	Bed vs. Cot	Crib vs. Cot
Offer nipple	$p = 0.38$	$p = 0.004$	$p = 0.10$
Feeding effort	$p = 0.37$	$p = 0.000$	$p = 0.04$
Successful feeds	$p = 0.33$	$p = 0.001$	$p = 0.05$

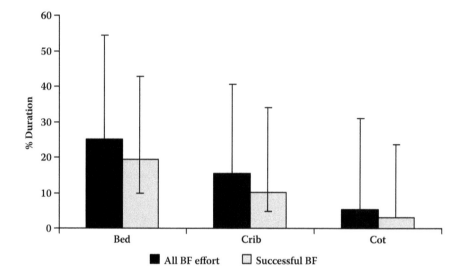

Figure 7.4   Median proportional duration per night (and interquartile range) of breast feeding activity.

feeding, bed versus cot, $p = 0.048$; all breast feeding effort, bed versus cot, $p = 0.016$, two-tailed Mann-Whitney U tests).

Breast Feeding Continuation

The continuation of breast feeding following discharge from the postnatal ward was ascertained via telephone interviews conducted at 2, 4, 8, and 16 postnatal weeks. The proportion of mothers and infants complying with their allocated sleep condition on at least one night of the trial and who were still engaged in exclusive or any breast feeding at each of the contact points is shown in figure 7.5.

Although all mothers initiated breast feeding prior to discharge from hospital, it is clear that the proportion of infants being exclusively breast fed fell rapidly in the standard-care (cot) group, while those in the intervention groups (bed and crib) showed a slower (and very similar) decline in exclusive breast feeding. Likewise for any breast feeding: the proportion

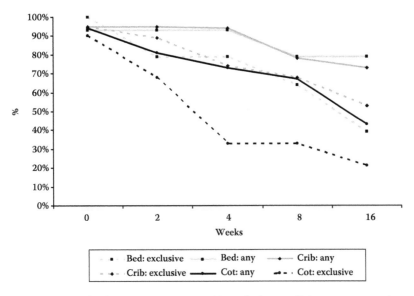

Figure 7.5 Breast feeding continuation to 16 weeks by condition on postnatal ward.

of those allocated to the cot condition who were still engaged in any breast feeding at 16 weeks was almost half that of the two intervention groups (cot = 43%, crib = 73%, bed = 79%). In testing the prediction that close sleeping proximity in the immediate postnatal period would be associated with greater breast feeding duration, we found significant differences between the bed and cot conditions for the duration (to at least 16 weeks) of any breast feeding (bed versus cot, $p = 0.031$, one-tailed t test) and exclusive breast feeding (bed versus cot, $p = 0.022$, one-tailed test), and between the crib and cot conditions for exclusive breast feeding (crib versus cot, $p = 0.015$, one-tailed t test). It is also important to note that no significant differences were found between the bed and crib conditions; figure 7.5 demonstrates that the decline in any and exclusive breast feeding up to 16 weeks in these two groups tracked one another closely. It should be noted that these breast feeding duration data are truncated at 16 weeks, and so the mean figures do not reflect absolute breast feeding duration for those who continued after this time.

Maternal and Infant Sleep

Maternal and infant sleep states were visually determined as "active awake," "passive awake," "appears asleep," and "indeterminate." Duration of "appears asleep" was calculated for all mothers and infants and expressed as a proportion of the time each was visible during the observation period. Mothers and infants achieved very similar proportions of sleep

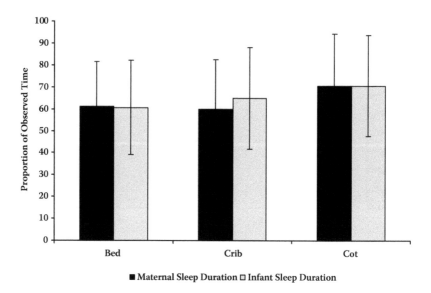

Figure 7.6   Maternal and infant mean (and SD) sleep duration.

in each of the three conditions (figure 7.6), and pair-wise two-tailed t tests revealed no significant difference for any intercondition comparison.

Sleep Proximity in the Home Environment

In order to assess whether intervening in the postnatal sleep location of infants affected their sleep location once they were at home, we asked mothers about their infant's sleep locations during each of the follow-up tele-phone interviews at 2, 4, 8, and 16 weeks. Up to the 16-week cutoff point the mean number of weeks during which infants bed shared at home at least once per week was: bed, 11.5 weeks; crib, 11.3 weeks; cot, 10.1 weeks. Pair-wise tests revealed no significant differences between the three groups in terms of bed-sharing behaviour at home.

Discussion

The results of this trial support hypotheses derived from an evolutionary view of mother-infant behaviour in the days immediately following birth and an understanding of lactation physiology. We predicted that enhanced nighttime proximity would be beneficial to breast feeding initiation. The three sleep conditions to which mothers and infants were randomly allo-cated in this trial are consistent with different amounts of physical contact (figure 7.2); therefore, this trial provides a valid test of the effects of night-time proximity in the immediate postpartum period. Mothers and infants with unhindered access to one another attempted to feed, fed successfully, and fed more frequently than those separated by a physical barrier in the form of the cot wall. This study, then, provides evidence that prolonged close nighttime contact between mothers and infants in the postpartum period is an effective mechanism for the successful establishment of breast feeding. The 20% discrepancy between the proportion of cot versus bed and crib mothers performing any breast feeding 2 weeks after the birth of their infant attests to the impact of this intervention. In the UK Infant Feeding Survey for the year 2000,[82] 23% of mothers who initiated breast feeding in hospital were no longer breast feeding 2 weeks later. It has long been recognised that feeding frequency in the early postnatal period is a key factor in establishing milk production and in learning how to suckle,[83–86] with the frequency of nighttime feeds being of particular significance.[76] The benefits of unhindered nighttime contact on the establishment of breast feeding by increasing feeding effort, raising prolactin levels, and encouraging mothers and infants to practice feeding technique appear to be effective in ameliorating breast feeding failure at an early stage.

We further predicted, based on current knowledge of lactation physiology, that prolonged close contact in the early postnatal period would increase breast feeding duration. The relationship between frequent feeding attempts (both successful and unsuccessful) and increased levels of prolactin is crucial to both the establishment of milk production and the maintenance of lactation. The significantly earlier termination of both exclusive and all breast feeding by cot mothers compared to crib and bed mothers by the 16 week cutoff point supports our prediction and reinforces the long-term consequences of unhindered contact in early infancy. The proportion of cot mothers engaged in any breast feeding at 16 weeks (50%) was consistent with the finding of the UK Infant Feeding Survey 2000[82] that 50% of mothers who initiated breast feeding were still doing so 4 months later. In contrast, approximately 80% of the mothers in the bed and crib conditions in this trial were still breast feeding at 16 weeks.

The evidence presented here, that prolonged close contact between mother and infant on the initial postnatal nights results in increased breast feeding frequency and enhanced long-term duration, enables us to link behavioural practices with both short-term physiological mechanisms and long-term survival consequences that have evolutionary implications. These findings challenge the current medical model of appropriate postnatal care, and provide support for the importance of mother-infant sleep contact. Data on infant safety in the three sleep locations are presented in our clinical report of this trial and yield no surprising findings.[81]

In view of the fact that both side-car crib and bed conditions were equally effective at promoting increased feeding frequency, and the side-car crib presented no difference from the stand-alone cot in terms of safety, while the bed condition showed a slightly increased potential risk, the side-car crib appears to be the most effective and safest means of maintaining extended mother-infant contact for the duration of the postpartum hospital stay.[81] Although the lack of individualised breast feeding support documented by Dykes[87,88] undoubtedly plays an important role in undermining mothers' efforts to initiate breast feeding on the postnatal ward, the current maternity "production line" is not the only aspect of postnatal care that could be improved in order to facilitate breast feeding initiation and establishment. As documented here, one non-labour-intensive change would be to provide facilities for mothers and infants to experience unhindered access to one another's bodies 24 hours a day.

Incorporating Evolutionary Perspectives into Policy and Practice

For most of human evolutionary history, successful mother-infant interaction in the immediate postnatal period was a critical component in our species' survival. Natural selection has produced human neonates equipped with a suite of innate characteristics that serve to stimulate and elicit maternal care, and thereby enhance their own survival chances. Human mothers, primed by a powerful mixture of neurochemicals and hormones, are programmed to respond both behaviourally and physiologically to their infant's cues and signals. Under the appropriate environmental conditions, the evolved system of interaction, feedback, and exchange between a newborn infant and its mother unfolds spontaneously according to a predictable sequence. Under inappropriate environmental conditions, the process can be so severely disrupted that maternal and infant health, and even infant survival, is compromised. Only in the last 100 years has the development of medical technologies made it possible for large numbers of neonates to survive the immediate postnatal period without a mother. This is a significant achievement of medical science, but one that came with a price. So successful was medical technology in promoting infant survival when mothers died or were unavoidably absent, that half a century or so ago, in the maternity wards of industrialised societies, maternal presence in the earliest period of an infant's life began to be considered superfluous, with mothers' increasingly passive role in the birth process ending with the removal of their infants to the care of experts. Half a century later we now observe the consequences of what we have lost.

Our research demonstrates the importance of maintaining physical contact between mothers and newborn infants, and the iatrogenic consequences of the disruption caused to the evolved maternal-infant relationship by their separation on the hospital postnatal ward, as tested by an experiment in evolutionary paediatrics. Throughout the course of this project, and our previous research, we have challenged clinicians, midwifery ward staff, community health care professionals, managers, health care policy makers, and parents to consider our evolved propensities as humans, primates, and mammals when contemplating care strategies for infants and proposing care interventions, and to work toward patterns of care that are congruent with the evolved characteristics of our species' behaviour and physiology.[89-93] This challenge has been accepted by many and found resonance with many more, and over the course of the past decade we have been invited to contribute, or have found our research being used to support, a variety of policy documents, position statements, and guidance papers for health professionals on issues surrounding mother-infant sleep contact.[94-102] Those working in the fields of childbirth, infant care, and breast feeding support have direct knowledge and experience of the capabilities and limitations of maternal and infant bodies and behaviour, and recognise many innate and instinctive features in mother-infant biology. For this reason, perhaps, in discussions and correspondence we have found that these health professionals are particularly receptive to an evolutionary approach (and many report experiencing a light-bulb moment during talks where the evolutionary explanation for the helplessness of human neonates is explained). The perspectives of evolutionary medicine on infant care have been embraced particularly strongly by those working to effect change within maternity provision and infant health in the UK—especially by those taking a position of advocacy for the rights of mothers and infants for physical access to one another's bodies, such as the UNICEF UK Baby Friendly Hospital Initiative, La Leche League, National Childbirth Trust, and breast feeding support organisations such as Baby Café, Sure Start, and Breastfeeding Network. However, the perspectives of evolutionary paediatrics have made minimal headway in convincing those organisations who hold steadfast to beliefs that maternal bodies are hazardous for infants. In the UK, the major cot death charities and the coroners represent this position most strongly, basing their arguments on the outcomes of epidemiological case control trials that have demonstrated detrimental outcomes of mother-infant sleep contact. In order for an evolutionary perspective to carry equivalent weight in an era when "evidence" is judged via meta-analyses and systematic reviews, we have found that it is necessary to subject the hypotheses of evolutionary medicine to rigorous testing via randomised trials, to report those trials in clinical journals, and to publicise the results widely. It is our experience that health professionals working in postnatal care are willing and able to implement lessons learned from an evolutionary perspective—and are even willing to directly challenge the establishment view if we, as researchers and proponents of the value of an evolutionary approach to health, are willing to invest the time and effort to make these perspectives relevant and accessible to their work, and to provide them with both encouragement and ammunition to initiate change in their health care domain.

References

1. Trevathan, W. R., Smith, E. O., McKenna, J. J., eds. 1999. *Evolutionary medicine.* New York: Oxford University Press.

2. Nesse, R. M., and Williams, G. C. 1995. *Evolution and healing: The new science of Darwinian medicine.* London: Phoenix.
3. Small, M. F. 1998. *Our babies, ourselves—How biology and culture shape the way we parent.* New York: Doubleday Dell Publishing Group.
4. Trevathan, W. R. 1993. The evolutionary history of childbirth. *Hum. Nat.* 4:337.
5. Dettwyler, K. 1995. A time to wean: The hominid blueprint for the natural age of weaning in modern human populations. In *Breastfeeding: Biocultural Perspectives,* ed. P. D. Stuart-Macadam and K. Dettwyler, 39–74. New York: Aldine de Gruyter.
6. McKenna, J. J., and McDade, T. 2005. Why babies should never sleep alone: A review of the co-sleeping controversy in relation to SIDS, bed-sharing and breastfeeding. *Ped. Resp. Rev.* 6:134.
7. Unziker, U. A., and Barr, R. G. 1986. Increased carrying reduces infant crying: A randomized controlled trial. *Pediatrics* 77:641.
8. Washburn, S. L. 1960. Tools and human evolution. *Sci. Am.* 203:3.
9. Rosenberg, K. R. 1992. The evolution of modern childbirth. *Yrbk. Phys. Anthropol.* 35:89.
10. Rosenberg, K. R., and Trevathan, W. R. 1996. Bipedalism and human birth: The obstetrical dilemma revisited. *Evol. Anthropol.* 4:61.
11. Martin, R. D. 1992. Primate reproduction. In *The Cambridge encyclopedia of human evolution,* ed. S. Jones, R. D. Martin, and D. Pilbeam, 86–90. Cambridge: Cambridge University Press.
12. Hrdy, S. B. 1992. Fitness tradeoffs in the history and evolution of delegated mothering with special reference to wet-nursing, abandonment, and infanticide. *Ethol. Sociobiol.* 13:409.
13. Christensson, K., Cabera, T., Christensson, E., Uvnas-Moberg, K., and Winberg, J. 1995. Separation distress call in the human neonate in the absence of maternal body contact. *Acta Paediatr.* 84:468.
14. Eaton, S. B., Shostak, M., and Konner, M. 1988. *The paleolithic prescription.* New York: Harper & Row.
15. BBC News. 2006. Shock after baby killed by dogs. September 25. http://news.bbc.co.uk/1/hi/england/leicestershire/5376788.stm (accessed November 17, 2006).
16. Christensson, K., Siles, C., Moreno, L., Belaustequi, A., et al. 1992. Temperature, metabolic adaptation and crying in healthy full-term newborns cared for skin-to-skin or in a cot. *Acta Paediatr.* 81:488.
17. Fransson, A. L., Karlsson, H., and Nilsson, K. 2005. Temperature variation in newborn babies: Importance of physical contact with the mother. *Arch. Dis. Child.* 90:500.
18. Goldstein Ferber, S., and Makhoul, I. R. 2004. The effect of skin-to-skin contact (kangaroo care) shortly after birth on the neurobehavioral responses of the term newborn: A randomised, controlled trial. *Pediatrics* 113:858–65.
19. Hock, E., McBride, S., and Gnezda, M. T. 1989. Maternal separation anxiety: Mother-infant separation from the maternal perspective. *Child Dev.* 60:793.
20. McKenna, J. J. 2002. Cultural influences on infant and childhood sleep biology, and the science that studies it: Toward a more inclusive paradigm. In *Sleep and breathing in children: A developmental approach,* ed. G. M. Loughlin, J. L. Carroll, and C. L. Marcus. New York: Marcel Dekker.
21. Valentin, S. R. 2005. Sleep in German infants—The "cult" of independence. *Pediatrics* 115:269.
22. Jenni, O., and O'Connor, B. 2005. Children's sleep: An interplay between culture and biology. *Pediatrics* 115:204.

23. Morelli, G. A., Rogoff, B., Oppenheim, D., and Goldsmith, D. 1992. Cultural variations in infants' sleeping arrangements: Questions of independence. *Dev. Psychol.* 28:604.
24. Abbott, S. 1992. Holding on and pushing away. *Ethos* 20:33.
25. Barry, H., and Paxson, L. M. 1971. Infancy and early childhood: Cross-cultural codes 2. *Ethnology* 10:466.
26. Javo, C., Ronning, J. A., and Heyerdahl, S. 2004. Child-rearing in an indigenous Sami population in Norway: A cross-cultural comparison of parental attitudes and expectations. *Scand. J. Psychol.* 45:67.
27. Reimao, R., Pires de Souza, J. C. R., Medeiros, M. M., and Almirao, R. 1998. Sleep habits in native Brazilian Terena children in the state of Mato Grosso do Sul, Brazil. *Arq. Neuropsiquiatr.* 56:703.
28. Hardyment, C. 1983. *Dream babies: Child care from Locke to Spock*. London: Jonathan Cape Ltd.
29. Hulbert, A. 2003. *Raising America: Experts, parents and a century of advice about children*. New York: Knopf Publishing.
30. Ulrich, L. T. 1990. *A midwife's tale: The life of Martha Ballard, based on her diary 1785–1812*. New York: Vintage Books.
31. Ehrenreich, B., and English, D. 1979. *For her own good: 150 years of the experts' advice to women*. London: Pluto Press.
32. Loudon, I. 1993. *An international study of maternal care and maternal mortality 1800–1950*. New York: OUP.
33. Tew, M. 1995. *Safer childbirth? A critical history of maternity care*. London: Chapman & Hall.
34. Pitcock, C. D., and Clark, R. B. 1992. From Fanny to Fernand: The development of consumerism in pain control during the birth process. *Am. J. Obstet. Gynecol.* 167:581.
35. Baker, P. A. 1989. Illustrations from the Wellcome Institute Library: The National Birthday Trust Fund records in the contemporary Medical Archives Centre. *Med. Hist.* 33:489.
36. Feldhusen, A. E. 2000. The history of midwifery and childbirth in America: A time line. *Midwifery Today*. www.midwiferytoday.com/Articles/Timeline.Asp (accessed May 12, 2006).
37. Nusche, J. 2002. Lying in. *Can. Med. Assoc. J.* 167:675.
38. McKenna, J. J. 1986. An anthropological perspective on the sudden infant death syndrome (SIDS): The role of parental breathing cues and speech breathing adaptations. *Med. Anthropol.* 10:9.
39. Konner, M., and Super, C. 1987. Sudden infant death syndrome: An anthropological hypothesis. In *The role of culture in developmental disorder*, ed. S. Harkness and C. Super, 95–108. New York: Academic Press.
40. Barr, R. 1998. Colic and crying syndrome in infants. *Pediatrics* 102:1282.
41. Hagen, E. H. 1999. The functions of postpartum depression. *Evol. Hum. Behav.* 20:325.
42. Barr, R. G., Trent, R. B., and Cross, J. 2006. Age-related incidence curve of hospitalized shaken baby syndrome cases: Convergent evidence for crying as a trigger to shaking. *Child Abuse Neglect* 30:7.
43. Gerhardt, S. 2004. *Why love matters: How affection shapes a baby's brain*. New York: Brunner-Routledge.
44. Blum, D. 2002. *Love at Goon Park: Harry Harlow and the science of affection*. Cambridge, MA: Perseus.
45. Harlow, H. F. 1959. Love in infant monkeys. *Sci. Am.* 200:68.

46. Righard, L., and Alade, M. O. 1990. Effect of delivery room routines on success of first breast-feed. *Lancet* 336:1105.

47. Varendi, H., and Porter, R. 2001. Breast odour as the only maternal stimulus elicits crawling towards the odour source. *Acta Paediatr.* 90:372.

48. Anderson, G. C., Moore, E., Hepworth, J., and Bergman, N. 2003. Early skin-to-skin contact for mothers and their healthy newborn infants. *Birth* 30:206.

49. De Chateau, P., and Wiberg, B. 1977. Long-term effect on mother-infant behaviours of extra contact during the first hour post partum. II. A follow up at three months. *Acta Paediatr. Scand.* 66, 145.

50. Wright, A. L. 2001. The rise of breastfeeding in the United States. In *Breastfeeding 2001*. Part 1. *The evidence for breastfeeding*, ed. R. J. Schanler. 1–12. Philadelphia: WB Saunders.

51. Cunningham, A. S., Jellife, D. B., and Jellife, E. E. P. 1991. Breastfeeding and health in the 1980s: A global epidemiologic review. *J. Pediatr.* 118:656.

52. Howie, P. W., Forsyth, J. S., Ogston, S. A., Clark, A., et al. 1990. Protective effect of breastfeeding against infection. *BMJ* 300:11.

53. Dewey, K. G., Heinig, M., and Nommsen-Rivers, L. A. 1995. Differences in morbidity between breastfed and formula fed infants. *J. Pediatr.* 126:696.

54. Nissen, E., Lilja, G., Matthiesen, A. S., Ransjo-Arvidsson, A. B., et al. 1995. Effects of maternal pethidine on infants' developing breast feeding behaviour. *Acta Paediatr.* 84:140.

55. Varendi, H., Porter, R. H., and Winberg, J. 1994. Does the newborn baby find the nipple by smell? *Lancet* 8:989.

56. Ransjo-Arvidson, A. B. Matthiesen, A. S., Lilja, G., Nissen, E., et al. 2001. Maternal analgesia during labor disturbs newborn behavior: Effects on breastfeeding, temperature, and crying. *Birth* 28:5.

57. Matthiesen, A., Ransjo-Arvidson, A., Nissen, E., and Uvnas-Moberg, K. 2001. Postpartum maternal oxytocin release by newborns: Effects of infant hand massage and sucking. *Birth* 28:13.

58. Johnston, J. M., and Amico, J. A. 1986. A prospective longitudinal study of the release of oxytocin and prolactin in response to infant suckling in long term lactation. *J. Clin. Endocrinol. Metab.* 62:653.

59. Uvnas-Moberg, K., Widstrom, A.-M., Werner, S., Matthiesen, A.-S., and Winberg, J. 1990. Oxytocin and prolactin levels in breast-feeding women. Correlation with milk yield and duration of breast-feeding. *Acta Obstet. Gyn. Scand.* 69:301.

60. Young, D. 2005. Rooming-in at night for mothers and babies: Sweden shows the way. *Birth* 32:161.

61. Yamauchi, Y., and Yamanouchi, I. 1990. The relationship between rooming-in/ not rooming-in and breast-feeding variables. *Acta Paediatr. Scand.* 79:1017.

62. Waldenstrom, U., and Swenson, A. 1991. Rooming-in at night in the postpartum ward. *Midwifery* 7:82.

63. Keefe, M. R. 1988. The impact of infant rooming-in on maternal sleep at night. *J. Obstet. Gynecol. Neonatal Nurs.* 17:122.

64. Keefe, M. 1987. Comparison of neonatal night time sleep-wake patterns in nursery versus rooming environments. *Nurs. Res.* 36:140.

65. UNICEF UK Baby Friendly Initiative. 1999. Why we recommend rooming-in. *Baby Friendly News* 4:3.

66. Svensson, K., Matthiesen, A. S., and Widstrom, A. M. 2005. Night rooming-in: Who decides? An example of staff influence on mother's attitude. *Birth* 32:99.

67. Gokcay, G., Uzel, N., Kayaturk, F., and Neyzi, O. 1997. Ten steps for success-ful breast-feeding: Assessment of hospital performance, its determinants and planning for improvement. *Child Care Health Dev.* 23:187.
68. Rice, P. L. 2000. Rooming-in and cultural practices: Choice or constraint? *J. Reprod. Inf. Psychol.* 18:21.
69. Svensson, K., Matthiesen, A.-S., and Widstrom, A.-M. 2005. Night rooming-in: Who decides? An example of staff influence on mother's attitude. *Birth* 32:99.
70. World Health Organisation. 1999. Postpartum care of the mother and new-born: A practical guide. Technical Working Group, World Health Organisa-tion. *Birth* 26:255.
71. DiGirolamo, A. M., Grummer-Strawn, L. M., and Fein, S. B. 2001. Maternity care practices: Implications for breastfeeding. *Birth* 28:94.
72. UNICEF. 2000. *Implementing the Baby Friendly best practice standards.* London: UNICEF UK Baby Friendly Initiative.
73. Perez-Escamilla, R., Pollitt, E., Lonnerdal, B., and Dewey, K. 1994. Infant feed-ing policies in maternity wards and their effect on breast-feeding success: An analytical overview. *Am. J. Pub. Health* 84:89.
74. Riordan, J. 2005. *Breastfeeding and human lactation.* 3rd ed. Boston: Jones and Bartlett.
75. Neville, M. C., Morton, J., and Umemura, S. 2001. Lactogenesis. The transition from pregnancy to lactation. *Pediatr. Clin. North Am.* 48:35.
76. Tennekoon, K. H., Arulambalam, P. D., Karunanayake, E. H., and Seneviratne, H. R. 1994. Prolactin response to suckling in a group of fully breastfeeding women during the early postpartum period. *Asia Oceania J. Obstet. Gyn.* 20:311.
77. Woolridge, M. W. 1994. Baby-controlled breastfeeding. In *Breastfeeding: Biocul-tural perspectives*, ed. P. Stuart-Macadam and K. Dettwyler, 217–42. New York: Aldine de Gruyter.
78. Marasco, L., and Barger, J. 1999. Cue feeding: Wisdom and science. *Breastfeeding Abstr.* 18:28.
79. Lozoff, B., and Brittenham, G. 1979. Infant care: Cache or carry? *J. Pediatr.* 95:478.
80. Lozoff, B. 1983. Birth and "bonding" in non-industrial societies. *Dev. Med. Child Neurol.* 25:595.
81. Ball, H. L., Ward-Platt, M. P., Heslop, E., Leech, S. J., and Brown, K. 2006. Ran-domised trial of mother-infant sleep proximity on the post-natal ward: Impli-cations for breastfeeding initiation and infant safety. *Arch. Dis. Child.* 91:1005.
82. Hamlyn, B., Brooker, S., Oleinikova, K., and Wands, S. 2002. *Infant feeding 2000.* London: Department of Health, TSO.
83. Salariya, E. M., Easton, P. M., and Cater, J. I. 1978. Duration of breast-feeding after early initiation and frequent feeding. *Lancet* 25:1141.
84. Hornell, A., Aarts, C., Kylberg, E., Hofvander,Y., and Gebre-Medhin, M. 1999. Breastfeeding patterns in exclusively breastfed infants: A longitudinal pro-spective study in Uppsala, Sweden. *Acta Paediatr.* 88:203.
85. Auerbach, K. 2000. Evidence-based care and the breastfeeding couple: Key con-cerns. *Journal of Midwifery and Women's Health* 45:205.
86. Inch, S. 1990. Postnatal care relating to breastfeeding. In *Midwifery practice post-natal care. A research based approach*, ed. J. Alexander, V. Levy, and S. Roch, 19-44. New York: Macmillan.
87. Dykes, F. 2005. A critical ethnographic study of encounters between mid-wives and breast-feeding women in postnatal wards in England. *Midwifery* 21:241–52.
88. Dykes, F. 2006. *Breastfeeding in hospital.* London: Routledge.

89. Ball, H. L. 2006. Parent-infant bed-sharing behavior: Effects of feeding type, and presence of father. *Hum. Nat.* 17:301.
90. Ball, H. L. 2003. Breastfeeding, bed-sharing and infant sleep. *Birth* 30:181.
91. Ball, H. L. 2002. Reasons to bed-share: Why parents sleep with their infants. *J. Reprod. Inf. Psychol.* 20:207.
92. Wailoo, M., Ball, H. L., Fleming, P. J., and Ward-Platt, M. P. 2004. Infants bed-sharing with mothers: Helpful, harmful or don't we know? *Arch. Dis. Child.* 89:1082.
93. Ball, H. L., Hooker, E., and Kelly, P. J. 1999. Where will the baby sleep? Attitudes and practices of new and experienced parents regarding cosleeping with their new-born infants. *Am. Anthropol.* 101:143.
94. UNICEF UK Baby Friendly Initiative. 2004. Babies sharing their mothers' bed while in hospital: A sample policy. www.babyfriendly.org.uk/pdfs/bedsharingpolicy.pdf (accessed May 12, 2006).
95. National Institute of Clinical Excellence (NICE). 2006. Clinical guidelines and evidence review for post natal care: Routine post natal care of recently delivered women and their babies. www.nice.org.uk/download.aspx?o=cg37fullguideline (accessed May 12, 2006).
96. International Lactation Consultant Association. 2005. Clinical guidelines for the establishment of exclusive breastfeeding. www.ilca.org/education/2005clinicalguidelines.php (accessed May 12, 2006).
97. Royal College of Midwives. 2004. National bed-sharing audit. www.rcm.org.uk/info/docs/040505184601-357-1.pdf (accessed May 12, 2006).
98. International Childbirth Education Association (ICEA). 2006. Position statement on postnatal education. www.icea.org/images/articles/0301.pdf (accessed May 12, 2006).
99. Encyclopedia of Early Childhood Development. 2006. Breastfeeding education and support. La Leche League Canada. www.excellence-earlychildhood.ca/documents/Pitman-Ayre-JaschkeANG.pdf (accessed May 12, 2006).
100. Royal Australasian College of Physicians. 2006. Breastfeeding paediatric policy statement. www.racp.edu.au/hpu/paed/Breastfeeding.pdf (accessed May 12, 2006).
101. Amicus/CPHVA (Community Practitioner & Health Visitor's Association). 2006. Fact sheet on co-sleeping. https://www.amicustheunion.org/docs/RD55020Co-Sleeping20(2).doc (accessed May 12, 2006).
102. Royal College of Midwives. 2004. Bed-sharing and cosleeping. Position Statement 8. www.rcm.org.uk/professional/docs/PS20820Bed20sharing.doc (accessed May 12, 2006).

8

The Role of Helminthes in Human Evolution
Implications for Global Health in the 21st Century

A. Magdalena Hurtado
School of Human Evolution and Social Change,
Arizona State University

M. Anderson Frey
Department of Anthropology, University of New Mexico

Inés Hurtado
Department of Anthropology, University of New Mexico
and Instituto Venezolana de Investigaciones Científicas

Kim Hill
School of Human Evolution and Social Change,
Arizona State University

Jack Baker
Bureau of Business and Economic Research,
University of New Mexico

Contents

Introduction

Although physiological and behavioral responses are often reported in animals with high worm burdens, or helminthes,[1]* the adaptive significance of these responses is unknown.[3,4] Similarly, although the field of biomedicine and medical anthropology is replete with publications on the short-term physiological and health effects of helminthes on their hosts, the evolutionary causes and consequences of those responses that might shed light on present-day human global health patterns are also unknown.

Our ignorance is due in part to assumptions that are pervasive in the fields of human biology and biomedicine. The first is the assumption that hunting and gathering humans have throughout history been much less infected with helminthes than their horticultural or agricultural counterparts. A recent review of the literature suggests that relatively few long-term co-evolutionary relationships between hominins and helminthes emerged early in our history, and for the most part, hominins would have contracted new parasites by coming into contact with or ingesting raw plants, insects, meat, and fish.[5] And the second is that even among humans infested with helminthes, such infections are primarily a nuisance rather than a serious threat to health and life, and are therefore unlikely to exert strong selective pressures on human physiological and immunological responses. The justification for this second assumption is that in most human populations, the primary causes of death are attributed to viral and bacterial infections, trauma, conspecific violence, and chronic health conditions, not helminthes (for examples of main causes of deaths among hunter-gatherers, see Hill and colleagues[6,7]). However, in populations with high worm burdens, such primary causes of death could be in many instances secondary or tertiary to helminthic infection. Epidemiologists have long debated over what constitutes a true cause of death—the heart attack or the viral infection that weakened that individual's heart, the automobile accident or the chronic onchocercosis infections of the eyes that blurred that individual's vision, and so on. Similarly, we will argue here that in populations with high worm burdens, rates of death and illness due to

---

* Helminthes are worms that are divided into three types that parasitize human hosts: Digenenan flukes, tapeworms (cestodes), and roundworms (nematodes).[2]

bacterial or viral infection are not independent of the levels of helminthic infection that they harbor.

As we grapple for answers to why infectious diseases now contribute more to human mortality across the globe than does any other health condition, in spite of huge technological and medical advances, researchers have found in recent years, in many different contexts and human populations, that helminthic comorbidity greatly increases the probability of infection and death due to HIV, tuberculosis, and malaria. Therefore, it is now clear that helminthes can exert strong selective pressures on human hosts by increasing rates of morbidity and mortality directly or by increasing the probability of mortality and morbidity that is more proximally induced by co-occurring bacteria and viruses in the same hosts. Much less is known about the effects of helminth comorbidity on rates of death due to chronic diseases such as asthma, diabetes, cancer, and cardiovascular disease. Could this have been true as well throughout most of our history? What energetic trade-offs might have been involved? And how is the human immune response implicated?

Thus, we need to approach the study of helminthes in past environments with rigor and ask what the potential role of helminthes may have been in shaping the mortality profiles of our ancestors, what selective pressures they may have exerted on ancestral genotypes that modern humans have now inherited, and what might be the implications of this helminth-focused view of human history for understanding global health patterns in the twenty-first century.

In this chapter, we show that the phylogeny of helminth infection in primates is such that nonhuman primates and extinct and extant nomadic and horticulturalist human groups are heavily parasitized. The very high prevalence of helminthes in nonhuman primates and humans suggests that throughout most of our history, helminthes may have exerted important selective pressures on physiological and immunological phenotypes by imposing substantial and chronic energetic costs on their hosts. Helminthes directly impose costs by draining hosts of nutrients essential to physiological performance. They indirectly impose costs by promoting the rapid loss of host nutrients through diarrhea or vomiting, two of the better-known strategies that helminthes use to ensure transmission between susceptible hosts. Moreover, helminthes cause energetically costly upregulation of host immune defense mechanisms. These direct and indirect costs add up to significant foregone investments in growth, reproduction, and maintenance (that is, opportunity costs).

In order to begin to systematically identify helminth-related selective pressures in past human environments, we minimally need answers to two questions: (1) Is chronic helminthic infection prevalent in our closest mammalian ancestors (nonhuman primates), prehistoric human populations, and extant human populations whose disease ecologies most closely resemble those of our past—indigenous populations? (2) What might have been some of the energetic costs of helminth infection among humans, and do they ebb and wane across the life course?

Figure 8.1 provides an overview of the chapter. In this figure, we show the disease ecology, physiological and life history components of interest, and the sections in which they will be discussed. We first present data on the distribution of intestinal helminthes in nonhuman primates, prehistoric populations, and indigenous peoples ("Helminthes and Early Humans"). We then summarize some of the immune system–related effects of helminth infection observed in animals and humans ("Costs of Defense against Helminthes"). The implications of the findings for global health in the twenty-first century are then considered ("Implications for Global Health").

Data Compilation Methods

Data compilation for this chapter was conducted through keyword searches in PubMed[8] for roundworm, whipworm, hookworm, helminthes, primates, South American Indians and prevalence. The PubMed Mesh system extracts records by using terms that are manually assigned to every article published in journals indexed by PubMed. For example, if an article includes information on any South American indigenous group, regardless of the term that the authors use to refer to that group, members of the PubMed staff assign to it the term "South American Indians." The PubMed Mesh system is the only electronic search engine that works in this way. We chose these key words because we were specifically interested in prevalence rates of soil-transmitted helminthes in South American indigenous populations and non-human primate populations.[9] Because there are at least 64 species of endohelminthes,[10] we focused our search for evidence of helminthes in prehistoric and extant populations on four frequently studied species of soil-transmit-

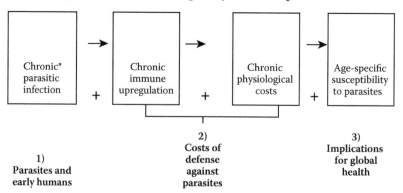

*Chronic refers to persistent presence in hosts from birth to death

Figure 8.1 Immune system–mediated effects of parasites on health and global health.

ted helminthes: *Ascaris lumbricoides* (roundworm), *Trichuris trichiura* (whipworm), and two species of hookworm—*Necator americanus* and *Anclyostoma duodenale*. Goncalves and colleagues[11] provided a comprehensive review of helminthes discovered in archeological settings, and that information was utilized to form the paleoparasitological tables. For the modern indigenous groups keyword searches of PubMed and Google Scholar include but are not limited to "soil-transmitted helminthes," "macrohelminthes," "prevalence," "hookworm," "ascaris," "paleaoparasitology," and "indigenous." Data on prevalence rates of roundworm, whipworm, and hookworm were systematically recorded into an Excel database. Two samples were generated using data on populations of South America: 125 and 38 studies of indigenous and nonindigenous groups, respectively.

Helminthes and Early Humans

Molecular genetics and paleoparasitology now allow us to state with some degree of confidence that helminthes have been an essential feature of hominin disease ecologies. Phylogenetic analysis of tapeworm divergence patterns indicates that the human taeniid forms split from the hyaenid, canid, and felid forms during a period of time when hominins appear to have become more dependent on hunting and scavenging, between 0.78 and 1.71 mya.[12] Other paleoparasitological studies, and biomedical studies of nonhuman primates and indigenous peoples also suggest that helminthes were persistent features of past ecologies (see below).

Because there are at least sixty-four species of endohelminthes,[10] we focused our search for evidence of helminthes in prehistoric and extant populations on four frequently studied species of soil-transmitted helminthes: *Ascaris lumbricoides* (roundworm), *Trichuris trichiura* (whipworm), and two species of hookworm—*Necator americanus* and *Anclyostoma duodenale*.

Infection with roundworm, whipworm, and occasionally *A. duodenale* (a hookworm species) originates through ingestion of fully developed eggs in contaminated food or water. Unlike roundworm and whipworm, the two hookworm species enter subcutaneous vessels when human skin contacts contaminated soil. After entrance into the host, the roundworm and whipworm larva burrow through the lining of the stomach, enter the circulatory system, and migrate to the lungs. In the respiratory system the larva pass over the epiglottis to reenter the gastrointestinal tract, where they mature into egg-laying adults.[13–15]

Nonhuman Primates

Helminth infection prevalence rates (i.e., number of individuals infested with helminthes divided by the total number of individuals sampled times 100) are high among nonhuman primates (table 8.1). Based on rates published in three studies by Murray,[16] Michaud,[17] Gillespie[18] and colleagues, it appears

Table 8.1　Prevalence of soil-transmitted helminthes in nonhuman primates

Strata	Primate species	Common name	Parasite species	Prevalence	Country	Reference
<10% prevalence						
	Saimiri sciureus macrodon	Common squirrel monkey	Trichuris	1.8	Peru	17
	Pan troglodytes	Chimpanzee	Trichuris	5.0	Tanzania	16
	Lagothrix lagotricha	Woolly monkey	Hookworm	8.3	Peru	17
	Lagothrix lagotricha	Woolly monkey	Ascaris	8.3	Peru	17
	Pan troglodytes	Chimpanzee	Hookworm	9.0	Tanzania	16
>10% prevalence						
	Cacajao calvus rubicundis	Uakari	Hookworm	20.0	Peru	17
	Cercopithecus ascanius	Redtail monkey	Trichuris	21.0	Uganda	18
	Colobus badius tephrosceles	Red Colobus	Trichuris	36.0	Uganda	18
	Colobus badius tephrosceles	Red Colobus	Trichuris	40.0	Uganda	18
	Papio cynocephalus	Baboon	Hookworm	44.0	Tanzania	16
	Cercopithecus ascanius	Redtail monkey	Trichuris	63.0	Uganda	18
	Papio cynocephalus	Baboon	Trichuris	66.0	Tanzania	16
	Colobus guereza	Abyssinian Colobus	Trichuris	79.0	Uganda	18
	Colobus guereza	Abyssinian Colobus	Trichuris	84.0	Uganda	18

that Old and New World wild nonhuman primates are similarly affected. Values range from 8.1 to 84% for one of three species, whipworm, hookworm, or roundworm. These are probably underestimates of true infection rates since it is very difficult to find worms and eggs in feces in any species and particularly in highly mobile wild communities.

The total number of prevalence rates reported in these studies is 14, and of those, 5 fell below 10% prevalence and 9 above 10%. New World and Old World nonhuman primates are similarly represented in both groups, suggesting that endohelminthes are a persistent feature of nonhuman primate disease ecology. Perhaps this was also true in hominin species that predated the emergence of *Homo sapiens*.

Prehistoric Populations

In an extensive review of the human paleoparasitological literature, Goncalves and colleagues[11] conclude that "almost all human specific helminthes have been found in ancient feces (p. 103)." With this, in combination with the possibility that hominin species that predated *H. sapiens* were at risk of helminthic infection as the molecular genetics and the nonhuman primate data suggest, it appears that endohelminthes may have been an unvarying feature of *H. sapiens* ecologies until recently.

Table 8.2 summarizes a subset of key published findings on roundworm, whipworm, and hookworm in prehistoric samples. To our knowledge, the oldest evidence of roundworm was found in fecal samples of early humans over 30,000 years ago in France. All other data are more recent, with the oldest sample dating back to 10,000 years ago and the most recent dating back to 2,700 years ago (see Goncalves et al.[11] for an extensive review of the paleoparasitological data on all helminthes).

The probable early presence of roundworm in France, at over 30,000 years ago, combined with the presence of both roundworm and whipworm at Kruger Cave before the advent of agricultural intensification in sub-Saharan Africa, means these helminthes could survive, if not thrive, in presedentary environments. Hence, soil-transmitted helminthes were likely present in humans before the advent of densely populated, sedentary populations in both South Africa and France.[19,20] Also, in the New World, soil-transmitted helminthes are present from 8,000–7,000 B.P. The pre-Columbian presence of roundworm, whipworm, and hookworm suggests these helminthes were major contributors to the disease load of the Americas since the initial migration to the New World over 12,000 years ago.[11,21]

In spite of the limitations of methods used to detect whole worms, eggs, or their genes in rehydrated coprolites, latrines, or mummified human remains using visual or molecular methods,[30] paleoparasitologists have found strong evidence that roundworm, whipworm, and hookworm were prevalent in prehistoric populations of *H. sapiens*. As methods improve, anthropologists

Table 8.2 Presence of soil transmitted helminthes in prehistoric human coprolites

Strata	Species	Archeological site	Country	Continent	Oldest approximate date (years ago)	References
Preagriculture	A. lumbricoides	Grand Grotte	France	Europe	30,160	19
Postagriculture	A. lumbricoides	Kruger Cave	South Africa	Africa	10,000	20
	T. trichiura	Kruger Cave	South Africa	Africa	10,000	20
	T. trichiura	Minas Gerais	Brazil	Americas	8,000	11
	E. vermicularis	Utah	United States	Americas	7,837	22
	Ancylostomid	Pedra Furada	Brazil	Americas	7,230	21
	E. vermicularis	Dirty Shame, Oregon	United States	Americas	6,300	23
	Ancylostomid	Tiliviches	Chile	Americas	4,100	11
	Ancylostomid	Clairvaux, Jura	France	Europe	3,600	24
	A. lumbricoides	Clairvaux, Jura	France	Europe	3,600	24
	T. trichiura	Clairvaux, Jura	France	Europe	3,600	24
	T. trichiura	Swifterbant	Netherlands	Europe	5,400	25
	A. lumbricoides	Arbon, Thurga	Switzerland	Europe	3,384	24
	T. trichiura	Arbon, Thurga	Switzerland	Europe	3,384	24
	T. trichiura	Otzal, Tyrol	Austria	Europe	3,300	26
	Ancylostomid	Minas Gerais	Brazil	Americas	4,905	27
	A. lumbricoides	Somerset	England	Europe	4,100	28
	A. lumbricoides	Chalain, Jura	France	Europe	2,700	24
	Ancylostomid	Chalain, Jura	France	Europe	2,700	24
	A. lumbricoides	Huarmey Valley	Peru	Americas	2,277	29

are likely to find higher prevalences of helminthes in prehistoric populations than those documented to date.

Indigenous Peoples

The disease ecologies of indigenous peoples residing in remote areas of the world are the best examples that we have of past human ecologies. Here we summarize data on population and age-specific prevalence rates of roundworm, whipworm, and hookworm among indigenous groups of South America. We use nonindigenous groups residing in similar regions as proxies of later humans. This allows us to compare rates of helminthic infection between populations residing in environments that represent our past and our present.

The foci of this section are two sets of figures (figures 8.2 and 8.3) with data that allow us to answer the following questions: (1) Are the prevalence rates of roundworm, whipworm, and hookworm significantly different between indigenous and their nonindigenous counterparts (figure 8.2)? (2) What are the shapes of the age-specific curves of roundworm, whipworm, and hookworm in South American indigenous groups and their nonindigenous counterparts (figure 8.3)?

Prevalence and Intensity

Prevalence rates of roundworm, whipworm, and hookworm are considerably higher among indigenous than among their nonindigenous counterparts. Figure 8.2a shows that although the distributions of roundworm, hookworm, and whipworm overlap between the two ethnic groups; in fact, we found two outlier values for urban groups that were as high as the highest prevalence rates for indigenous groups (close to 100% prevalence). However, the median values are consistently higher for roundworm, whipworm, and hookworm in indigenous groups (n = 125; 29.6–72%) when compared to nonindigenous groups (n = 38; 16.1–36.8%). For hookworm, the difference is considerable; 72 versus 13.9 for indigenous versus nonindigenous. It is less so for roundworm (54.1 versus 36.8%), and much less so for whipworm (29.6 versus 16.1%).

Another way to evaluate differences in helminthic infection between indigenous and other groups is to measure prevalence rates in groups of individuals with light, moderate, and heavy worm burdens. Figure 8.2b shows prevalence rates stratified by group, and then by group and degree of infection intensity.[38] Interestingly, in the indigenous sample, 88% of those surveyed were infected with roundworm, compared to 12.5% of the urban population. Within the urban sample, a larger percentage (7.3) had light infections, with lower percentages in the moderate (5.8) and heavy (2.9) infection groups. This Poisson distribution is typical in nonindigenous communities.[49] However, in Scolari's study,[38] 44% of the individuals under study

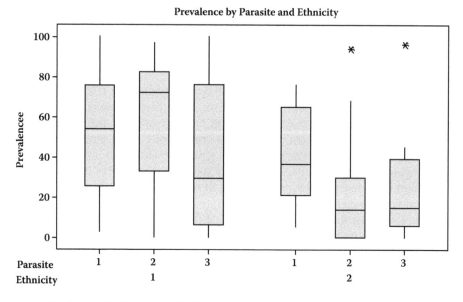

Prevalence by Parasite and Ethnicity

For Parasite 1 = Ascaris, 2 = Hookworm, 3 = Trichuris
For Ethnicity 1 = Indigenous, 2 = Non-Indigenous

a

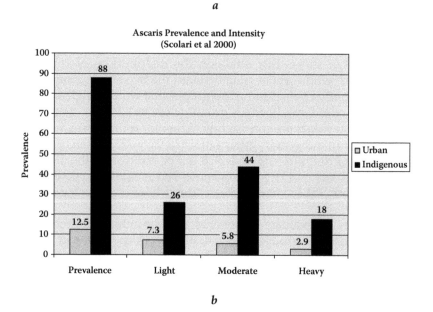

b

Figure 8.2 (a) Prevalence rates stratified by parasite species and ethnicity.[9,17,31–96] (b) Prevalence of roundworm stratified by intensity and ethnic group (urban = nonindigenous and admixed urban populations) (based on Scolari et al.[38]).

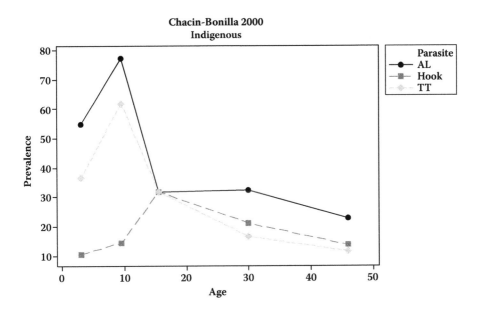

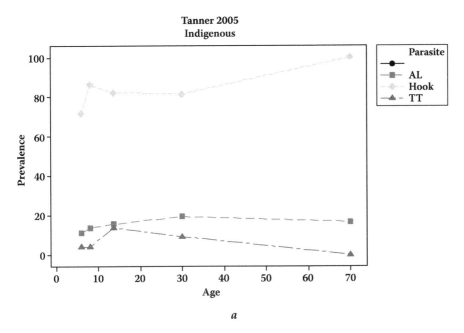

a

Figure 8.3   (a) Age-specific curves of roundworm, hookworm, and whipworm in two South American indigenous groups (based on Chacin-Bonilla and Sanchez-Chavez[37] and Tanner[45]).

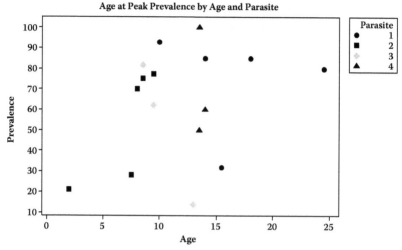

1 = Hookworm, 2 = Ascaris, 3 = Trichuris, and 4 = All parasites

b

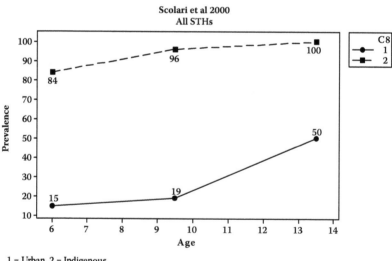

1 = Urban, 2 = Indigenous

c

Figure 8.3 (b) Age at peak prevalence among indigenous groups of South Americans stratified by parasite.[34,36–38,40,45–48] (c) Age-specific prevalence of parasite infection stratified by ethnic group (urban = nonindigenous and admixed urban populations). Based on Scolari et al.[38]

had moderate intensity infections, compared to 26% with light and 18% with heavy infections. This bell-shaped distribution in intensity of infection is unexpected and shows that relative to the urban population, when indigenous groups are infected with roundworm, not only do they have higher overall prevalence but also a higher percentage of individuals harbor larger worm loads.

Age-Specific Curves

If prevalence rates of roundworm, whipworm, and hookworm are generally higher among indigenous groups than the nonindigenous, then it must also be the case that in every age group indigenous individuals also have higher rates.

Figure 8.3a shows two panels with data on indigenous groups of Venezuela and Bolivia. The first panel shows the age distribution of prevalence rates of roundworm, whipworm, and hookworm among 433 indigenous individuals from two villages of Southern Venezuela.[37] For both roundworm and whipworm the prevalence rates increase throughout childhood and eventually reach a peak between 7 and 12 years of age at 77.3 and 62%, respectively. Then, prevalence of infection decreases during adolescence and tapers down to 28% for roundworm and 15% for whipworm in adulthood. Hookworm prevalence peaks later in life at 31% between the ages of 13 and 18 before decreasing in adulthood.

The second panel shows the age distribution for a sample of 317 Tsimané, an indigenous population of lowland Bolivia. The hookworm prevalence peaks at 86.2% between the ages of 6 and 9 years and remains at a steady high level until old age. In this study, 100% of elderly individuals (70 years old or older) were infected with hookworm (n = 6). In contrast, roundworm and whipworm were much less prevalent across all ages, although the rates are not biologically insignificant. Between 10 and 17 years of age, the Tsimane in the sample had a prevalence of 13.7%. And infants had infection rates of 21% for roundworm and 16% for hookworm.

When we compare the two studies, we find that in both groups at least one soil-transmitted helminth has a prevalence rate of over 70% at one or more stages of the life course. We also find that in Venezuela, a dramatic increase occurs from infancy to adolescence, followed by a long period of resistance into old age. In contrast, in Bolivia susceptibility to roundworm, whipworm, and hookworm appears to be high at all ages, particularly in the case of hookworm.

From these two studies, it appears that risk of helminth infection increases in indigenous groups during periods of brain and body growth, mainly from infancy to adolescence. Other studies support this conclusion. Figure 8.3b shows the distribution of age at peak prevalence for hookworm, roundworm, and whipworm among indigenous groups of South America. Most values fall between 5 and 17 years of age, and within this age range, the prevalence rates are greater than 50%. The Tsimané of Bolivia are an exception to this pattern; by 2 years of age, individuals reach a peak of 21% prevalence of roundworm. In addition, in groups of Itagua, Paraguay, and American-inhas of Brazil, the age at peak prevalence for hookworm is delayed into the 20s.[36,45,46] We also learn from figure 8.3b that roundworm and whipworm tend to peak in prevalence at earlier ages than does hookworm.

In summary, the age distributions of prevalence rates of roundworm, whipworm, and hookworm are similar in some ways but different in others when compared across indigenous groups. But how much do they differ from the age curves of indigenous groups?

Figure 8.3c shows that they can be extremely different in both magnitude and shape between the ages of 5 and 15. A study on the prevalence of round-worm, whipworm, and hookworm infection among indigenous and urban schoolchildren in Ortigueira, Brazil,[38] shows that the overall prevalence of those infections was significantly higher in the 100 indigenous children compared to the 136 urban children in the sample. The main culprits of infection in the urban and indigenous samples were roundworm (88 versus 12.5% for indigenous and urban) and hookworm (38 versus 5.8%). In both samples, collected by the same team of researchers using the same methods in two locations, the prevalence rate increased throughout childhood and peaked between 12 and 15 years of age. However, the urban prevalence peak was only 50%, compared to 100% among indigenous children between 12 and 15 years of age. While in the indigenous sample, susceptibles are close to being completely exhausted before the age of 10 (i.e., 96% of the children were already infected with at least one helminth), in the urban sample, only 19% of the children were infected.

Costs of Immune Defense against Helminthes

If endohelminthes, as the previous sections on nonhuman primates, prehis-toric populations, and indigenous peoples suggest, were a major feature of the disease ecologies of early humans, what costs did they incur across the life course, and across generations?

Based on a review of the literature on immune system-related effects on health, we learned that immune upregulation due to helminthic infection has at least three short-term effects and five long-term effects (figure 8.4). Over the short and long term, helminth-induced immune upregulation is associated with hypermetabolism, an exaggerated increase in metabolic rate, and catabolism, or destructive processes by which cells convert complex molecules into simpler compounds. These include lipolysis, protolysis, and glycolysis, or the decomposition of lipids, proteins, and carbohydrates.

Over the short term, these physiological responses in turn lead to increases in body temperature and fever, weight loss, and loss of skeletal muscle (fig-ure 8.4). And over the long term, as these physiological responses become chronic, hosts continually invest in the synthesis of new proteins; suffer from immunopathologies, impaired neurological function, and viral and bacterial disease; and develop morphological traits that are asymmetrical (figure 8.4).

The predictor variable in the figure is helminth-induced immune system upregulation, which in humans involves significant increases in the production of immunoglobulin E (IgE) and eosinophils. In helminth-endemic regions of the world, high IgE and eosinophil levels are associated with resistance to reinfection.[50–52] In helminthic infection, IgE, the immunoglobulin with the lowest circulating concentration, relies on the high affinity of mast cells for IgE via their e-heavy-chain Fc receptors (FcERs) to clear endohelminthes. Mast cells degranulate upon binding to IgE, and release histamine, which in turn activates eosinophils. The latter then release cationic granule proteins with activities known to be cytotoxic, or toxic to the cells of endohelminthes, including eosinophil cationic protein, eosinophil-derived neurotoxin, and eosinophil peroxidase.[53]

The upregulation of IgE and eosinophils is significantly greater in populations whose disease ecologies most closely resemble those of early humans, mainly extant indigenous groups residing in remote regions of the world.[54] Moreover, IgE and eosinophils are not the only cells that are upregulated during helminthic infection. The concentration of other immune cells such as T-lymphocytes is also elevated in tribal populations of West Africa and Papua New Guinea.[55–58]

We did not find any studies that measured the effects of hyper-IgE, eosinophil production, or any other immune cell on the health of indigenous peoples. Instead, we found studies in the animal literature and the medical literature on nonindigenous peoples that systematically measured some of the relationships of interest summarized in figure 8.4. This published work is a useful starting point for thinking about how helminth-induced mechanisms can exert strong selective pressures on the expression of genes that regulate differential energetic investments into helminth clearance across the life course.

Short-Term Effects

Over the short term, the initial immune upregulation associated with helminthic infection requires mobilization of protein and energy to combat infection.[58] Even brief infections are sufficient to instill protein malnutrition within a few days in hospital patients with sepsis.[59] In bumblebees and caterpillars, these protein and energy requirements have especially devastating effects during periods of starvation because hosts have to rely on their own tissue for nutrients, and thus frequently succumb to death.[60,61] In humans, similar phenomena are observed in patients suffering from trauma or sepsis, a health condition that also elevates the production of IgE and eosinophils.[62] The high glucose and glutamine requirements of immune cells like IgE and eosinophils require the breakdown of protein (proteolysis), carbohydrate (glycolysis), and lipid (lipolysis) reserves in septic hospital patients.[63,64] And

(A) SHORT TERM

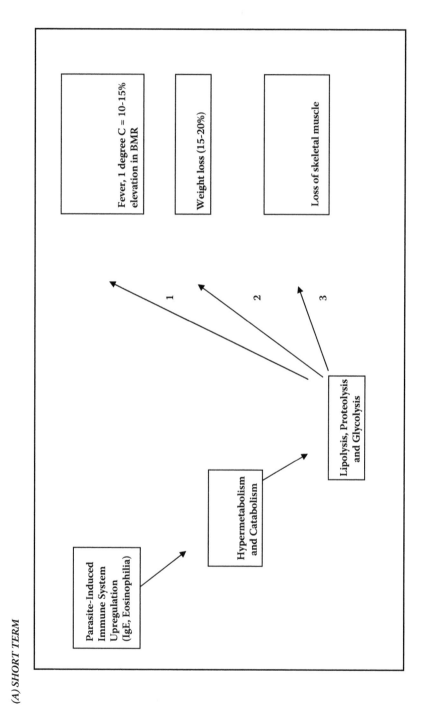

Figure 8.4a The costs of parasite-induced immune system upregulation on short-term physiological funciton, morphology, and health outcomes.

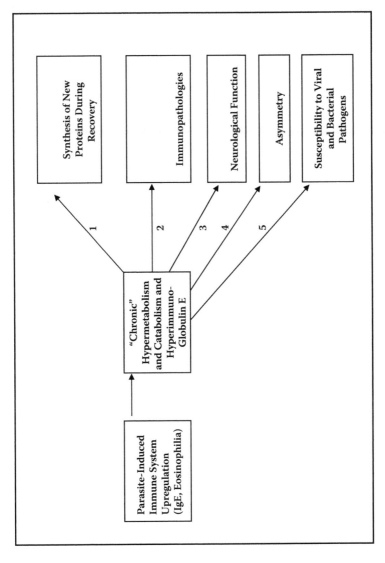

(B) LONG TERM

Figure 8.4b The costs of parasite-induced immune system upregulation on long-term physiological function, morphology, and health outcomes.

lipolysis, proteolysis, and glycolysis have been found to reach exaggerated levels in trauma patients and must be quelled quickly with therapeutic interventions to prevent disability and death.[65–67]

Lypolysis, proteolysis, and glycolysis are also associated with an increase in body temperature in hospital settings.[66] In at least one study of patients suffering from major trauma, researchers found that with each 1°C in fever, basal metabolic rate (BMR) increased by 10–15%.[68] Even a typhoid vaccination can raise the metabolic rate of human hosts between 15 and 30%,[69] and metabolic rate can increase 20–25% during septic infections.[70] Moreover, an increase in metabolic rate leads to weight loss and loss of skeletal muscle such that in some studies researchers found that humans suffering from infections or severe trauma, and birds infested with ectohelminthes, experienced a dramatic decrease in weight.[65,71,72] In humans, severe sepsis infections can lead to a loss of 15–30% of overall body weight.[59]

Few studies have measured the energetic costs of infection by specific groups of pathogens. That is, for example, are the costs similar or different for bacteria, viruses, or helminthes? This is a difficult question to answer because it requires experimentation that requires controlling for factors other than "type of pathogen," such as pathogen virulence factors, host susceptibility, pathogen load, and replication rates within hosts. Thus at present we are unable to provide a summary of differences in costs between helminthes and other microbial agents such as viruses and bacteria. For the time being, we can only guess that the energetic costs of helminth infection must be significant for individuals who are infected because, unlike many of the better known pathogenic viruses and bacteria that afflict humans, helminth infections are never completely cleared, reinfection is chronic throughout the life course, and the complex interplay of cells involved in humoral immunity that is associated with helminth clearance is less effective than is the more specialized cell-mediated immunity responsible for the defense against viruses and bacteria.[73]

Long-Term Effects

Extended periods of hypermetabolism require the synthesis of new proteins to make up for their loss in injured humans as well as septic infections in humans and rats.[74,75] The cost of synthesizing new proteins can be substantial, requiring an additional 24 kcal intake in human hosts to replace each gram of protein lost during the period of infection.[76] Under conditions of limited or fluctuating food availability, hosts may never be able to make up for lost tissue even after elimination of the infection.

In addition, chronic immune attacks against helminthes throughout the life course increase the probability that cells of the immune system will be less able to discriminate between self and nonself (i.e., immunopathology) for reasons such as (1) the likelihood that some of the molecular arrangements of pathogen cells and host cells will match increases with persistent infection,

(2) mechanisms that downregulate immune cell production become less efficient, leading to unintended attacks on host cells as the immune response continues to fight pathogens unabated, and (3) the immune response becomes increasingly more exaggerated and less discriminating between host and pathogen cells with each subsequent bout of infection, i.e., a superantigenic response (see also [78–80]).

Moreover, chronic helminthic infection throughout the life course may force hosts to compromise investment in important tissues such as brain matter. There is suggestive evidence of reduced gray matter in conjunction with malaria, and some evidence suggests that in humans, chronic helminthic infection is associated with a decline in the volume of gray matter and in the ability to consolidate memory to long-term storage as well.[81,82]

Other tissues are also compromised when helminthic infection is persistent and chronic early in life. This compromise is associated with asymmetrical development of morphological traits and growth faltering. Research on birds shows that in some species the feathers of immune-challenged individuals grow more asymmetrically than in controls.[83–87] In addition, in at least one study of a tribal population, the growth rates of children with higher levels of C-reactive protein, an indicator of systemic acute inflammation, was lower than in other children.[88]

Last, chronic immune defense against helminthes can influence susceptibility to viral and bacterial pathogens. This is because helminthes skew immune defense toward Th2 and compromise the effectiveness of Th1 response, which is essential to clearing viral and bacterial infections.[54,84–94]

In summary, if endohelminthes were a major feature of the disease ecologies of early humans, what mortality costs did they incur? And can the mortality profiles of indigenous groups today provide a glimpse into the past? Possibly. Currently we find rates of mortality among indigenous groups that far exceed those of their their nonindigenous neighbours and other humans globally who have lower or negligible helminth infestation rates or no exposure to helminthes. High rates of mortality among indigenous populations, including those of South America, have been documented in many books and articles. Ribeiro[95] reports that 87 of 230 native groups in Brazil went extinct between 1900 and 1957, and studies estimate that the current native population in Brazil is only 5% of the estimated size for the 1500s. Contrary to the received wisdom, this dismal picture did not end with the period of conquest. Indigenous groups continue to be vulnerable to all sorts of diseases, and much of this vulnerability could be due to, by current public health standards, high and sometimes extraordinary prevalence rates of helminth loads.

Without the luxury of controlled experiments, it will be extremely challenging to determine in future decades the extent to which excess indigenous mortality is primarily driven by the costs that helminthes exert on their hosts as opposed to other factors. But for the time being, we can at least propose that helminthes may be an important contributing factor, just as researchers in Africa have suggested that the HIV/AIDS and tuberculosis epidemics could be more easily contained if helminthes were eradicated from that continent.[96] At least at present it is important to note that that the rates of helminth infestations among indigenous peoples of South America are among the highest in the world, and that even Highland groups that reside in colder climates are not spared.[9] We also know that bacterial epidemics such as tuberculosis take huge tolls on these populations after decades of first contact, even when they are no longer virgin soil epidemics, and the number of susceptibles should be exhausted, at least in principle.[97] In the 1950s, groups like the Southern Kayapo lost 99% of their original population in 50 years. They did not lose the majority of the population at contact: Losses continued through five decades. And, in the 1950s and 1970s, some Xikrin and Yanomamo communities that had had contact with outsiders in previous years declined in numbers by more than 20% within 10–15 years. Equally baffling is the finding that in the 1990s the infant mortality rate among the Xavante was three times higher than among other nonnative Brazilians, decades after first contact, and that life expectancy at birth in 2000 was frequently 20+ years lower among indigenous groups when compared to their nonindigenous counterparts. In fact, the life expectancy of indigenous peoples in Brazil and Venezuela is lower than that for the U.S. population in 1900 and lower than in Sierra Leone in 2000, which has the lowest reported national life expectancy in the world today.[9]

Implications for Global Health in the Twenty-First Century

A review of the literature on the antiquity and on population and age-specific prevalence rates of helminthic infection in humans leads us to conclude that helminthes in general and helminthes in particular were probably the group of pathogens that exerted the strongest disease-related selective pressures on immune system and life history traits in most past human environments. Helminthes appear to be ubiquitous in paleoarcheological remains in spite of our inability to easily detect the presence of worms, eggs, or their genes in ancient feces or guts. Second, they also appear to be ubiquitous among nonhuman primates in the New and Old Worlds, again, in spite of our limited capacity to collect fecal samples from highly mobile, wild primates. And third, they are more prevalent in extant, genetically homogenous, and small human populations that most closely resemble those of early humans, that is, indigenous peoples, than in contemporaneous, genetically heterogenous, and large populations with longer exposure to public health programs and germ theory.

Complementary reviews suggest that helminthe-induced immune upregulation is costly over the short and long term during the life of an individual, and that many of these costs are related through metabolic pathways. Thus, if the helminth fauna of prehistoric human populations is similar to that of nonhuman primates and indigenous peoples, then immune upregulation and associated costs, including mortality, was probably the norm in early humans. Thus, we argue that although indigenous groups have lower overall mortality than do nonhuman primates,[98,99] the higher disease-related mortality that we observe in presanitation, or sanitation absent human environments, compared to sanitized environments[6] is probably primarily caused by chronic and multiple helminthic infection, rather than viral and bacterial agents. Unlike helminthes, these agents wax and wane through virulent and avirulent states over periods that can exceed many human generations.[100]

Taken together, the data on helminth infection suggest that, in the hominin lineage, one of the greatest feats of natural selection was to produce a gracile primate with the largest brain and highest fertility for its body size and the lowest mortality at all ages in spite of multiple chronic, immunologically and metabolically costly helminthic infections throughout the life course. These observations suggest that selection against individuals who could not simultaneously grow a large brain, reproduce at short intervals, and mount effective immune responses against helminthes would have been great during early human evolution (and up to the present time in populations that continue to be heavily parasitized). Moreover, selection against individuals who did not help their kin or establish reciprocally altruistic alliances that would allow them to simultaneously grow, reproduce, keep their children healthy, and mount effective defenses against helminthes must also have been great, particularly during periods of the life course that are most reproductively and immunologically demanding, mainly during childhood and peri-adolescent periods when the probability of helminthic infection is oftentimes highest, the neocortex is not yet fully developed, and reproductive costs increase exponentially (particularly in females).

If our conjectures are correct, then one of the tragic side effects of helminthic infection in hominins is that the physiological and immunological genotypes that we have inherited from our ancestors evolved in hosts with rich, diverse, and chronic helminth faunas. In present-day, highly sanitized environments, these genotypes may be highly maladaptive and may play a causative role in the emergence of a wide range of puzzling conditions, including asthma, diabetes, and acne.[101–104] These conditions are expressed early in life, and as our environments become even more sanitized, the age of onset of these conditions continues to decline.[105] It is possible that immune activity was "programmed" into a developmental schedule millions of years or hundreds of thousands of years ago in response to the age profile of helminthic exposure and defense. The human immunological system may be primed to respond to helminthes early in life, and in their absence, it triggers responses to innocuous substances that damage the host and become chronic. There is no *a priori* reason to believe why responses to innocuous substances should not be acute, and cease to cause morbidity as do responses to other substances, like man-made and natural vaccines. In the future it might be useful to systematically identify differences and similarities between the immunological mechanisms involved in helminth defense, childhood asthma, and childhood diabetes I and II, particularly in light of the exponential growth of research on the relationship between immunological phenotypes that are expressed early in life and health outcomes later in life.[106,107]

Again, if our conjectures are correct, another tragic side effect of helminthic infection in hominins is that the worms that we harbor today have been with us for thousands if not millions of years. This does not mean that maintaining large worm burdens is not costly to hosts, but rather that the majority of humans have lost, and continue to lose, this host-pathogen arms race. It would be interesting to come up with an approximate number for the total volume of the helminth biomass that humans harbor in their intestines in the world today, to divide that number by the total volume of the biomass of human hosts, and then to do the same calculation for indigenous versus nonindigenous rural and urban groups in developed and underdeveloped countries. This would give us a better appreciation for the energy that helminthes are extracting from humans on a macroscale.

The evolutionary implications of this helminth-centered view of human biology in the twenty-first century are considerable. How much of the energy that humans now invest directly or indirectly in defense against helminthes could be diverted into other fitness- and health-enhancing physiological functions across the life course if helminthes could be eradicated from the planet? How would that affect global health patterns? In addition to the more obvious effects, such as healthier overall developmental trajectories for children, and a longer, healthier life span, the impact of helminth eradication on the rates of emergence of new viral and bacterial infections, and the rates of transmission of their older counterparts, could be significant. If in fact helminthes make hosts more susceptible to pathogenic viruses and bacteria through mechanisms that downregulate the cells involved in the Th1 response, then, on a global scale, eradication of helminthes would in effect lead to a reduction of the number of susceptibles to new and old viral and bacterial infections. Such effects could be modeled mathematically and compared to the effects of immunization campaigns. Second, we are not aware of research on the effects of helminth-induced immunosuppresion on viral and bacterial genetic mutation rates. It would be important to know whether pathogens mutate at faster rates in populations of hosts with a reduced battery of immune attacks than in populations with a more extensive armory. If viral and bacterial mutation rates are higher in populations with high worm burdens, then the eradication of helminthes not only would globally reduce infectious disease morbidity and mortality, but also would potentially slow down the emergence of new diseases for centuries to come.

References

1. Grove, D. 1990. *A history of human helminthology*. Wallingford, UK: CAB International.

2. University of Cambridge, Department of Pathology. 2007. *Helminth infections of man.* http://www.path.cam.ac.uk/~schisto/General_Parasitology/Hm.helminths.html. Accessed 18 Feb 2008.
3. Minchella, D. J. 1985. Host life history variation in response to parasitism. *Parasitology* 90:205–16.
4. Poulin, R., Morand, S., and Skorping, A. 2000. *Evolutionary biology of host-parasite relationships: Theory meets reality.* New York: Elsevier.
5. Barrett, R., Kuzawa, C., McDade, T., and Armelagos, G. J. 1998. Emerging and re-emerging infectious diseases: The third epidemiologic transition. *Ann. Rev. Anthropol.* 27:247–71.
6. Hill, K., Hurtado, A. M., and Walker, R. S. 2007. High adult mortality among Hiwi hunter-gatherers: Implications for human evolution. *J. Hum. Evol.* 52:443–54.
7. Hill, K., and Hurtado, A. M. 1996. *Ache life history: The ecology and demography of a foraging people.* New York: Aldine Press.
8. http://www.pubmed.org.
9. Hurtado, A. M., Lambourne, C. A., James, P., Hill, K., Cheman, K., and Baca, K. 2005. Human rights, biomedical science, and infectious diseases among South American indigenous groups. *Annu. Rev. Anthropol.* 34:639–65.
10. Centers for Disease Control and Prevention. 2007. *Parasites of the intestinal tract.* Atlanta, GA: Infectious Disease Department. http://www.dpd.cdc.gov/DPDx/HTML/Para_Health.htm. Accessed 18 Feb 2008.
11. Goncalves, M. L., Araujo, A., and Ferreira, L. F. 2003. Human intestinal parasites in the past: New findings and a review. *Mem. Inst. Oswaldo Cruz* 98:103–18.
12. Hoberg, E. P., Alkire, N. L., de Queiroz, A., and Jones, A. 2001. Out of Africa: Origins of the *Taenia* tapeworm in humans. *Proc. R. Soc. B* 268:781–87.
13. Hotez, P. J., Brooker, S., Bethony, J. M., Bottazzi, M. E., Loukas, A., and Xiao, S. 2004. Hookworm infection. *N. Engl. J. Med.* 351:799–807.
14. Matthews, B. E. 2005. *An introduction to parasitology.* Cambridge: Cambridge University Press.
15. Bethony, J., Brooker, S., Albonico, M., Geiger, S. M., Loukas, A., Diemert, D., and Hotez, P. J. 2006. Soil-transmitted helminth infections: Ascariasis, trichuriasis, and hookworm. *Lancet* 367:1521–32.
16. Murray, S., Stem, C., Boudreau, B., and Goodall, J. 2000. Intestinal parasites of baboons (*Papio cynocephalus anubis*) and chimpanzees (*Pan troglodytes*) in Gombe National Park. *J. Zool. Wild Med.* 31:176–78.
17. Michaud, C., Tantalean, M., Ique, C., Montoya, E., and Gozalo, A. 2003. A survey for helminth parasites in feral New World non-human primate populations and its comparison with parasitological data from man in the region. *J. Med. Primatol.* 32:341–45.
18. Gillespie, T. R., Chapman, C. A., and Greiner, E. C. 2005. Effects of logging on gastrointestinal parasite infections and infection risk in African primates. *J. Appl. Ecol.* 42:699–707.
19. Bouchet, F., Baffier, D., Girard, M., Morel, P. H., Paicheler, J. C., and David, F. 1996. Paleoparasitology in a Pleistocene context: Initial observations in the Grande Grotte at Arcy-sur-Cure. *C.R. Acad. Sci. Ser.* 319:147–51.
20. Evans, A. C., Markus, M. B., Mason, R. J., and Steel, R. 1996. Late stone-age coprolite reveals evidence of prehistoric parasitism. *SAMJ* 86:846–49.
21. Ferreira, L. F., Araujo, A., Confalonieri, U., Chame, M., and Ribeiro Filho, B. 1987. The finding of hookworm eggs in human coprolites from 7230 ± 80 years BP, from Piaui, Brazil. *An. Acad. Bras. Cienc.* 59:280–81.

22. Fry, G. F., and Hall, H. J. 1969. Parasitological examination of prehistoric human coprolites from Utah. *Proc. Utah Acad. Sci. Art. Lett.* 46 (Part 2):102–5.
23. Hall, H. J. 1976. Untitled notes. *Paleopathol. News* 13:9.
24. Dommelier-Espejo, S. 2001. Contribution a l'etude paleoparasitologique des sites Neolithiques en environnement lacustre dans les domaines Jurassien et Peri-alpin. Thesis, Universite de Reims.
25. Roever-Bonnet, H., Rijpstra, C., van Renesse, M. A., and Peen, C. H. 1979. Helminth eggs and gregarines from coprolites from the excavations at Swifterbant. *Helinium* 19:7–12.
26. Aspock, H., Auer, H., and Picher, O. 1996. *Trichuris trichiura* eggs in the neolithic glacier mummy from the Alps. *Parasitol. Today* 12:255–56.
27. Ferreira, L. F., Araujo, A., and Confalonieri, U. 1982. Untitled note. *Paleopathol. News* 38:5.
28. Jones, A. K., and Nicholson, C. 1988. Recent finds of *Trichuris* and *Ascaris* ova from Britain. *Paleopathol. News* 62:5–6.
29. Patrucco, R., Tello, R., and Bonavia, D. 1983. Parasitological studies of coprolites of pre-Hispanic Peruvian populations. *Curr. Anthropol.* 24:393–94.
30. Iniguez, A. M., Reinhard, K., Goncalves, M. L., Ferreira, L. F., Araujo, A., and Vicente, A. C. 2006. SL1 RNA gene recovery from *Enterobius vermicularis* ancient DNA in pre-Columbian human coprolites. *Int. J. Parasitol.* 36:1419–25.
31. Udonsi, J. K., and Ogan, V. N. 1993. Assesment of the effectiveness of primary health care interventions in the control of three intestinal nematode infections in rural communities. *Pub. Health* 107:53–60.
32. Ferreira, C. S., Ferreira, M. U., and Nogueira, M. R. 1994. The prevalence of infection by intestinal parasites in an urban slum in Sao Paulo, Brazil. *J. Trop. Med. Hyg.* 97:121–27.
33. Kobayashi, J., Hasegawa, H., Forli, A. A., Nishimura, N. F., Yamanaka, A., Shimabukuro, T., and Sato, Y. 1995. Prevalence of intestinal parasitic infection in five farms in Holambra, Sao Paulo, Brazil. *Rev. Inst. Med. Trop. Sao Paulo* 37:13–18.
34. Hopkins, R. M., Gracey, M. S., Hobbs, R. P., Spargo, R. M., Yates, M., and Thompson, R. C. 1997. The prevalence of hookworm infection, iron deficiency and anaemia in an aboriginal community in north-west Australia. *Med. J. Aust.* 166:241–44.
35. Stoltzfus, R. L., Chwaya, H. M., Tielsch, J. M., Schilze, K. J., Albonico, M., and Savioli, L. 1997. Epidemiology of iron deficiency anemia in Zanzibari schoolchildren: The importance of hookworms. *Am. J. Clin. Nutr.* 65:153–59.
36. Labiano-Abello, N., Canese, J., Velazquez, M. E., Hawdow, J. M., Wilson, M. L., and Hotez, P. J. 1999. Epidemiology of hookworm infection in Itagua, Paraguay: A cross sectional study. *Mem. Inst. Oswaldo Cruz* 94:583–86.
37. Chacin-Bonilla, L., and Sanchez-Chavez, Y. 2000. Intestinal parasitic infections, with a special emphasis on cryptosporidiosis, in Amerindians from western Venezuela. *Am. J. Trop. Med. Hyg.* 62:347–52.
38. Scolari, C., Torti, C., Beltrame, A., Matteelli, A., Castelli, F., Gulletta, M., Ribas, M., Morana, S., and Urbani, C. 2000. Prevalence and distribution of soil-transmitted helminth (STH) infections in urban and indigenous schoolchildren in Ortigueira, state of Parana, Brasil: Implications for control. *Trop. Med. Int. Health* 5:302–7.
39. Andrade, C., Alava, T., Palacio, I. A., Poggio, P. D., Jamoletti, C., Gulletta, M., and Montresor, A. 2001. Prevalence and intensity of soil-transmitted helminthiasis in the city of Portoviejo (Ecuador). *Mem. Inst. Oswaldo Cruz* 96:1075–79.

40. Flores, A., Esteban, J. G., Angles, R., and Mas-Coma, S. 2001. Soil-transmitted helminth infections at a very high altitude in Bolivia. *Trans. R. Soc. Trop. Med. Hyg.* 95:272–77.
41. Carme, B., Motard, A., Bau, P., Day, C., Aznar, C., and Moreau, B. 2002. Intestinal parasitoses among Wayampi Indians from French Guiana. *Parasite* 9:167–74.
42. Taranto, N. J., Cajal, S. P., De Marzi, M. C., Fernandez, M. M., Frank, F. M., Bru, A. M., Minvielle, M. C., Basualdo, J. A., and Malchiodi, E. L. 2003. Clinical status and parasitic infection in a Wichi Aboriginal community in Salta, Argentina. *Trans. R. Soc. Trop. Med. Hyg.* 97:554–58.
43. De Quadros, R. M., Marques, S., Arruda, A. A., Delfes, P. S., and Medeiros, I. A. 2004. Intestinal parasites in nursery schools of Lages, southern Brazil. *Rev. Soc. Bras. Med. Trop.* 37(5).
44. Araujo, C. F., and Fernandez, C. L. 2005. Prevalence of intestinal parasitosis in the city of Eirunepe, Amazon. *Rev. Soc. Bras. Med. Trop.* 38(1).
45. Tanner, S. N. 2005. A population in transition: Health, cultural change, and intestinal parasitism among the Tsimane' of lowland Bolivia. Unpublished dissertation, University of Michigan.
46. Fleming, F. M., Brooker, S., Geiger, S. M., Calda, I. R., Correa-Oliveira, R., Hotez, P. J., and Bethony, J. M. 2006. Synergistic associations between hookworm and other helminth species in a rural community in Brazil. *Top. Med. Int. Health* 11:56–64.
47. Ferrari, J. O., Ferreira, M. U., Camargo, L. M. A., and Ferreira, C. S. 1992. Intestinal parasites among Karitiana Indians from Rondonia State, Brazil. *Rev. Inst. Med. Trop. Sao Paulo* 34:223–25.
48. Ferreira, M. R., Souza, W., Perez, E. P., Lapa, T., Carvalho, A. B., Furtado, A., Coutinho, H. B., and Wakelin, D. 1998. Intestinal helminthiasis and anemia in youngsters from Matriz da Luz, district of Sao Lourenco da Mata, state of Pernambuco, Brazil. *Mem. Inst. Oswaldo Cruz* 93:289–93.
49. Anderson, R. M., and May, R. M. 1991. *Infectious diseases of humans: Dynamics and control.* Oxford: Oxford University Press.
50. Hagan, P., Blumenthal, U. J., Dunn, D., Simpson, A. J., and Wilkins, H. A. 1991. Human IgE, IgG4 and resistance to reinfection with *Schistosoma haematobium*. *Nature* 349:243.
51. Rihet, P., Demeure, C. E., Bourgois, A., Prata, A., and Dessein, A. J. 1991. Evidence for an association between human resistance to *Schistosoma mansoni* and high anti-larval IgE levels. *Eur. J. Immunol.* 21:2679.
52. Dunne, D. W., Butterworth, A. E., Fulford, A. J., Kariuki, H. C., Langley, J. G., Ouma, J. H., Capron, A., Pierce, R. J., and Sturrock, R. F. 1992. Immunity after treatment of human schistosomiasis: Association between IgE antibodies to adult worm antigens and resistance to reinfection. *Eur. J. Immunol.* 22:1483.
53. Ackerman, S. J. 1993. Characterization and function of eosinophil granule proteins. In *Eosinophils: Biological and clinical aspects,* ed. S. Mkino and T. Fukuda, 33–71. Boca Raton, FL: CRC Press.
54. Hurtado, A. M., Hurtado, I., and Hill, K. 2004. Public health and adaptive immunity among natives of South America. In *Lost paradises and the ethics of research and publication,* ed. F. M. Salzano and A. M. Hurtado, 164–90. New York: Oxford University Press.
55. McDade, T. W. 2005. The ecologies of human immune function. *Annu. Rev. Anthropol.* 34:495–521.

56. Lisse, I. M., Aaby, P., Whittle, H., Jensen, H., Engelmann, M., and Christensen, L. B. 1997. T-lymphocyte subsets in West African children: Impact of age, sex, season. *J. Pediatr.* 130:77–85.

57. Witt, C. S., and Alpers, M. P. 1991. Lymphocyte subsets in Eastern Highlanders of Papua New Guinea. *PNG Med. J.* 34:98–103.

58. Muehlenbein, M. P., Bribiescas, R. G. 2005. Testosterone-mediated immune functions and male life histories. *Am. J. Hum. Biol.* 17:527–558.

59. Long, C. L. 1977. Energy balance and carbohydrate metabolism in infection and sepsis. *Am. J. Clin. Nutr.* 30:1301–10.

60. Moret, Y., and Schmid-Hempel, P. 2000. Survival for immunity: The price of immune system activation for bumblebee workers. *Science* 290:1166–68.

61. Lee, K. P., Cory, J. S., Wilson, K., Raubenheimer, D., and Simpson, S. J. 2006. Flexible diet choice offsets protein costs of pathogen resistance in a caterpillar. *Proc. R. Soc. B* 273:823–29.

62. DiPiro, J. T., Hamilton, R. G., Howdieshell, T. R., Adkinson, N. F., and Mansberg, A. R. 1992. Total IgE in plasma is elevated after traumatic injury and is associated with sepsis syndrome. *Ann. Surg.* 215:460–66.

63. Crouser, E. D., and Dorinsky, P. M. 1996. Metabolic consequences of sepsis: Correlation with altered intracellular calcium homeostasis. *Clin. Chest Med.* 17:249–61.

64. Michie, H. R. 1996. Metabolism of sepsis and multiple organ failure. *World J. Surg.* 20:460–64.

65. Chiolero, R., Revelly, J. P., and Tappy, L. 1997. Energy metabolism in sepsis and injury. *Nutrition* 13:45–51.

66. Patiño, J. F., de Pimiento, S. E., Vergara, A., Savino, P., Rodriguez, M., and Escallón, J. 1999. Hypocaloric support in the critically ill. *World J. Surg.* 23:553–59.

67. Tredget, E. E., and Yu, Y. M. 1992. The metabolic effects of thermal injury. *World J. Surg.* 16:68–79.

68. Roe, C., and Kinney, J. 1965. The caloric equivalent of fever: Influence of major trauma. *Ann. Surg.* 161:140–48.

69. Cooper, A. L., Horan, M. A., Little, R. A., and Rothwell, N. J. 1992. Metabolic and febrile responses to typhoid vaccine in humans: Effect of B-adrenergie blockage. *J. Appl. Physiol.* 72:2322–28.

70. Kreymann, G., Grosser, S., Buggisch, P., Gottschall, C., Matthaei, S., and Greten, H. 1993. Oxygen consumption and resting metabolic rate in sepsis, sepsis syndrome and septic shock. *Crit. Care Med.* 21:1012–19.

71. Christe, P. 2002. Intraseasonal variation in immune defence, body mass, and hematocrit in adult house martins *Delichon urbica*. *J. Avian Biol.* 33:321–25.

72. Shell-Duncan, B., and Wood, J. W. 1997. The evaluation of delayed-type hypersensitivity responsiveness and nutritional status as predictors of gastro-intestinal and acute respiratory infection: A prospective field study among traditional nomadic Kenyan children. *J. Trop. Pediatr.* 43:25–32.

73. Ulvestad, E. 2007. *Defending Life: The Nature of Host-Parasite Relations.* The Netherlands: Springer.

74. Biolo, G., Toigo, G., Ciocchi, B., Situlin, R., Iscra, F., Gullo, A., and Guarnieri, G. 1997. Metabolic response to injury and sepsis: Changes in protein metabolism. *Nutrition* 13:52–57.

75. Hobler, S. C., Williams, A. B., Fischer, J. E., and Hasselgren, P. O. 1998. IGF-I stimulates protein synthesis but does not inhibit protein breakdown in muscle from septic rats. *Am. J. Physiol. Reg. Integ. Comp. Physiol.* 274:R571–76.

76. Scrimshaw, N. S. 1991. Effect of infection on nutrient requirements. *J. Parent. Enter. Nutr.* 15:589–600.
77. Nelson, R. J., and Demas, G. E. 1996. Seasonal changes in immune function. *Q. Rev. Biol.* 71:511–48.
78. Zuk, M., and Stoehr, A. M. 2002. Immune defense and host life history. *Am. Nat.* 160:S9–22.
79. Soler, J. J., de Neve, L., Perez-Contreras, T., Soler, M., and Sorci, G. 2003. Trade-off between immunocompetence and growth in magpies: An experimental study. *Proc. Biol. Sci.* 270:241–48.
80. Roy, B. A., and Kirchner, J. W. 2000. Evolutionary dynamics of pathogen resistance and tolerance. *Evolution* 54:51–63.
81. Lekander, M. 2002. Ecological immunology: The role of the immune system in psychology and neuroscience. *Eur. Psychol.* 7:98–115.
82. Steen, R. G., Emudianughe, T., Hunte, M., Glass, J., Wu, S., Xiong, X., and Reddick, W. E. 2005. Brain volume in pediatric patients with sickle cell disease: Evidence of volumetric growth delay? *Am. J. Neuroradiol.* 26:455–62.
83. Amat, J. A., Aguilera, E., and Visser, G. H. 2007. Energetic and developmental costs of mounting an immune response in greenfinches (*Carduelis chloris*). *Ecol. Res.* 22:282–87.
84. Fair, J. M., and Myers, O. B. 2002. The ecological and physiological costs of lead shot and immunological challenge to developing western bluebirds. *Ecotoxicology* 11:199–208.
85. Fair, J. M., and Ricklefs, R. E. 2002. Physiological, growth, and immune responses of Japanese quail chicks to the multiple stressors of immunological challenge and lead shot. *Arch. Environ. Cont. Toxicol.* 42:77–87.
86. Sanz, J. J., Moreno, J., Merino, S., and Tomas, G. 2004. A trade-off between two resource demanding functions: Post-nuptial moult and immunity during reproduction in male pied flycatchers. *J. Anim. Ecol.* 73:441–47.
87. Whitaker, S., and Fair, J. 2002. The costs of immunological challenge to developing mountain chickadees, *Poecile gambeli*, in the wild. *Oikos* 99:161–65.
88. Panter-Brick, C., Lunn, P. G., Baker, R., and Todd, A. 2000. Elevated acute-phase protein in stunted Nepali children reporting low morbidity: Different rural and urban profiles. *Br. J. Nutr.* 85:1–8.
89. Hurtado A. M., Hill, K., Kaplan, H., and Lancaster, J. 2001. The epidemiology of infectious diseases among South American Indians. *Curr. Anthropol.* 42:425–32.
90. Hurtado, A. M., Hill, K. R., and Hurtado, I. 2001. The compromised health of South American Indians: The need for study under ethical rules. *Interciencia* 26:166–70.
91. Nacher, M., Singhasivanon, P., Yimsamran, S., Manibunyong, W., Thanyavanich, N., Wuthisen, P., and Looareesuwan, S. 2002. Intestinal helminth infections are associated with increased incidence of *Plasmodium falciparum* malaria in Thailand. *J. Parasitol.* 88:55–58.
92. Bentwich, Z., Kalinkovich, A., Weisman, Z., Borkow, G., Beyers, N., and Beyers, A. D. 1999. Can eradication of helminthic infections change the face of AIDS and tuberculosis? *Immunol. Today* 20:485–87.
93. Christensen, N. O., Nansen, P., Fagbemi, B. O., and Monrad, J. 1987. Heterologous antagonistic and synergistic interactions between helminthes and between helminthes and protozoans in concurrent experimental infection of mammalian hosts. *Parasitol. Res.* 73:387–410.

94. Long, K. Z., Nanthakumar, N. 2004. Energetic and nutritional regulation of the adaptive immune response and trade-offs in ecological immunology. *Am. J. Hum. Biol.* 16: 499–507.
95. Ribeiro, D. 1971. *The Americas and Civilization.* London: Allen and Unwin.
96. Bentwich, Z., Kalinkovich, A., Weisman, Z., Borkowa, G., Beyers, N., Beyers, A. D. 1999. Can eradication of helminthic infections change the face of AIDS and tuberculosis? *Immunol. Today* 20: 485–487.
97. Hurtado, A. M., Hill, K. R., Rosenblatt, W., Scharmen, T. 2003. Longitudinal study of tuberculosis outcomes among immunologically naïve Ache natives of Paraguay. *Am. J. Phys. Anthropol.* 121:134–150.
98. Hill, K., Boesch, C., Goodall, J., Pusey, A., Williams, J., and Wrangham, R. 2001. Mortality rates among wild chimpanzees. *J. Hum. Evol.* 40:437–50.
99. Montenegro, R.A., and Stephens, C. 2006. Indigenous health in Latin America and the Caribbean. *Lancet* 367:1859–69.
100. Ewald, P. 1996. *Evolution of infectious disease.* Oxford: Oxford University Press.
101. Stearns, S.C. 1998. *Evolution in health and disease.* Oxford: Oxford University Press.
102. Hurtado, A. M., Arenas de Hurtado, I., Hill, K., and Rodriguez, S. 1997. The evolutionary ecology of chronic allergic conditions: The Hiwi of Venezuela. *Hum. Nat.* 8:51–75.
103. Hurtado, A. M., Hurtado, I., Sapien, R., and Hill, K. 1999. The evolutionary ecology of childhood asthma. In *Evolutionary medicine,* ed. W. Trevathan, J. McKenna, and E.O. Smith, 101–34. Oxford: Oxford University Press.
104. Wickelgren, I. 2004. Immunotherapy: can worms tame the immune system? *Science* 305: 170–171.
105. Graham, A. L., Allen, J. E., and Read, A. F. 2005. Evolutionary causes and consequences of immunopathology. *Annu. Rev. Ecol. Evol. Syst.* 36:373–97.
106. Gluckman, P. D., Cutfield, W., Hofman, P., and Hanson, M. A. 2005. The fetal, neonatal, and infant environments—The long-term consequences for disease risk. *Early Hum. Dev.* 81:51–59.
107. Morley, R. 2006. Fetal origins of adult disease. *Semin. Fetal Neonatal Med.* 11:73–78.

9

Nonbizarre Delusions as Strategic Deception

Edward H. Hagen
Department of Anthropology, Washington State University

Contents

Introduction

Strategic analysis has yielded several surprising insights into animal behavior. Costly and seemingly harmful traits, such as the large, cumbersome tail of the peacock, or the exhausting bouts of roaring by red deer, are now understood to credibly signal aspects of quality to potential mates. Yet psychiatry, despite its focus on costly and therefore seemingly harmful behavior, has made almost no use of this powerful conceptual tool (the work of Nesse and a few others being notable exceptions). Seen through a strategic lens, it is conceivable that some behaviors currently thought to indicate madness might have a method to them, an idea that appears, surprisingly, in the early work one of psychiatry's harshest critics, Thomas Szasz.

Szasz is well known both for his biting critique of the mental illness concept[1] and his vehement condemnation of what he views as the coercive nature of modern psychiatry.[2] Less well known is that in addition to these philosophical and social critiques, Szasz has offered a constructive theory of mental illness, namely, that so-called mental illnesses are really strategies in the social games in which we are all engaged. Hysteria and all other phenomena called mental illnesses are

> made to happen by sentient, intelligent human beings and can be understood best, in my opinion, in the framework of games. "Mental illnesses" thus differ fundamentally from ordinary diseases and are similar, rather, to certain moves or techniques in playing games. Suffering from hysteria is thus far from being sick and could more accurately be thought of a playing a game, correctly or incorrectly, skillfully or clumsily, successfully or unsuccessfully, as the case might be.[1] (p. 225)

According to Szasz, these strategies are incorrectly labeled illnesses because they often involve socially undesirable behaviors like lying, cheating, and deception. The illness label then justifies the social control of these behaviors.[3] Psychiatry, however, is strictly prohibited from considering a strategic view of mental illness as, for example, a type of lie:

> For the contemporary psychiatrist to speak of lying in connection with so-called mental illness is anathema. Once a person is called a "patient" his psychiatrist is no longer even permitted to consider such a thing as lying. The prohibition placed on this term and all it connotes has been at least as strong as that on sex in Victorian society, and perhaps even greater. Anyone who speaks of lying in connection with psychiatric

problems, tends ipso facto to be identified as "antipsy-
chiatric" and "antihumanitarian," meaning thereby
that he is both wrong and bad. I believe this is most
regrettable, and merely signifies the contemporary
psychiatrist's (and lay person's) sentimentalizing attitude toward the so-called mentally ill. Such an attitude
toward mental illness is harmful to science and has no
place in it.[1] (p. 272)

Using concepts from modern evolutionary biology, Szasz's strategic theory
of mental illness can be framed as testable hypotheses, at least for a restricted
range of psychiatric symptoms. Given this framing, much evidence collected
using the illness model actually supports Szasz. The argument I will develop,
however, differs in important ways from Szasz's. First, I am not advancing
a social critique of psychiatry; instead, I am interested in whether the illness model is the correct scientific model of some psychiatric symptoms or
whether other models fare equally well or better. Second, I am not proposing
that Szasz is correct about all mental illnesses—I strongly suspect that he is
not, especially for conditions like autism and schizophrenia. Here I will only
be investigating a single psychiatric symptom: nonbizarre delusions. Finally,
unlike Szasz, I will specify in detail the special social circumstances that
should elicit deceptive strategies and the benefits such strategies can deliver
in the types of social environments in which humans evolved.

The Mystery of Delusions

Delusions are tenaciously held false beliefs that are unresponsive to the presentation of evidence contrary to the belief. The individual is preoccupied
with the belief, finds it difficult to avoid thinking or talking about it, and
does not report subjective efforts to resist it (in contrast to patients with
obsessional ideas). The belief involves personal reference, rather than unconventional religious, scientific, or political conviction.[4,5]

Delusions are generally divided into two categories, bizarre and nonbizarre.
Bizarre delusions are beliefs that are inconsistent with a person's culture, for
example, an American's tenaciously held belief that insects were living in his
brain.* Nonbizarre delusions are tenaciously held false beliefs that nonetheless could be accepted as true in that individual's culture, for example, an
American's tenaciously held false belief that he knew of an assassination plot
against the president. Nonbizarre delusions are often systematized, with the

---

* A meta-analysis by Bell et al.[6] concluded that although delusions in general can be reliably diagnosed, the diagnosis of bizarre delusions was unreliable. They noted, however,
  that many of the studies reviewed were poorly designed or suffered significant confounds. Because the distinction between bizarre and nonbizarre delusions plays a key
  role in the DSM, they suggested several criteria for adequate future studies.

delusional system forming a logical and coherent whole. Recent events may be incorporated into the system, or used as supporting evidence. This chapter will be solely concerned with nonbizarre delusions, which are accompanied by the preservation of clear and orderly thinking, and whose etiology, as discussed in more detail below, almost certainly involves severe social problems and is therefore distinct from that of bizarre delusions.

A final criterion for delusions is that the beliefs are not shared by others.[5] It is critical to the thesis advanced here, however, that delusions *are* believed by healthy members of the wider community. There is evidence that this is the case. As Bell et al.[7] showed in a study of mind control experiences reported on the Internet, even these bizarre delusions (albeit ones with a distinctly persecutory flavor, e.g., reports of police using brain implants) attract adherents. Although Bell et al. admit that some adherents were likely also psychotic, they found that the authors of mind control reports were often actively engaged with a nonpsychotic community who had thematically similar concerns.

Enormously disruptive to sufferers and their families, delusions are among the most difficult psychiatric conditions to treat. After more than a century of research, however, no compelling explanation of delusions has emerged. Delusions have been attributed to disturbances in affect and thinking, deficits in perception, deficits in the psyche, projections or externalizations of personal wishes, conflicts, or fears, altered views of the self, susceptible personality types, existential conflicts, avoidance responses, unsuccessful social interactions, and cybernetic regulation of the self and others. Most theories can be characterized by two major themes: delusions are either motivational (individuals are motivated to explain unusual perceptions, or they are motivated to reduce or ameliorate uncomfortable emotional or psychic states) or a sign of an underlying cognitive deficit (see Winters and Neale[8] for references and critique).

Cognitive deficit models of delusions appear to be attracting the most research attention. This research has revealed numerous deficits in cognition that distinguish individuals with persecutory delusions from other psychiatric patients as well as normal controls. These deficits are typically grouped into a limited set of categories, such as attentional biases, attributional biases, jumping-to-conclusion biases, and theory-of-mind deficits. For example, compared to nondelusional psychiatric patients and controls, individuals with persecutory delusions preferentially attend to threat-related stimuli, preferentially recall threatening episodes, spend less time reappraising potential threats in ambiguous pictures, take more credit for successes, more strongly deny responsibility for failures, tend to attribute failures to active malevolence on the part of others, draw conclusions based on less information and are more confident in these conclusions, and are less able to correctly infer the mental states of others (for reviews, see references 9–11).

But do these findings reveal cognitive deficits, or do they simply reveal cognitive *differences*? Imagine, for the sake of argument, that a person with persecutory fears had real enemies. It would not be surprising that this person preferentially attended to threat-related stimuli, preferentially recalled threatening episodes, tended to attribute failures to the malevolence of others, and so forth. Because none of these studies controlled for individuals' social circumstances, it is impossible to conclude that these cognitive differences are evidence of genuine cognitive deficits, that is, of mental illness. The evidence is equally consistent with a strategic interpretation that views delusions as an adaptive response to certain kinds of real social threats.

Further, most, if not all, of these differences are state differences, not trait differences: cognitive differences covary with delusional symptoms. Correlation is not causation, so it could be that cognitive differences are the cause of delusions, that delusions are the cause of cognitive differences, or, as I will argue here, that both are correlated with a third factor: genuine social problems.

There is excellent evidence that delusions are caused by changes in brain biochemistry—most antipsychotic drugs work by blocking dopamine and serotonin receptors, such as the D_2 and $5HT_{2A}$ receptors—but this is evidence in support of materialism, not of dysfunction. The brain is an electrobiochemical machine, so every difference in psychological state is caused by changes in electrobiochemistry. A person who is in love has brain levels of dopamine and norepinephrine that are different from a person who is not in love, and yet medicine would not say that a person in love is suffering from an excess of dopamine or norepinephrine, nor would it say that a person who is not in love is suffering from dopamine or norepinephrine deficits. By blocking or activating various receptors in the brain, it should be possible to suppress or activate just about any brain function, including the formation of memories, rational thought, language, emotions, and laughter.

Regarding the many other theories of delusions, a comprehensive review concluded, "In sum, despite large numbers of explanation and theories on delusional thinking, there is no agreed upon conceptualization or general model concerning their nature and very few theories enjoy empirical support."[8]

Should a Function for Delusions Be Considered?

Genuine brain dysfunctions such as Alzheimer's disease and stroke-related brain damage are best understood within mainstream psychiatry's illness model. It is less apparent, however, whether the same is true of other distressful psychiatric states like depression, anxiety, and delusions. As numerous critics of the Western concepts of "normal" and "abnormal" psychology have pointed out, labeling undesirable behaviors and emotions as abnormal allows them to be "treated," often with powerful drugs, and allows persons exhibiting them to be committed to institutions.[1,3,12] According to these critics, psychiatry then ceases to be medicine and instead becomes a form of social control.

Wakefield's concept of mental disorders as *harmful dysfunctions*[13-15] provides a compelling resolution to the debate between psychiatry and its critics. Traits that evolved to serve some function—adaptations—are not ill-nesses, even if they are deemed harmful by society. If aggression is an adap-tation, for example, it is not an illness, even if it causes social harm. On this view, aggression is not then a medical problem but a social problem. Con-versely, traits that are dysfunctional but cause no harm are also not illnesses. A vasectomy sterilizes a man, but this reproductive dysfunction is exactly what he desires. Only mental or physical conditions that are both harmful and dysfunctions, like Alzheimer's disease, are illnesses. Disentangling bio-logical function from judgments of harm permits the latter to be more easily debated and critiqued.

Establishing that a psychological phenomenon is an adaptation, and therefore not an illness, requires that (1) some important reproductive prob-lem posed by the physical or social environment be identified (the *selection pressure*), and (2) the psychological phenomenon in question be shown to effectively solve that problem. I will argue that severe social failure was an important selection pressure on the evolution of human psychology. I will then argue that certain types of deception would have effectively mitigated the costs of severe social failure. Finally, I will argue that delusions, for three reasons, are exactly these types of deception. The first reason is that severe social problems appear to be an important cause of delusions; the second reason is that delusions seem well designed to elicit benefits from others; and the third reason is that, in small-scale societies at least, delusions do elicit benefits from others.

Inquiring whether delusions are functional is especially urgent. The long-term use of older "typical," and even the newer and safer "atypical," antipsychotic drugs used to treat delusions is particularly dangerous. In a significant fraction of patients these treatments cause serious side effects like parkinsonism, and even irreversible brain damage, such as tardive dyskine-sia: repetitive, involuntary, purposeless movements.[16] If delusions are func-tional, they are not illnesses, so the use of antipsychotic drugs to suppress them would require additional ethical considerations, and new approaches to alleviate suffering would be conceivable.

The Selection Pressure: Severe Social Failure

Identifying an evolved function—an adaptation—always requires positing an associated ancestral environmental context, and a fitness benefit.* Vision, for example, requires an ancestral environment with sunlight, and the fit-ness benefit of seeing; hemoglobin an ancestral environment containing oxy-gen, and the fitness benefit of this oxidant for metabolism; and the immune system an ancestral environment with pathogens, and the fitness benefit of eliminating them from the body.

---

* Fitness is an individual's capability to reproduce; fitness benefits and costs increase and decrease this capability, respectively.

The hypothesis I will develop here makes two basic assumptions about ancestral human communities. First, that they comprised small, interdependent groups. And second, that, on occasion, conflicts arose in these groups that restricted a particular individual's access to benefits provided by group members, imposing a severe fitness cost on him or her.

The first assumption is supported by four lines of evidence:

1. Many primate species, including our closest relatives, the chimpanzees and bonobos, live in small groups and obtain important benefits from group members.[17,18]
2. There is clear archaeological evidence from the Late Middle Pleistocene on showing that *Homo* hunted big game, returning large packages of meat to caves and other central sites where it was processed and consumed by multiple individuals.[19]
3. Members of the African and Eurasian predator guilds to which Pleistocene *Homo* appears to have belonged, such as lions, hyenas, African dogs, and wolves, also live in small groups that provide important benefits to members.[19]
4. All known modern hunter-gatherers live in relatively small groups.[20]

In such small-scale and traditional societies, individuals receive very important benefits from their relations with others, including food, protection, health care, and mates.[20-27] Conflicts and ruptures in social relationships obviously put these benefits at risk.

Regarding the second assumption, the rate of social problems in ancestral human communities is an open question, of course, but social rejection, exclusion, shunning, and ostracism have been documented among wild chimpanzees,[28,29] baboons,[30] lemurs,[31] a number of other primate species,[32] as well as numerous cultures, most institutions (e.g., government, military, education), all types of relationships (formal and informal), and among children, adolescents, and adults.[33] More generally, it is hard to imagine that there is a single member of the human species who has not experienced threats to his or her relationships. A spouse can fall in love with another, a parent can die, a friend can betray, and so forth.

The pain induced by problems in social relationships, termed social pain, is intense. The pain and distress experienced when recollecting a socially painful event, for instance, especially ostracism, are substantially higher than when recollecting a physically painful event, with levels of social pain comparable to chronic back pain and childbirth.[33] Social pain might actually have evolved from physical pain,[34,35] and it appears that similar brain regions are involved in social and physical pain.[36] Leary et al.[37] found that 99% of recollected instances of social pain involved relational devaluation, usually by someone close.

Social pain is believed by most specialists to be the product of a long evolutionary history, especially among mammals, of heavy reliance on social relationships, including mother-infant bonds, and the disastrous consequences when such bonds are weakened or severed.[33–35,38–40] Infants of less socially integrated baboon mothers, for example, are less likely to survive than infants of more socially integrated mothers, even after controlling for dominance rank, group membership, and environmental conditions.[41]

Social problems such as loss of close kin, a failure to form friendships, poor relationships, few benefits provided by social partners, renegotiation of relationships on less favorable terms, termination of one or more relationships, loss of social status, or hostile individuals impeding one's attempt to socialize with others or ostracizing one from the group would have decreased or eliminated access to essential resources, critically reducing one's biological fitness. Social failures would have greatly increased the difficulties in finding or keeping a mate, children would have received less care and investment, and close kin may have suffered as well. For a comprehensive review of the negative consequences of one type of social problem, stigmatization, see Crocker et al.[42] For an evolutionary analysis of stigmatization, see Kurzban and Leary.[43]

The price that humans pay for their almost unprecedented reliance on social relationships is the serious fitness cost that attends social failure.

The Adaptations: Vigilance and Exploitative Deception

In the rest of this chapter, I will argue a simple proposition: that when individuals are in what would have been, in ancestral environments, a bad social situation, they will increase their vigilance and, in some cases, will lie. When they are in a disastrous social situation, they will experience a strong compulsion to lie, and they will believe their own lies to increase the odds that others believe them too. What I am adding to this prosaic idea is simply that there are specialized psychological adaptations to increase vigilance and to lie. The lies are compulsory and completely unconscious, and are tightly focused on themes that garnered social benefits in ancestral social environments. Because these lies are often (but not always) implausible in modern states, they succeed less often than they would have in ancestral environments, and have therefore been misidentified as a psychopathology termed nonbizarre delusions.

In this section I outline what I think adaptations for increased vigilance and deception would look like. In subsequent sections, I show how delusions correspond to this outline, I establish the important role of social problems in the etiology of delusions, and I document that, in small-scale and traditional societies, delusions elicit benefits.

Increased Vigilance

There is substantial evidence of specialized neural mechanisms for the detection and recollection of social threats. In particular, there are attentional and memory biases for threat-related stimuli. People more rapidly detect angry faces than faces expressing other emotions; visual scanning of threat-related faces, compared to faces exhibiting other emotions, is characterized by distinct visual scan path strategies, such as increased fixations on feature areas (e.g., eyes, mouth) that appear particularly important for recognizing anger and fear. Evidence from behavioral, lesion, and imaging studies indicates these functions appear to depend critically on the amygdala and prefrontal cortices.[44] Because delusional and delusion-prone individuals show consistent differences in performance on threat-related tasks relative to controls, Green and Phillips[44] suggest that clinical levels of paranoia represent the dysfunctioning of evolved social threat detection mechanisms. As I will document below, however, real social problems are precursors to, and probable causes of, delusions. Hence, the documented neurocognitive differences could equally well represent an adaptively heightened vigilance to real threats.

Individuals with real social problems should show increased vigilance, i.e., paranoia, toward social threats. Individuals suffering severe social failure have, by definition, few social partners or weak social bonds. They have few people to take care of them if they are injured or fall ill, they have few people to provide critical resources like food, and they have few allies to help defend them in conflicts with others. Consequently, they are far more vulnerable than others to illness, injury, resource shortages, and social conflicts.[27] Given this increased vulnerability, it becomes increasingly necessary to avoid such costly circumstances. To do so, socially vulnerable individuals must increase their vigilance, at the expense both of devoting more time and effort to other tasks and of mistaking benign situations for dangerous ones, what Nesse[45] refers to as the smoke detector principle (smoke detector thresholds must be set low so that there are very few false negatives, at a cost of higher frequencies of false positives).

Signaling and Deception: General Theory

From an evolutionary perspective, adaptations for communicating information or sending signals evolved because they benefited the sender, and not necessarily the receiver.[46] Organisms may communicate either true or false information when it is in their fitness interest to do so. Because conflicts between organisms are common, deception should be rife in nature, and it is. Mimicry and crypsis are extremely widespread in vertebrates, arthropods, and opisthobranch gastropods.[47] Myrmecomorphy—morphological and behavioral mimicry of ants—has evolved at least 70 times, for example, in a total of more than 2,000 species belonging to 200 genera in 54 families, including 15 times in spiders, 10 times in plant bugs, and 7 times in staphylinid beetles.[48] And these do not even include the many species that mimic ant chemical signals! (I briefly discuss these below.) In nature, bluff and deception are often the rule rather than the exception.

In cooperative social relationships, however, where communication enhances the effectiveness of cooperation and future interactions are likely, outright exploitation of receivers should be rare.[49] In fact, in cooperative social systems, signals should be cheap, reliable, and easy to send because this reduces the cost of cooperation, thus increasing its net fitness benefit. Receivers in cooperative relationships are nonetheless susceptible to third-party parasites, termed *social parasites*,* that mimic the sender's signal, since discrimination against the parasite's signals may jeopardize the benefits obtained by communicating and cooperating with the sender.[50]

Social parasites are known in a wide variety of vertebrate and invertebrate species. In vertebrates, well-studied examples include avian brood parasites, such as cuckoos and cowbirds, that lay eggs in the nests of other species to avoid the costs of brood care.[51] Ants, however, might provide a better point of comparison because, like humans, they have an elaborate system of cooperation based on "cheap" signals,† in this case chemical and behavioral rather than linguistic. To defend against social parasites, ants have evolved a sophisticated chemical recognition system, probably based on cuticle hydrocarbons, enabling them to behave altruistically toward nestmates and reject non-nestmates.[52]

Despite their recognition system, ant species are parasitized by a number of arthropods, including butterflies, beetles, and even other ants. In fact, of the 10,000 or so known ant species, more than 200, or about 2%, parasitize other, often closely related, species, and in the unusually well-characterized ant fauna of Switzerland, about 1/3 of the species are parasitic.[53]

Ants are exploited by parasites in a number of ways, including enslavement and the takeover of nests by foreign queens. Penetration of the nest by social parasites is believed to involve either chemical mimicry, where the parasite synthesizes chemical signals similar or identical to host signals, or chemical camouflage, where the parasite acquires the requisite chemicals from the host. Chemical mimicry has now been confirmed for several parasitic species, including species of beetles, flies, and butterflies.[52] Larvae of the lycaenid butterfly *Meculinea rebeli*, for example, engage in a particularly impressive form of parasitism using evolved chemical signals to break the communication and recognition codes of the ant host *Myrmica schencki*. *Meculinea* caterpillars chemically masquerade as ant larvae, causing them to be transported into the ant nest brood by foraging ant workers. There, the caterpillars are fed by the ants.[54]

---

* More precisely, social parasites exploit some aspect of the the social behavior of their hosts.

† Cheap signals are those that can be sent with only a small cost to fitness, and are therefore more easily faked.

Human Social Parasites

Humans, too, facilitate the exchange of extremely valuable benefits using a communication system (language) that relies on "cheap" signals, and so are vulnerable to exploitation by social parasites, in this case other humans, that can mimic these signals.

In biological theory, one of the principal mechanisms to deter deception in cheap signaling systems is to punish false or deceptive signals by defecting from repeated future cooperative interactions to the deficit of the deceiver (e.g., Silk et al.[55] and references therein). An important cost that humans face for deceiving other group members, in other words, is the termination of social relationships. This consequence of "cheating" is predicted by virtually all models of the evolution of cooperation based on social exchange.[56,57] However, an individual who is already suffering severe social failure—that is, one with few or no profitable social relationships and little access to future social benefits—cannot be deterred by such threats. This individual has nothing to lose and much to gain from successful deception that elicits social benefits they otherwise have no access to. An adaptation to deceive and exploit social partners should be present in all individuals, but only activated in those for whom the benefits of deception outweigh the costs. Among individuals already suffering severe social failure, the benefits of deception and exploitation will almost always outweigh the costs because there are few or no costs.

What would such a deceptive, exploitative adaptation look like? First, it should cause individuals suffering severe social failure to signal to others that they need social benefits, and that they can provide social benefits in return. These individuals should behave in ways that are difficult to consciously imitate, like displaying intense fear or excitement,[58] because such behavior may be more likely to convince others. They should be able to give reasons for their behavior that are difficult to independently verify, at least immediately. Examples include the claim that one possesses important information or has an intimate relationship with a high-status individual. The deceptive signals, like cues of need and distress, should be supported by explanations or additional information that provide a plausible basis for the signals. Individuals attempting to extract social benefits from others via deception will be plausible recipients of the intended benefits, and they should feel compelled to communicate their deceptions to others. The adaptation should deactivate if and when social partnerships are established.

There is evidence, discussed below, that delusions satisfy every hypothesis, and conversely, that these hypotheses account for most of the significant clinical, etiological, and demographic aspects of delusions, a psychotic psychiatric symptom. The only previous (brief) suggestion that I have encountered that psychoses function to mitigate social exclusion is that of Wallace.[59] He presents no rationale for this function, however. As I discussed earlier, Szasz[1] has argued that "mental illness" in general is often a form of deception. Henderson[60,61] has carefully investigated the hypothesis that neuroses, though not psychoses, function to elicit care, and Sullivan[62] is well known for his interpersonal approach to psychiatry. The exploitative deception hypothesis of delusions is consistent with the argument that self-deception functions to facilitate the receipt of social benefits.[40,63–65]

Domains of Deception

There are three domains where humans receive substantial social benefits: social exchange, defense, and mating. Each of these should consequently be the target of individuals wishing to extract social benefits via deception.

Social Exchange

Individuals prefer to cooperate with individuals who have valuable benefits to offer.[66,67] Deceptive cues of access to important information, people, or of possessing valuable skills should therefore increase one's social value to others, increasing access to social benefits.

Additionally, individuals help others when they can provide large benefits to others at low cost to themselves (throwing a rope to a drowning man, for example) because they are then eligible for a return on this investment when the benefited individuals reciprocate.[68,69] Humans give off numerous cues of distress, like crying and expressions of fear,[70,71] indicating they are eligible for receiving these kinds of social investments. Social norms also often dictate providing assistance to needy group members. Deceptive cues of illness, fear, or distress should therefore elicit social investments from unsuspecting fellow group members.

Defense

Belief that there is an external threat provides a very strong impetus for cooperation among humans,[72] and it has been argued that external threats were a significant selection pressure for the initial evolution of cooperation among hominids.[73] Because a high level of within-group cooperation among a large number of individuals is essential to successful defense, external threats provide an extremely strong incentive to suppress internal political conflicts. Further, in the face of an external threat, each healthy group member has considerable value to other group members as a defender. Group members should readily cooperate against possible external threats because the costs of responding to a false threat are lower than the costs of not responding to a real threat. Deceptive claims of external threats should therefore elicit social benefits by reducing internal political conflicts that might threaten those with few allies, and by increasing one's social value as a provider of important information about enemies, and as a defender.

Mating

A mating relationship is usually a close and intimate relationship, in which partners have considerable influence on one another. Deceptive claims of a romantic relationship with a high-status person would be difficult to disprove, and they imply that one has influence on that person, as well as access to his or her power and resources. It should be possible to trade on one's perceived relationship with a person of status and power to increase one's own status and power.

Delusions as Exploitative Deception

Mental Illness as Adaptation

Several authors have suggested that certain psychiatric symptoms and syndromes may be adaptations.[39,74–84] Unpleasant experiences like nausea, vomiting, and fever are healthy, functional physiological responses to toxins and infections. Analogously, intense, negative psychological experiences like delusions and hypochondriasis may be healthy, functional responses to certain types of social failure. If so, under Wakefield's illness concept they are not illnesses, however distressing or harmful they might be.

Delusional Disorder

To avoid confounding the etiology of delusions with the etiology of depression, hallucinations, brain damage, substance use, or catatonic behavior, all of which can be associated with delusions,[85] I will restrict my focus to delusions in the absence of any other symptom, that is, to the distinct nosological entity delusional disorder (DD). (Although there is still some debate whether DD is a valid and distinct psychiatric entity, it has been accepted as such in the *Diagnostic and Statistical Manual of Mental Disorders* [DSM-IV]; see, e.g., references 86–94 for work on the nosological validity of DD and related delusional psychoses.)

DD is defined by the presence of nonbizarre delusions of at least one month's duration, and by the absence of hallucinations, disorganized speech, disorganized or catatonic behavior, flattening of affect, markedly impaired functioning, odd or bizarre behavior, underlying medical condition, or physiological effects of a substance (i.e., drug use).[95] Paranoid disorder (DSM-III) is an older term for DD that included only persecutory or jealous delusions.* In other words, individuals with DD are cognitively, emotionally, and physically unimpaired, and their only symptom is a nonbizarre delusional framework.

Paranoid schizophrenia (DSM-IV) is similar to DD except that prominent auditory or visual hallucinations are present in addition to delusions. This

---

* These older DSM-III criteria are still commonly encountered in the research literature.

chapter will not propose an adaptive function for paranoid, catatonic, or any other type of schizophrenia. Unfortunately, studies of delusions often include individuals who might be diagnosed as schizophrenic or for whom a diagnosis of DD is excluded due to the presence of prominent hallucinations or other psychotic symptoms. Besides delusions and hallucinations, psychotic symptoms include disorganized speech and grossly disorganized or catatonic behavior. The use of data including any such individuals will be noted.

Although DD is rare (with a prevalence of approximately 0.01–0.03%), delusions in concert with other symptoms like depression and auditory hallucinations are not. One population survey found the prevalence of delusions to be 3.3%.[96] Another large (n = 18,980) cross-cultural survey found the prevalence of delusions to be 1.9%.[97] Though delusions can be associated with a variety of other conditions, individuals with DD have delusions and nothing else. Identifying the cause of DD might therefore reveal the specific cause of nonbizarre delusions, a cause that could then explain the association of delusions with other disorders. Let us call this unknown cause X. The association of delusions with, e.g., brain damage, hallucinations, catatonia, substance use, or depression might be via the association of brain damage, hallucinations, catatonia, substance use, or depression with X. For example, brain damage could cause X, which then causes delusions. Seen from this perspective, the prevalence of delusions is expected to be much higher than DD. In the "Social Problems Cause Delusions" section I will discuss the considerable evidence that X, the specific cause of nonbizarre delusions, is severe social problems.

Where possible, findings for DD will be contrasted with those for schizophrenia. Schizophrenia provides an excellent control case for DD since it is also a psychotic disorder whose symptoms include both bizarre and nonbizarre delusions, as well as the more disabling psychotic symptoms. As will be seen below, DD has a social "fingerprint" quite distinct from schizophrenia. When delusions are separated from other symptoms and conditions, an etiology of social exclusion and isolation emerges.

Paranoia as Increased Vigilance

Paranoid personality disorder (PPD, DSM-IV) is not considered to be a psychotic disorder; individuals are not delusional—they do not cling tenaciously to an elaborated false belief—nor have they experienced other psychotic symptoms. They are, however, very distrustful and suspicious of others, whose motives are interpreted as malevolent. PPD may be an adaptation to social problems that employs vigilance rather than deception. Socially threatened individuals must be on the constant lookout for attempts to deprive them of material, social, or reproductive resources. Because they do not have social partners that would help them, they must also be more vigilant in avoiding injury and disease. PPD, anxiety, obsessive-compulsive, and certain somatoform "disorders" may therefore be vigilance-type adaptations to social and physical threats (see also Green and Phillips[44] and Boyer and Lienard[98]). PPD appears to be more common than persecutory delusions and, if an adaptation, may be used instead of deception for less severe social threats. Socially threatened individuals who fear members of their in-group may be increasing their vigilance toward likely internal adversaries rather than attempting to exploit them.

Delusions as Adaptations for Exploitative Deception

DD is characterized by the presence of a full-blown delusional framework. Delusional themes are not random or arbitrary. In principle, delusional themes could orbit any domain of human cognition involving belief formation, including any aspect of the physical environment (e.g., beliefs about the location of streets and buildings), biological environment (e.g., beliefs about apples and lettuce), material culture (e.g., beliefs about how to open a car door or put on a pair of pants), or even numerous aspects of the social environment (e.g., beliefs about the meaning of English words). But they do not. Cross-culturally, the vast majority of delusions can be characterized by a tiny subset of all conceivable themes: grandiose, persecutory, erotomanic, somatic, and jealous.[95] These themes almost exactly match the domains of deception that are most likely to garner social benefits, as discussed above in the "Domains of Deception" section: social exchange, defense, and mating. Grandiosity deceptively increases one's social value; somatic delusions deceptively indicate that one is sick and therefore deserving of aid; paranoia deceptively indicates a need for protection from an external threat, a threat that could increase group cohesion and one's value as a provider of information about enemies and as a defender; and erotomania deceptively indicates a relationship with a high-status individual that could be traded on to increase one's own status. Jealous delusions represent increased vigilance, not deception. See table 9.1 for a summary of the deceptive or vigilant functions proposed for each delusional theme.

If delusions are to effectively deceive others, delusional individuals must act in accordance with their delusions. Importantly, most do. Wessely et al.[99] found that 60% of their sample of deluded individuals* reported at least one action based on delusion; third-party informants reported that 52% of the sample probably or definitely acted on delusions. Persecutory delusions were significantly more likely to be acted upon than other beliefs. In a sample of patients with DD who were being supervised by a forensic psychiatric service after violent or threatening acts, Kennedy et al.[100] similarly found that 80% of the acts were related to the delusion. Other actions, such as fleeing or barricading to avoid delusional persecutors, were also consistent with the delusion.

For delusions to be a universal psychological adaptation, they must be found in all cultures. That appears to be the case.[101,102] Westermeyer,[102] relying

---

* Sample included individuals diagnosed with schizophrenia and affective psychosis.

Table 9.1 The Five Nonbizarre Delusional Themes according to DSM-IV, and Their Possible Functions

Delusional Theme	Hypothesized Function
Grandiose: Individuals are convinced they possess important information, have a special relationship with a very important person, or have some great (but unrecognized) talent or insight.	**Deception:** Individuals are presenting themselves as highly valuable social partners in order to gain friends, allies, and other social benefits.
Persecutory: Individuals believe that they are threatened by powerful others. These are the most common type of delusions.[9] Individuals with these delusions can give very convincing accounts of the reputed threat, behave consistently with the delusion,[99] and give cues of genuine fear and distress.[100]	**Vigilance:** Socially threatened individuals need to greatly increase their vigilance toward the social environment to prevent further harm. **Deception:** Belief in an external threat provides a very strong impetus for cooperation among members of the same group, especially those living in small, autonomous bands with real enemies. These delusions exploit the willingness of others to cooperate in mutual defense, decreasing internal conflicts and increasing the mutual value of all group members.
Erotomanic: Individuals believe that another person, usually of high status, is in love with them. Males with erotomanic delusions often attempt to rescue females from some imagined danger.[95] Note that the delusional person does not necessarily claim to be in love with the target.	**Deception:** Individuals that are highly valued by, and have an important connection with, a high status individual have higher value themselves. Claims of sexual relationships may have been particularly difficult for others to disprove because even when such relationships exist individuals often deny them. Males falsely claiming to offer defensive benefits to females are probably attempting to obtain both social and sexual benefits.

Delusional Theme	Hypothesized Function
Somatic: Individuals with somatic delusions, which are often difficult to distinguish from hypochondriasis,[95] are preoccupied with the fear or idea that they have a serious disease based on a misinterpretation of one or more bodily signs or symptoms. The fear persists despite medical reassurance.	**Vigilance:** Socially threatened individuals need to be particularly concerned about falling ill because of the uncertainty that others will care for them. **Deception:** Group members are tricked into providing care under the assumption that they are helping a seriously ill person (who might then return the favor in the future). Social norms may also dictate providing assistance to those who appear in need.
Jealous: Individuals believe their mate to be unfaithful.	**Vigilance:** Socially threatened individuals are likely at greater risk for losing their mates. Jealous delusions are therefore not examples of exploitative deception, but are simply a greatly increased form of normal jealousy.

on a review of the literature, 4 years of field work in Asia, 15 years at an international clinic at the University of Minnesota Hospitals and Clinics, and several studies of culture and psychopathology conducted in the United States, makes the following cross-cultural generalizations about delusions: delusional themes (e.g., grandiose, persecutory) vary little, if any, across cultures, whereas the specific content may be influenced by culture; culture-bound (e.g., persecution by hekura spirits) and secular (e.g., persecution by political enemies) delusional content are not mutually exclusive, but may coexist in the same individual; and delusional content can be quite etic, or secular, and yet still give rise to behaviors that are highly culture bound or emic (such as building a religious shrine or undertaking amok-type violence).

The hypothesis, in sum, is that individuals facing severe social threats developed powerful delusional systems. These caused them to unconsciously deceive their fellow group members in order to receive social benefits that they had lost or been unable to obtain. For example, an individual experiencing a persecutory delusion—"Group Z is trying to kill me"—would display very convincing signs of fear and distress and be able to cite evidence of the truth of his or her claims. In a small, somewhat isolated band with genuinely hostile Group Z neighbors, such a display could be convincing enough that fellow group members would cooperate with this individual against Group Z, a common enemy. Indeed, it is difficult to see why an otherwise normal individual displaying convincing, culturally consistent fear toward a known enemy would not be believed at least some of the time. And if he or she were believed, it is difficult to see why he or she would not at least occasionally obtain protection and other social benefits. On this view, delusions are a protective response to social problems.

Because only a tiny fraction of the world's population currently lives in small, isolated communities with hostile neighbors, delusions, even if they are adaptations, will often fail to elicit benefits. Citizens of industrial societies live in large communities with extensive police and military forces, and have access to many sources of information. Since external attacks are unlikely and exaggerated fears are often easy to disprove in these contexts, delusional displays of persecution have little chance of success and, in fact, are usually maladaptive—tragically, they tend to intensify social isolation rather than mitigate it. Although social problems should cause delusions in all societies, delusions would usually provide social benefits only in the now rare small, kin-based societies.

Social Problems Cause Delusions

Delusions are strongly associated with social problems. In the *social selection* hypothesis, this is attributed to the delusions themselves: delusions, it is claimed, prevent people from forming and maintaining social relationships. Alternatively, in the *social causation* hypothesis, severe social problems cause delusions in otherwise healthy individuals. If delusions are adaptations to severe social problems, then social problems should cause delusions. Several lines of evidence indicate that otherwise healthy individuals first suffer severe social problems, and then suffer delusions.

Psychiatric Populations

Cameron[103] was among the first to explicitly locate the genesis of delusional systems in the social arena. He identified the importance of social isolation and lack of social communication in the development of a delusional framework, noting that paranoiac attitudes and actions grow out of a breakdown in the machinery of social cooperation. Cameron, however, felt that isolation from the community was only the final outcome of a process that led the delusional individual to act detrimentally on his environment. Interestingly, he, too, recognized that delusional behavior may occasionally make an individual a distinguished person and, rarely, a leader of men.

In contrast to Cameron, Lemert[104] found strong evidence for the causal role of social exclusion in paranoia. He retrospectively studied eight cases of persons with "prominent paranoid characteristics." Four cases involved persons admitted to the state hospital at Napa, California, with diagnoses of paranoid schizophrenia. The lack of any history or evidence of hallucinations or intellectual impairment, however, excludes schizophrenia as a likely diagnosis for these cases. The others involved persons admitted to hospitals, involved with the law, or having chronic job difficulties. One case resembled paranoid personality disorder.

Lemert spent as much as 200 hours per case collecting data from anyone who played a significant role in the life of the person involved, attempting to establish the order in which delusions and social exclusion occurred. He found that

> the paranoid process begins with persistent interpersonal difficulties between the individual and his family, or his work associates and superiors, or neighbors, or other persons in the community. These frequently or even typically arise out of bona fide or recognizable issues centering upon some actual or threatened loss of status for the individual. This is related to such things as the death of relatives, loss of a position, loss of professional certification, failure to be promoted, age and physiological life cycle changes, mutilations, and changes in family and marital relationships. The status changes are distinguished by the fact that they leave no alternative acceptable to the individual, from whence comes their "intolerable" or "unendurable" quality. For example: the man trained to be a teacher who loses his certificate, which means he can never teach; or the man of 50 years of age who is faced with loss of a promotion which is a regular order of upward mobility in an organization, who knows that he can't "start over;" or the wife undergoing hysterectomy, which mutilates her image as a woman. (p. 7)

Lemert concluded that it is this process of exclusion and isolation that leads to the development of the delusional framework and not the converse. He notes that paranoia emerges in situations where "the goals of the individual can be reached only through cooperation from particular others, and in which the ends held by others are realizable if cooperation is forthcoming from ego."

In another retrospective study, this one of a group of thirty-four individuals with DD (DSM-III paranoid disorder), Kaffman[105] (p. 7) found that in every case there was a clear and realistic connection between paranoid premises and facts and events in the patient's life. He also found that authentic past and current interpersonal transactions play a dominant role in generating and activating the paranoid beliefs. From the case studies presented, these transactions appear to have involved isolation and rejection.

Kendler[88] argues that DD is distinguished from schizophrenia by low rates of psychiatric illness among family members of patients with DD, and the fact that environmental factors look to be more etiologically important than do genetic-constitutional ones. Several lines of evidence support the hypothesis that these environmental factors are social problems. Principal among them are case-control studies of DD versus schizophrenia. Because the symptoms of DD are less disabling than those of schizophrenia, social selection theory would predict that DD will be associated with fewer social problems than will schizophrenia. Several studies, two of which are described here, show just the opposite: DD, the less severe syndrome, is associated with more social problems than schizophrenia, supporting a social causation theory of DD.

Based on an analysis of case notes and follow-up interviews, Retterstöl's retrospective/case-control study of 301 first-admission psychiatric patients with paranoid and paranoiac symptoms[106] found that 100% of paranoid psychoses were caused by an event that "provokes the insecurity of the individual," i.e., those that tended to isolate the individual and make him feel an outsider, either by making him unpopular within his own group or by transplanting him to new and strange surroundings (p. 133). This was true of only 54% of cases diagnosed with schizophrenia.

Kay et al.[107] conducted a case-control study between psychiatric patients diagnosed with either paranoid psychosis (n = 54) or affective psychosis (n = 57). A minority of the paranoid patients were diagnosed as schizophrenic. Before the onset of the illness, paranoid patients were found to have had more difficulty than affective patients in forming and maintaining satisfactory interpersonal relationships, and had been more solitary, shy, reserved, and suspicious, and less able to display sympathy or emotion. At the onset of illness, the following features distinguished the paranoid group from the affective group: low social class, having few or no surviving children, living alone, and social deafness. All of these indicate an increased likelihood of social problems. Kay et al. conclude that their data support a multifactorial hypothesis where various adverse circumstances, especially in combination, such as being unmarried, having few close relatives, belonging to lower social class groups, or becoming deaf, increase the chances of hardship, insecurity, and loneliness in later life. The accumulated sense of deprivation and injustice is conducive to paranoid illness. Because socially impaired personalities were not associated with low social position, they disfavor downward social drift as an explanation for the correlation of social problems with paranoid illness.

Longitudinal Population Surveys

The causal role of social problems in delusion formation is also strongly indicated by recent longitudinal studies that assessed various types of social problems at time 1 in large samples of the general population and then found high rates of delusions at time 2 among those who suffered severe social problems at time 1 (screening out, or controlling for, individuals with a history of psychotic symptoms at time 1).

A large (7,076) random sample of members of the Dutch population (all fluent Dutch speakers), for example, was screened for a 3-year longitudinal study.[108] Individuals with any history or evidence of psychotic symptoms (or psychosis-like experiences) at the initial interview were excluded from the study. Individuals who experience discrimination based on ethnicity, sex, sexual orientation, age, disability, or appearance are at increased risk for social problems. Perceived discrimination reported during the initial interview in one domain (e.g., skin color) was associated with a near doubling of the rate of delusional ideation found at the final interview 3 years later, relative to those who reported no discrimination. Perceived discrimination reported in multiple domains (e.g., skin color plus sexual orientation) was associated with a more than fivefold increase in the rate of delusional ideation found 3 years later. These associations remained after adjustment for variables measured at the initial interview, like employment status, marital status, and education level, nonpsychotic DSM-III-R diagnosis, indicators of premorbid social adjustment, and personality measures of neuroticism, self-esteem, and locus of control. Interestingly, no association was found between discrimination and onset of hallucinatory experiences, suggesting that discrimination increases risk for delusions, and not psychotic symptoms per se.

In a similar study,[109] 2,524 adolescents aged 14–24 years provided self-reports at time 1 of lifetime exposure to trauma, including physical threats, rape, sexual abuse, and serious accidents. They were also assessed for psychotic symptoms, and potential confounds like psychosis-proneness, socioeconomic status, urbanicity, cannabis use, major depression, bipolar disorder, anxiety disorder, and hypomanic episode. At time 2, an average of 42 months later, participants were interviewed for presence of psychotic symptoms (eleven delusion items and four hallucination items), major depression, and bipolar disorder. Controlling for the aforementioned confounds, the odds ratio for the association between experiencing any trauma and psychosis narrowly defined (i.e., three psychotic symptoms) was 1.89 (results were not reported separately for delusions and hallucinations). When trauma categories were inspected separately, all were significantly associated with psychosis except "other" and "serious accident," indicating that psychosis is not caused by trauma in general, but rather social trauma ("natural catastrophe" might be an exception). Trauma was also not associated with new cases of major depression or bipolar disorder at time 2, indicating that trauma was a risk factor specifically for psychotic symptoms, not psychopathology in general.

Immigrants and Refugees

Immigrants and refugees are quite likely to suffer social problems since they have often left family, friends, and other important social ties behind, and will face increased difficulties competing for social benefits in a foreign, and perhaps hostile, society. The successful formation of new social ties in the adopted country is far from assured. Tellingly, numerous studies have found extremely high rates of delusional and paranoid symptoms among immigrant and refugee populations.[88,110–115] Two studies show rates of DD among immigrants to be forty to fifty times that of the indigenous population,[113,115] compared to only a 3½-fold increase for schizophrenia.[113] Kendler[88] found rates of DD among the foreign born to greatly exceed rates of either schizophrenia or affective illness. DD clearly has a particular association with immigrant/refugee status.

In an attempt to resolve whether these results are best explained by social selection theory, social causation theory, or other factors, Westermeyer[115] conducted a careful study of paranoid symptoms and disorders among 100 Hmong refugees living in the United States. In six of nine cases (66%), no pre-emigration factors could be found, supporting social causation theory. His study indicates that successful acculturation, assessed in several ways, is associated with low paranoid symptoms. Chiu and Rimón[111] report that 56% of the paranoid immigrants in their study had no history of psychiatric treatment prior to immigration, again supporting social causation theory.* Social causation appears to contribute to the high prevalence of delusional symptoms among immigrants, although social selection is probably a factor as well.

Low Socioeconomic Status

DD is associated with poor social and economic standing, as is mental illness in general.[116] This association, however, is particularly strong in the case of DD. In a review of the demographics of DD, Kendler[88] found that patients with DD were more likely to come from poor economic backgrounds and to be more poorly educated than patients with either affective illness or (in most cases) schizophrenia. Kendler argues that this pattern speaks against the hypothesis that disabling symptoms alone are the cause of downward social drift. Because schizophrenia produces more disabling symptoms than DD, it should produce greater psychosocial disability and, therefore, more downward social drift. The fact that DD was, nevertheless, associated with lower SES suggests that low SES is a precursor of DD, rather than a consequence of disabling symptoms. Kay et al.[107] also found paranoid patients to be significantly associated with low social class compared to patients with affective disorders. They, too, disfavor the social selection hypothesis.

But is low SES associated with social problems of the kind hypothesized to cause delusions? Mirowsky and Ross,[117] using data on 463 individuals collected during a community mental health survey in El Paso, Texas, and Juarez, Mexico, found that low socioeconomic status together with belief in external locus of control—the expectation that outcomes are determined by

---

* Of these patients, 22% had a DSM-III paranoid disorder, while 61% were classified as paranoid schizophrenic.

forces external to one's self, such as powerful others, luck, fate, or chance—was strongly associated with mistrust, the feeling that it is safer to trust no one. Mistrust, in turn, was associated with paranoia (paranoia being determined by responses to four questions similar to diagnostic criteria for DSM-IV paranoid personality disorder). Mirowsky and Ross conclude that powerlessness, victimization, and exploitation were the causative factors of mistrust and thus paranoia.

Intuitively, severe social failure would seem to be a consequence of suffering delusions. The facts, however, strongly suggest the opposite: severe social problems both precede and significantly increase the risk for the onset of delusions, an increased risk that persists even after controlling for numerous confounds. This is compelling evidence that social problems cause delusions.

Delusions "Work" in Small-Scale Societies

If delusions function to alleviate social problems, then delusional individuals must (1) convince others to share their delusions and (2) garner social benefits as a consequence. There is strong evidence for (1) and a fair amount of evidence for (2).

Psychiatry recognizes that in most societies, including Western societies, delusional individuals can at least occasionally convince others to share their delusional framework, reifying the phenomenon as shared psychosis (Folie à Deux). According to DSM-IV:[95]

> The essential feature of Shared Psychotic Disorder is a delusion that develops in an individual involved in a close relationship with another person (sometimes termed the "inducer" or "the primary case") who already has a Psychotic Disorder with prominent delusions....The [secondary] individual comes to share the delusional beliefs of the primary case in whole or in part....Usually the primary case in Shared Psychotic Disorder is dominant in the relationship and gradually imposes the delusional system on the more passive and initially healthy second person. Individuals who come to share delusional beliefs are often related by blood or marriage and have lived together for a long time, sometimes in relative isolation. If the relationship with the primary case is interrupted, the delusional beliefs of the other individual usually diminish or disappear. Although most commonly seen in relationships of only two people, Shared Psychotic Disorder can occur among a larger number of individuals, especially in family situations. ... (p. 332–333)

Shared psychosis is labeled a disorder, but it appears to simply describe situations in which the delusions of a stronger personality are believed by weaker personalities. In Western societies, secondaries are often vulnerable individuals who may have a preexisting psychiatric disturbance or physical disability.[118] In traditional societies, however, this is not necessarily the case. There are a number of examples in the ethnographic record where social conflict is associated with delusions, which in turn are believed by fellow group members, eliciting benefits.

Ethnopsychiatrist Burton-Bradley worked among the diverse indigenous Papua New Guinea (PNG) population, including remote highland groups, from the late 1950s to the early 1970s. His observations of cargo cults provide compelling evidence that delusions are frequently believed, garnering social benefits. There is a vast literature on cargo cults, which arose in colonial Melanesia in response to rapid and disruptive social and cultural change. Burton-Bradley describes them as follows[119] (p. 12):

> A prophet, leader, or messiah emerges. He is often a mediocrity, as measured by different culture standards, and one who is not averse to the use, or threatened use, of sorcery in bringing dissidents into line, although recourse to this action is seldom necessary. He has a fantasy solution to offer his followers initiated by a revelation which may take the form of a dream or visual hallucination, both powerful agents in effecting conversion. He proclaims a great future event, or a millennium, and may even provide the specific date. Preparations are made to deal with the expected changes. Airstrips, wharves, or helipads are constructed to receive the ancestral spirits who bear the much-valued cargo. An iconoclastic contraculture may develop, and new social mores may be adopted. Money is destroyed, food gardens are neglected, and livestock killed on the theory that they will no longer be needed. When prophecy fails, the cult wanes and becomes latent.

Burton-Bradley approvingly noted that the early view of cargo cults as mere reflections of individual mental disorder had been discredited—current work rightly emphasized social rather than medical causes. But he goes on[120] (p. 124):

> An unfortunate and unanticipated by-product of this new interpretation is the implicit and occasionally explicit assumption in some quarters that psychotics are never leaders. This latter view is false.

Burton-Bradley presents several case studies from PNG in which the prophet was almost certainly schizophrenic. What is remarkable is that the prophet's grandiose delusions of the imminent arrival of cargo did not merely elicit minor social benefits, but actually catapulted the prophet to a leadership position. This despite the recognition by many of his followers that he was *longlong* (insane).* Although some of the prophet's closest followers might themselves have been suffering from psychiatric disturbances, the vast majority of followers were almost certainly in a state of good mental health.[119] Sharp, a medical officer who worked in the same area in the late 1970s, also described a movement where the principal prophet had paranoid schizophrenia. He concluded that "if the distinguishing feature of crisis movement leaders is mental disorder, then that part of human behaviour and experience we call mental disturbance or madness can play a far more significant role in our affairs than we generally admit"[121] (p. 119). In these examples, grandiose delusions appear to be protective against the social problems that are often caused by the other symptoms of schizophrenia.[122]

Stevens and Price[123] investigate cult phenomena from an evolutionary perspective as well. They provide numerous examples of delusional individuals gaining cult leadership positions and the attendant social benefits. Their thesis, however, differs significantly from that presented here. They mainly argue that schizoid traits evolved to facilitate group fissioning when resources were scarce: charismatic, often schizoid, cult leaders lead a subgroup to a new promised land. In contrast, I focus solely on the deceptive functions of nonbizarre delusions, which can occur alone or as one symptom of a psychiatric syndrome like schizophrenia or affective psychosis; I claim no evolved function for any variant of these syndromes as a whole. Further, gaining cult leadership status via grandiose delusions is not the only benefit that accrues to delusional individuals in small-scale societies. Paranoid delusions appear to deliver social benefits of a different sort, namely, increased solidarity with the group.

In a psychiatric survey of isolated groups of Australian Aboriginals who had only recently abandoned hunting and gathering, Eastwell[124–126] found that in a total population of 10,500, 57 were suffering from reactive psychosis, or fear-of-sorcery syndrome. This syndrome is characterized as an anxiety state with paranoid features magnified to psychotic proportions. The patient fears imminent death from the sorcery of a traditional enemy. According to Eastwell, sorcery in this population is thought to be directed toward the clan as a whole rather than one member alone. Fellow clan members believed delusions of enemy sorcery so much so that Eastwell often found multiple members of a family suffering psychotic episodes in reaction to the same or closely related event. Following the

---

* Hallucinations are not part of the exploitative deception hypothesis, but they appear to play a role in some of these cases.

DSM definition of shared psychosis, he termed these delusional episodes associative or identificatory illness. Eastwell observed that members of the clan closed ranks with the patient in indignation against the putative enemy sorcerers, exactly the outcome predicted by the exploitative deception hypothesis.

There are other similar accounts of delusions being taken seriously by family and community members. El-Islam,[127] for example, studied the remission of delusions among a group of deluded psychotics from the Arab Gulf states. The existence of traditionally shared beliefs in the family and community set the stage for remission. The patient often attributed the remission of his delusions to relatives dealing with the object of delusion through prayer or through traditional healers, or the delusion was "absorbed" into the cultural belief system and lost its force. El Sendiony[128] and Murphy,[122] cited in Westermeyer,[102] also note the phenomenon of relatives accepting an individual's delusional framework. Finally, the Internet study of delusional beliefs discussed earlier[7] shows that online communities form around websites devoted to these beliefs and that many participants are not themselves delusional.

Social Benefits and the Remission of Delusions

According to the exploitative deception hypotheses, delusions and persecutory fears should remit in individuals who receive sufficient social benefits. Jørgensen and Aagaard[130] studied the relationship of a number of social variables to impairment, remission, and relapse. They found that being married, living with others, having frequent social contacts, working full-time, and belonging to high-status social groups were important predictors of good outcome. Living alone, having few social contacts, and not working prior to admission, on the other hand, were by far the best predictors of poor outcome for this group of patients.

Jørgensen and Aagaard conclude that social variables like having social contacts and useful work are more valuable than any of the clinical variables in predicting outcome. Because they are correlations, the results presented by these researchers do not favor social causation over social selection theories, but they do demonstrate the strong and *necessary* association of positive social variables with the remission of DD. Finally, even patients who attributed their delusions to biological disease nonetheless stressed the importance of strong, supportive social environments to dispelling delusions.[131]

Detecting Exploitative Deception

Over evolutionary time, could individuals suffering severe social problems lie and get away with it? After all, why not just ignore anything said by an individual suffering severe social problems? The fact is that delusions are often believed.[7,119–121,124–129] Although I have emphasized the effectiveness of delusions in small-scale societies, cults with grandiose leaders thrive in Western societies, and large segments of the public believe things that closely resemble common delusional themes. They believe in conspiracy theories, UFOs, and that certain people, such as psychics and astrologers, have special powers and abilities.

The evolution of an adaptation to unconsciously lie in dire social circumstances does not seem out of the question, especially since individuals facing social failure need not change everyone's opinion of them—they only need to manipulate the social calculus of a few group members in their favor. The question then becomes, why are humans so gullible?

At the theoretical level, there are several factors that favor exploitative deceivers. As Hölldobler[50] has argued for social mimics among ants, individuals who evolve to successfully discriminate against exploitative deceivers risk inadvertently discriminating against real cooperators. Because the benefits received through cooperative signaling are so valuable, individuals may evolve to tolerate some exploitation rather than risk losing the benefits obtained from the far more common genuine cooperator. Additionally, because social failure was a deadly threat, whereas being exploited was likely a less-than-deadly threat, the selection pressure on adaptations for exploitative deception was stronger than it was on detection mechanisms. Exploitative deception adaptations can then be expected to outperform detection mechanisms as a consequence of this asymmetrical, intraspecific arms race.[132]

Further, exploiters may attempt to target individuals who have little or no information concerning the social status of the exploiter. These could include individuals from other groups, or individuals from competing factions within the group. Many known hunter-gatherers lived in fission-fusion societies. Group size fluctuated dramatically with season, with smaller foraging bands aggregating into much larger groups to participate in communal hunts.[20] This periodic aggregation and dispersal would have enhanced the opportunities for successful deception. Information transfer would have been slowed during times of dispersion, hindering the detection of deception by naive individuals during aggregations.

Another powerful argument in favor of social failures successfully employing exploitative deception is that it is very difficult to identify complete social failures—those who are not valued by anyone. In order to detect individuals who are not valued by anyone, one must track the entire social network. For even modestly sized groups, the time and effort required are high and possibly prohibitive, growing quadratically with group size. Estimates are that ancestral hunter-gatherers may often have lived in groups ranging in size from 25 to 150 individuals.[20,133] Tracking how everyone felt about everyone in a medium-sized group of 50 would have required 2,450 different assessments, a considerable, and probably impossible, undertaking.* In sum, in the high-stakes game of relationship formation and maintenance, there would have occasionally been an odd man out. Identifying him may not have been trivial, improving his chances of deceptively exploiting others.

Delusions with Other Symptoms

Delusions commonly occur with other psychiatric symptoms and conditions like depression, auditory hallucinations, the (nondelusional) symptoms of schizophrenia, brain injury, and substance use.[85] One population survey found, for example, that 4.1% of individuals suffering depressive symptoms also had delusions.[97] Another found an approximately 0.7% prevalence of delusions with auditory hallucinations in the general population.[134]

The association of depressive symptoms and delusions is clearly consistent with the hypothesis explored here. Individuals suffering a loss of social standing sufficient to trigger delusions would obviously be vulnerable to depression as well. The association of brain injury with delusions is also consistent. If a brain injury or other neurological deficit causes individuals to lose their social relationships, then delusions would, under the hypothesis, be an adaptive response to the loss of social relationships, not to the brain injury per se. Interestingly, two studies found extremely high rates of delusions following brain injury,[135,136] but in 42 and 66% of the cases the delusions onset more than 10 years after the injury. This long delay suggests that delusions might have been caused by the social consequences of the injury rather than the brain injury itself.

If the nondelusional symptoms of schizophrenia cause a loss of social relationships then, again, delusions could be seen as an adaptive response to the loss of social relationships, explaining the association of delusions with other schizophrenic symptoms. Speculatively, given that a large fraction of individuals in most societies believe in supernatural agents or powers,[21,137] auditory hallucinations, a prominent feature of schizophrenia, may not have interfered significantly with the deceptive function proposed for delusions. Burton-Bradley, in his observation of cult leaders, noted that hallucinations and dreams were important agents in effecting conversions.[119]

---

* The costs of tracking the entire social network might be reduced by gossiping, yet there are reasons why individuals would not want to readily advertise their valuation of others. When circumstances change, valuations can change dramatically. If one discovers, for example, that a low-valued person is a relative of a highly desired potential mate (and could therefore facilitate a marriage), their social value to an individual might well skyrocket. But, if the previously low-valued person knew that an individual had spoken disparagingly of them, they would be much less likely to be cooperative.

Conclusion

Social systems that rely on cheap signals for the exchange of substantial benefits, like those of ants and humans, are susceptible to exploitative deception. For humans, exploitative deceivers should often be individuals facing severe social failure, because in these circumstances there is little downside to lying and potentially a huge upside.

Decades of research in industrialized societies have shown that severe social problems precede, and probably cause, delusions in otherwise healthy individuals. Paranoia (i.e., heightened vigilance), at the very least, is an understandable reaction to real social problems. Full-blown delusions, however, cannot be explained simply as increased vigilance. Instead, delusions have all the features of a mimetic, or deceptive, signaling system. Individuals with DD are cognitively, emotionally, and physically unimpaired, and their only symptom is a nonbizarre delusional framework. Of the entire universe of conceivable false beliefs, delusions comprise only a tiny set of themes that, not coincidentally, I argue, generate cues that would have elicited benefits from others: possession of important information and abilities, fears of external threat, illness, and intimate relations with high-status individuals. Each of these situations would have been difficult for others to verify, at least in the short term, making them ideal candidates for exploitative deception.

But do delusions actually elicit benefits? With a few notable exceptions (such as research on shared psychosis), studies in Western societies rarely explore the social consequences of delusions. For that, we must turn to ethnographic research in small, kin-based societies, where studies show that delusions *are* believed and garner social benefits. Assuming delusions in industrialized societies are essentially the same phenomenon as delusions in small-scale societies, the etiological findings in Western societies and the ethnographic findings in small, non-Western societies together strongly imply that severe social problems cause delusions that, in turn, mitigate the problems by eliciting benefits from others.

Although considerably more evidence is needed that delusions generate enough benefits in small, kin-based societies to outweigh their costs, Szasz's argument that lies and deception are important aspects of what is usually termed mental illness, reframed here as an adaptationist account of delusions, is reasonably well supported by the available evidence.

This hypothesis, if proven, has some good news and some bad news for clinicians. Currently, powerful drugs are regularly used to suppress delusions, drugs that often fail to improve patients' lives yet cause dangerous side effects, including serious and sometimes irreversible brain damage.[16] A social cause for delusions implies that modifying the social environment in positive ways, instead of the patient's brain, could prevent delusions, or send them into full remission. The bad news is that, contrary to the illness model prevailing in psychiatry, the problem is the social environment—the patient's relationships with all his or her friends, relatives, colleagues, and acquaintances—not (necessarily) the patient's brain. The very term *patient*, in fact, would not really apply. The power of a clinician to convince all the members of a patient's social network to invest more in the patient—when they have already decided they do not want to—is extremely limited.

To make matters worse, in some situations the social exclusion of a particular individual might be well justified, or at least unavoidable. What could a clinician do to ameliorate such ostracism? Probably not much. Nevertheless, a correct scientific model of delusions would no doubt open up a variety of treatment options. Given that more than a century of research on delusions using the illness model has failed to explain them, it is time to rethink our approach to these deeply mysterious cognitive processes.

Acknowledgments

Many thanks to Don Symons, Nicole Hess, Andy Thomson, Paul Watson, Paul Andrews, and members of the Institute for Theoretical Biology, and the Center for Evolutionary Psychology for numerous comments and suggestions. This chapter is based on a talk at the Human Behavior and Evolution Society Annual Meeting, Santa Barbara, 1995.

References

1. T. S. Szasz. 1961. *The myth of mental illness*. New York: Dell.
2. T. S. Szasz. 1970. *The manufacture of madness: A comparative study of the inquisition and the mental health movement*. New York: Harper and Row.
3. T. J. Scheff. 1999. *Being mentally ill: A sociological theory*. 3rd ed. New York: Aldine de Gruyter.
4. W. W. Meissner. 1987. The diagnosis of paranoid disorders. In *Diagnostics and psychopathology*, ed. F. Flach. Number 1 in Directions in Psychiatry monograph series. New York: W. W. Norton and Co.
5. T. F. Oltmanns. 1988. Approaches to the definition and study of delusions. In *Delusional beliefs*, ed. T. F. Oltmanns and B. A. Maher, 3–11. Wiley series on personality processes. New York: John Wiley & Sons.
6. V. Bell, P. W. Halligan, and H. D. Ellis. 2006. Diagnosing delusions: A review of inter-rater reliability. *Schizophrenia Research* 86:76–79.
7. V. Bell, C. Maiden, A. Muñoz-Solomando, and V. Reddy. 2006. "Mind control" experiences on the Internet: Implications for the psychiatric diagnosis of delusions. *Psychopathology* 39:87–91.
8. K. C. Winters and J. M. Neale. 1983. Delusions and delusional thinking in psychotics: A review of the literature. *Clinical Psychology Review* 3:227–53.
9. R. P. Bentall, R. Corcoran, R. Howard, N. Blackwood, and P. Kinderman. 2001. Persecutory delusions: A review and theoretical integration. *Clinical Psychiatry Review* 21:1143–92.
10. N. J. Blackwood, R. J. Howard, R. P. Bentall, and R. M. Murray. 2001. Cognitive neuropsychiatric models of persecutory delusions. *American Journal of Psychiatry* 158:527–39.
11. V. Bell, P. W. Halligan, and H. D. Ellis. 2006. Explaining delusions: A cognitive perspective. *Trends in Cognitive Science* 10:219–26.

12. M. Foucault. 1965. *Madness and civilization: A history of insanity in the age of reason.* New York: Pantheon.

13. J. C. Wakefield. 1992. The concept of mental disorder: On the boundary between biological facts and social values. *American Psychologist* 47:373–88.

14. J. C. Wakefield. 1992. Disorder as harmful dysfunction: A conceptual critique of DSM-III-R's definition of mental disorder. *Psychological Review* 99:232–47.

15. J. C. Wakefield. 1999. Evolutionary versus prototype analyses of the concept of disorder. *Journal of Abnormal Psychology* 108:374–99.

16. A. Bagnall, L. Jones, L. Ginnelly, R. Lewis, J. Glanville, S. Gilbody, L. Davies, D. Torgerson, and J. Kleijnen. 2003. A systematic review of atypical antipsychotic drugs in schizophrenia. *Health Technology Assessment* 7:1–502.

17. J. C. Mitani, D. A. Merriwether, and C. Zhang. 2000. Male affiliation, cooperation and kinship in wild chimpanzees. *Animal Behaviour* 59:885–93.

18. F. de Waal. 1998. *Chimpanzee politics: Power and sex among apes.* Baltimore: Johns Hopkins University Press.

19. M. C. Stiner. 2002. Carnivory, coevolution, and the geographic spread of the genus *Homo. Journal of Archaeological Research* 10:1–63.

20. R. L. Kelly. 1995. *The foraging spectrum.* Smithsonian Institute Press.

21. D. E. Brown. 1991. *Human universals.* New York: McGraw-Hill.

22. M. N. Cohen. 1977. *The food crisis in prehistory: Overpopulation and the origins of agriculture.* New Haven, CT: Yale University Press.

23. L. H. Keeley. 1996. *War before civilization.* Oxford: Oxford University Press.

24. R. B. Lee and I. DeVore. 1968. *Man the hunter.* Chicago: Aldine Publishing Company.

25. M. D. Sahlins. 1972. *Stone age economics.* Tavistock.

26. P. L. van den Berghe. 1990. *Human family systems: An evolutionary view.* Illinois: Waveland Press.

27. L. S. Sugiyama. 2004. Illness, injury, and disability among Shiwiar forager-horticulturalists: Implications of health-risk buffering for the evolution of human life history. *American Journal of Physical Anthropology* 123:371–89.

28. J. Goodall. 1986. Social rejection, exclusion, and shunning among the gombe chimpanzees. *Ethology and Sociobiology* 7:227–36.

29. T. Nishida, K. Hosaka, M. Nakamura, and M. Hamai. 1995. A within-group gang attack on a young adult male chimpanzee: Ostracism of an ill-mannered member? *Primates* 36:207–11.

30. A. H. Harcourt. 1978. Strategies of emigration and transfer by primates, with particular reference to gorillas. *Z. Tierpsychol.* 48:401–20.

31. I. Fornasieri and J. J. Roeder. 1992. Marking behaviour in two lemur species (*L. fulvus* and *L. macaco*): Relation to social status, reproduction, aggression and environmental change. *Folia Primatol* (Basel) 59:137–48.

32. J. Lancaster. 1986. Primate social behavior and ostracism. *Ethology and Sociobiology* 7:215–25.

33. K. D. Williams. 2007. Ostracism. *Annual Review of Psychology* 58:425–52.

34. J. Panksepp. 1998. *Affective neuroscience: The foundations of human and animal emotions.* Oxford: Oxford University Press.

35. G. MacDonald and M. R. Leary. 2005. Why does social exclusion hurt? The relationship between social and physical pain. *Psychological Bulletin* 131:202–23.

36. N. I. Eisenberger, M. D. Lieberman, and K. D. Williams. 2003. Does rejection hurt? An fMRI study of social exclusion. *Science* 302:290–292.

37. M. R. Leary, C. Springer, L. Negel, E. Ansell, and K. Evans. 1998. The causes, phenomenology, and consequences of hurt feelings. *Journal of Personality and Social Psychology* 74:1225–37.
38. J. Bowlby. 1969. *Attachment and Loss*. Vol. 1. Penguin.
39. N. Thornhill and R. Thornhill. 1990. An evolutionary analysis of psychological pain following rape. 1. The effects of victim's age and marital status. *Ethology and Sociobiology* 11:155–76.
40. R. Nesse. 1990. The evolutionary functions of repression and the ego defenses. *Journal of the American Academy of Psychoanalysis* 18:260–85.
41. J. B. Silk, S. C. Alberts, and J. Altmann. 2003. Social bonds of female baboons enhance infant survival. *Science* 302:1231–1234.
42. J. Crocker, B. Major, and C. Steele. 1998. Social stigma. In *The Handbook of Social Psychology*, 504–53. Vol. 2.
43. R. Kurzban and M. R. Leary. 2001. Evolutionary origins of stigmatization: The functions of social exclusion. *Psychological Bulletin* 127:187–208.
44. M. J. Green and M. L. Phillips. 2004. Social threat perception and the evolution of paranoia. *Neuroscience and Biobehavioral Reviews* 28:333–42.
45. R. M. Nesse. 2001. The smoke detector principle: Natural selection and the regulation of defensive responses. *Annals of the New York Academy of Sciences* 935:75–85.
46. R. Dawkins and J. R. Krebs. 1978. Animal signals: Information or manipulation? In *Behavioral ecology*, ed. J. R. Krebs and N. B. Davies, 282–309. Blackwell.
47. A. Starrett. 1993. Adaptive resemblance: A unifying concept for mimicry and crypsis. *Biological Journal of the Linnean Society* 48:299–317.
48. J. D. Mclver and G. Stonedahl. 1993. Myrmecomorphy: Morphological and behavioral mimicry of ants. *Annual Review of Entomology* 38:351–77.
49. H. Markl. 1985. Manipulation, modulation, information, cognition: Some of the riddles of communication. In *Experimental behavioral ecology and sociobiology*, ed. B Hölldobler and M Lindauer, 163–94. Gustav Fischer Verlag.
50. B. Hölldobler. 1977. Communication in social hymenoptera. In *How animals communicate*, ed. A. Sebeok, 418–71. Bloomington: Indiana University Press.
51. S. I. Rothstein and S. K. Robinson. 1998. *Parasitic birds and their hosts: Studies in coevolution*. Oxford: Oxford University Press.
52. A. Lenoir, P. D'Ettorre, C. Errard, and A. Hefetz. 2001. Chemical ecology and social parasitism in ants. *Annu. Rev. Entomol.* 46:573–99.
53. A. Buschinger. 1986. Evolution of social parasitism in ants. *Trends Ecol. Evol.* 1:155–60.
54. T. Akino, J. J. Knapp, J. A. Thomas, and G. W. Elmes. 1999. Chemical mimicry and host specificity in the butterfly *Maculinea rebeli*, a social parasite of *Myrmica* ant colonies. *Proceedings: Biological Sciences* 266:1419–26.
55. J. B. Silk, E. Kaldor, and R. Boyd. 2000. Cheap talk when interests conflict. *Animal Behaviour* 59:423–32.
56. R. Axelrod and D. Dion. 1988. The further evolution of cooperation. *Science* 242:1385–90.
57. R. Axelrod and W. D. Hamilton. 1984. The evolution of cooperation in biological systems. In *The evolution of cooperation*, ed. R. Axelrod, 88–105. New York: Basic Books.
58. P. Ekman, G. Roper, and J. C. Hager. 1980. Deliberate facial movement. *Child Development* 51:886–91.
59. A. F. C. Wallace. 1960. The biocultural theory of schizophrenia. *International Record of Medicine* 173:700–14.

60. S. Henderson. 1974. Care-eliciting behavior in man. *Journal of Nervous Mental Disorders* 159:172–81.
61. S. Henderson. 1981. Social relationships, adversity and neurosis: An analysis of prospective observations. *British Journal of Psychiatry* 138:391–98.
62. H. S. Sullivan. 1953. *The interpersonal theory of psychiatry*. New York: Norton.
63. R. D. Alexander. 1979. *Darwinism and human affairs*. Seattle: University of Washington Press.
64. M. O. Slavin. 1985. The origins of psychic conflict and the adaptive function of repression: An evolutionary biological view. *Psychoanalysis and Contemporary Thought* 8:407–40.
65. R. L. Trivers. 1985. *Social evolution*. Reading, MA: Addison-Wesley.
66. L. A. Dugatkin. 1995. Partner choice, game theory and social behavior. *Journal of Quantitative Anthropology* 5:3–14.
67. J. Henrich and F. Gil-White. 2001. The evolution of prestige: Freely conferred deference as a mechanism for enhancing the benefits of cultural transmission. *Evolution and Human Behavior* 22:165–96.
68. A. Gouldner. 1960. The norm of reciprocity: A preliminary statement. *American Sociology Review* 47:73–80.
69. R. L. Trivers. 1971. The evolution of reciprocal altruism. *Quarterly Review of Biology* 46:35–57.
70. C. Darwin. 1872. *The expression of the emotions in man and animals*. New York: D. Appleton and Company.
71. P. Ekman. 1989. The argument and evidence about universals in facial expressions of emotion. In *Handbook of social psychophysiology*, ed. H. Wagner and A. Manstead, 143–64. Chichester, England: Wiley.
72. R. A. LeVine and D. T. Campbell. 1972. *Ethnocentrism: Theories of conflict, ethnic attitudes, and group behavior*. New York: John Wiley & Sons.
73. R. D. Alexander. 1987. *The biology of moral systems*. Aldine.
74. C. Badcock. 1990. *Oedipus in evolution*. Oxford: Blackwell.
75. M. R. A. Chance, ed. 1988. *Social fabrics of the mind*. LEA.
76. R. J. Gardner. 1982. Mechanisms in major depressive disorder: An evolutionary model. *Archives of General Psychiatry* 39:1436–41.
77. P. Gilbert. 1989. *Human nature and suffering*. Philadelphia: LEA.
78. K. Glantz and J. Pearce. 1989. *Exiles from Eden: Psychotherapy from an evolutionary perspective*. New York: Norton.
79. E. H. Hagen. 1999. The functions of postpartum depression. *Evolution and Human Behavior* 20:325–59.
80. E. H. Hagen. 2003. The bargaining model of depression. In *Genetic and cultural evolution of cooperation*, ed. P. Hammerstein, 95–123. Cambridge, MA: MIT Press.
81. J. S. Huxley, E. Mayr, H. Osmond, and A. Hoffer. 1964. Schizophrenia as a genetic morphism. *Nature* 204:220–25.
82. R. Nesse. 1991. What good is feeling bad—The evolutionary benefits of psychic pain. *Sciences* 31:30–37.
83. J. S. Price. 1972. Genetic and phylogenetic aspects of mood variation. *International Journal of Mental Health* 1:124–44.
84. D. A. Wilson. 1993. Evolutionary epidemiology: Darwinian theory in the service of medicine and psychiatry. *Acta Biotheoretica* 41:205–19.
85. T. C. Manschreck. 1989. The paranoid syndrome and delusional (paranoid) disorders. In *Outpatient psychiatry: Diagnosis and treatment*, ed. A Lazare. Baltimore: Williams and Wilkins.

86. P. Jørgensen and J. Jensen. 1988. An attempt to operationalize reactive delusional psychosis. *Acta Psychiatrica Scandinavica* 78:627–31.
87. K. S. Kendler. 1980. The nosologic validity of paranoia (simple delusional disorder): A review. *Archives of General Psychiatry* 37:699–706.
88. K. S. Kendler. 1982. Demography of paranoid psychosis (delusional disorder): A review and comparison with schizophrenia and affective illness. *Archives of General Psychiatry* 39:890–902.
89. K. S. Kendler. 1984. Paranoia (delusional disorder): A valid psychiatric entity? *Trends in Neurosciences* 7:14–17.
90. K. S. Kendler. 1987. Paranoid disorders in DSM-III: A critical review. In *Diagnosis and classification in psychiatry: A critical appraisal of DSM-III*, ed. G. L. Tischler, 57–83. New York: Cambridge University Press.
91. K. S. Kendler and M. T. Tsuang. 1981. Nosology of paranoid schizophrenia and other paranoid psychoses. *Schizophrenia Bulletin* 7:594–610.
92. K. Koehler and C. Hornstein. 1986. 100 years of DSM-III paranoia: How stable a diagnosis over time? *European Archives of Psychiatry and Neurological Sciences* 235:255–58.
93. S. Opjordsmoen. 1987. Toward an operationalization of reactive paranoid psychoses (reactive delusional disorder). *Psychopathology* 20:72–78.
94. G. Winokur. 1977. Delusional disorder (paranoia). *Comprehensive Psychiatry* 18:511–21.
95. APA. 1994. *Diagnostic and statistical manual of mental disorders*. American Psychiatric Association.
96. J. Van Os, M. Hanssen, R. V. Bijl, and A. Ravelli. 2000. Strauss (1969) revisited: A psychosis continuum in the general population? *Schizophrenia Research* 45:11–20.
97. M. M. Ohayon and A. F. Schatzberg. 2002. Prevalence of depressive episodes with psychotic features in the general population. *American Journal of Psychiatry* 159:1855–61.
98. P. Boyer and P. Lienard. n.d. Why ritualized behavior? Precaution systems and action parsing in developmental, pathological and cultural rituals. *Behavioral and Brain Sciences*.
99. S. Wessely, A. Buchanan, A. Reed, J. Cutting, et al. 1993. Acting on delusions. I. Prevalence. *British Journal of Psychiatry* 163:69–76.
100. H. G. Kennedy, L. I. Kemp, and D. E. Dyer. 1992. Fear and anger in delusional (paranoid) disorder: The association with violence. *British Journal of Psychiatry* 160:488–92.
101. D. M. Ndetei and A. Vadher. 1984. Frequency and clinical significance of delusions across cultures. *Acta Psychiatrica Scandinavica* 70:73–76.
102. J. Westermeyer. 1988. Some cross-cultural aspects of delusions. In *Delusional beliefs*, ed. T. F. Oltmanns and B. A. Maher, 212–29. New York: John Wiley & Sons.
103. N. Cameron. 1943. The development of paranoiac thinking. *Psychological Review* 50:219–34.
104. E. M. Lemert. 1962. Paranoia and the dynamics of exclusion. *Sociometry* 25:2–20.
105. M. Kaffman. 1981. Paranoid disorders: The core of truth behind the delusional system. *International Journal of Family Therapy* 3:29–41.
106. N. Retterstöl. 1966. *Paranoid and paranoiac psychoses*. Springfield, IL: Charles C. Thomas.

107. D. W. Kay, A. F. Cooper, R. F. Garside, and M. Roth. 1976. The differentiation of paranoid from affective psychoses by patients' premorbid characteristics. *British Journal of Psychiatry* 129:207–15.
108. I. Janssen, M. Hanssen, M. Bak, R. V. Bijl, R. De Graaf, W. Vollebergh, K. Mckenzie, and J. Van Os. 2003. Discrimination and delusional ideation. *British Journal of Psychiatry* 182:71–76.
109. J. Spauwen, L. Krabbendam, R. Lieb, H. U. Wittchen, and J. Van Os. 2006. Impact of psychological trauma on the development of psychotic symptoms: Relationship with psychosis proneness. *British Journal of Psychiatry* 188:527–33.
110. L. Carpenter and I. F. Brockington. 1980. A study of mental illness in Asians, West Indians and Africans living in Manchester. *British Journal of Psychiatry* 137:201–5.
111. L. P. Chiu and R. Rimón. 1987. Relationship of migration to paranoid and somatoform symptoms in Chinese patients. *Psychopathology* 20:203–12.
112. L. Ettinger. 1959. The incidence of mental disease among refugees in Norway. *Journal of Mental Science* 105:326–38.
113. L. Ettinger. 1960. The symptomology of mental disease among refugees in Norway. *Journal of Mental Science* 106:947–66.
114. Ø. Ødegaard. 1932. Immigration and insanity. *Acta Psychiatrica Neurologica Scandinavia*, Suppl. 4.
115. J. Westermeyer. 1989. Paranoid symptoms and disorders among 100 Hmong refugees: A longitudinal study. *Acta Psychiatrica Scandinavica* 80:47–59.
116. L. N. Robins, B. Z. Locke, and D. A. Regier. 1991. An overview of psychiatric disorders in America. In *Psychiatric disorders in America: The epidemiologic catchment area study*, ed. L. N. Robins and D. A. Regier, 328–66. New York: Macmillan.
117. J. Mirowsky and C. E. Ross. 1983. Paranoia and the structure of powerlessness. *American Sociological Review* 48:228–39.
118. S. D. Soni and G. J. Rockley. 1974. Socio-clinical substrates of Folie à Deux. *British Journal of Psychiatry* 125:230–35.
119. B. G. Burton-Bradley. 1975. *Stone age crisis: A psychiatric appraisal.* Nashville, TN: Vanderbilt University Press.
120. B. G. Burton-Bradley. 1970. The New Guinea prophet: Is the cultist always normal? *Medical Journal of Australia* 1:124–29.
121. Peter T. Sharp. 1990. The searching sun: The Lyeime movement—Crisis, tragic events and Folie à Duex in the Papua New Guinea Highlands. *Papua New Guinea Medical Journal* 33:111–20.
122. J. M. Murphy. 1976. Psychiatric labeling in cross-cultural perspective. *Science* 191:1019–28.
123. A. Stevens and J. Price. 2000. *Prophets, cults and madness.* London: Duckworth.
124. H. D. Eastwell. 1976. Associative illness among Aboriginals. *Australian and New Zealand Journal of Psychiatry* 10:89–94.
125. H. D. Eastwell. 1977. Projective and identificatory illnesses among ex-hunter-gatherers: A seven year survey of a remote Australian Aboriginal community. *Psychiatry* 40:330–43.
126. H. D. Eastwell. 1982. Psychological disorders among the Australian Aboriginals. In *Extraordinary disorders of human behavior*, ed. C. T. H. Friedmann and R. A. Faguet, 229–57. New York: Plenum Press.
127. M. F. El-Islam. 1980. Symptom onset and involution of delusions. *Social Psychiatry* 15:157–60.
128. H. F. M. El Sendiony. 1976. Cultural aspects of delusions: A psychiatry study of Egypt. *Australia New Zealand Journal of Psychiatry* 10:201–7.

129. H. B. M. Murphy. 1967. Cultural aspects of delusion. *Stadium Generale* 20:684–92.
130. P. Jørgensen and J. Aagaard. 1988. A multivariate predictor analysis of course and outcome in delusional psychosis. *Acta Psychiatrica Scandinavica* 77:543–50.
131. B. Stanton and A. David. 2000. First-person accounts of delusions. *Psychiatric Bulletin* 24:333–36.
132. R. Dawkins and J. R. Krebs. 1979. Arms races between and within species. *Proceedings of the Royal Society of London B* 205:489–511.
133. R. Dunbar. 1993. Co-evolution of neocortex size, group size and language in humans. *Behavioral and Brain Sciences* 16:681–735.
134. L. N. Robins and D. A. Regier, ed. 1991. *Psychiatric disorders in America: The epidemiologic catchment area study.* New York: Macmillan.
135. K. A. Achte, E. Hillbom, and V. Aalberg. 1969. Psychoses following war brain injuries. *Acta Psychiatrica Scandinavica* 45:1–18.
136. S. Koponen, T. Taiminen, R. Portin, L. Himanen, H. Isoniemi, H. Heinonen, H. Hinkka, and O. Tenovuo. 2002. Axis I and II psychiatric disorders after traumatic brain injury: A 30-year follow-up study. *American Journal of Psychiatry* 159:1315–21.
137. P. Boyer. 1994. *The naturalness of religious ideas: A cognitive theory of religion.* Berkeley: University of California Press.

10

Practical Aspects of Evolutionary Medicine

Gillian R. Bentley
Department of Anthropology and Wolfson
Research Institute, Durham University

Robert Aunger
Hygiene Centre, London School of Hygiene and Tropical Medicine

Contents

Introduction

Since Randy Nesse's and George Williams's seminal 1991 article developing a paradigm for evolutionary medicine (EM) the field has been expanding rapidly, although the number of books focused on the general topic still number less than ten. One of the earliest problems recognised by Nesse and Williams was the need to convince clinicians that this emerging field was relevant to their sphere of work. As stated in their 1991 article, "Medicine is a practical enterprise, and it hasn't been immediately obvious how evolutionary explanations might help us prevent or treat disease."[1] Although the number of papers pointing out the utility of EM for clinical applications has

217

been growing, there has been no comprehensive discussion to our knowledge of what makes EM eminently practical or, perhaps more appropriately, "applied," and thus the problem for this field in acquiring recognition from practicing doctors will continue.

The purpose of this chapter is to write a short review covering five examples of topics that cut across both clinical medicine and evolutionary theory and that demonstrate the practical benefits of EM. The chapter differs from others in this book since it draws not only on published research but also on the first author's (practical) experience in teaching a class called Evolutionary Medicine at University College London (UCL) from 2002 through 2006. The class differed from the usual format of university lectures in that a series of guests were actually interviewed in class on their topics of speciality relevant to EM. For some of the classes two guests were invited (often from different disciplines or different perspectives) in order to generate debate. Students were encouraged to ask questions at any point during the interview in order to increase their active participation in learning. Most of the interviews from 2003 onwards were videotaped by Media Resources at UCL.[2]

The purpose of designing the class as a series of interviews and the teaching and learning benefits of this format will not be dealt with in this chapter. Instead, it will focus on one of the questions that was consistently presented to the guests, namely, whether they thought their own particular speciality had practical benefits for EM, or whether they considered evolutionary explanations for a particular pathology, or set of pathologies, might have practical applications. In a sense, then, these interviews represent an informal survey of clinicians, evolutionary biologists, anthropologists, and others of their opinions concerning the practical aspects of EM. This chapter also purposely attempts to cover some newer areas that have not, to our knowledge, been previously discussed in other books specifically about EM (such as emergency medicine and "racial" medicine), but will also cover a few topics that have been dealt with in a limited way in other books (obstetrics, public health, and various aspects of epigenetics). Topics such as host-parasite evolution and the development of drug-resistant pathogens have been dealt with extensively in other edited volumes and will not be covered here. In order to emphasize each topic's relevance to EM as defined by Nesse and Williams, we also attempt to place each topic within the framework that these authors devised; that is useful for thinking about human pathologies, such as whether they might be defences or historical legacies.[1] Given the limited space in the chapter, this is intended as a summary rather than an exhaustive review of the topic.

Racial Medicine

This topic is a contentious and relatively new area that embraces genetics, pharmacology, and social and environmental issues. It primarily concerns pharmaceutical applications that might be more efficacious for specific ethnic groups either because they differ in their susceptibility to specific diseases or because they differ in their response to particular medicines (the study of genetic variation in drug response is known as pharmocogenomics). Up until 2004, twenty-nine medicines and possibly more were indeed claimed to have different effects in different human groups, depending on the ethnic population targeted.[3] Susceptibility to pathologies may depend on genetic polymorphisms that have been selected for over long periods of time, as in sickle cell anaemia, thalassemia, or lactose tolerance, conditions that have well-known associations with geographical areas, and particular ethnic groups (see chapter 2). These pathologies are relatively easy to diagnose and their genetic and evolutionary origins are fairly clear.

Increasingly, however, researchers are uncovering more subtle genetic variation both within and between human populations that have epidemiological implications. For example, Jonathan Cohen and colleagues[4] have recently analysed allelic variation in the proprotein convertase subtilisin/kexin type 9 serine protease gene, or PCSK9, among both U.S. blacks and whites. Out of 3,363 self-reported African Americans, 0.8% carried $PCSK9^{142x}$ compared to only 0.02% of whites, while 1.8% of blacks had the allele $PCSK9^{679x}$ compared to 0.04% of whites. In contrast, 3.2% of white subjects carried $PCSK9^{46L}$ compared to 0.7% of blacks. Carriers of these alleles benefited from substantial reductions in low-density lipoprotein (LDL) cholesterol and risk for coronary heart disease (CHD). The origin of these allelic variations is unclear.

Studies like the above example offer the potential to tailor medicines along individual lines. There are, in fact, other specific examples where this has happened.[5] Indeed, the idea of individualising medicine is gaining increasing acceptance as the optimal form of clinical care. The time and expense involved in such an endeavour, however, so far outweigh any other considerations in most circumstances. However, for many diseases, genetic susceptibilities may be dwarfed by environmental issues. Even in the Cohen study, cited above, one black carrier of the beneficial allele did have CHD, but was described as obese with a body mass index of 34, with hypertension, and smoked. As stated by Francis Collins, "In many instances, the causes of health disparities…have little to do with genetics, but rather derive from differences in culture, diet, socioeconomic status, access to health care, education, environmental exposures, social marginalisation, discrimination, stress and other factors."[6] Collins went on to espouse a list of objectives in the search for those factors that underlie health risks, to include adequate consideration of these other factors quoted here, and is supported by many other advocates.[3,7]

One of the major problems, in fact, with racial medicine is that the genetic, or ethnic, focus has been on precisely those pathologies with very clear *environmental* components. Two notable examples are CHD and kidney disease, pathologies more common in individuals with a history of poor diets, low rates of exercise, and smoking.[8] Many of these characteristics also tend to be associated with low socioeconomic status, at least in Europe and the United

States. Social issues like poverty are often difficult to separate from ethnicity in certain groups depending on their history of immigration to northern and western countries, reinforcing the concerns expressed by Collins and others outlined above.

There is already a vast literature in anthropology and genetics that deals with the issue of race itself, and whether this is an artificial construct[9-12] (cf. chapter 13), as well as an equally large literature dealing with inequalities in health resulting from social factors correlated with ethnicity.[3,6,7,13] We therefore, advisedly, put inverted commas around the first use of this term to highlight the problems with this terminology, albeit recognising that this term appears within the relevant literature whether deservedly or not. Suffice it to say here, however, that our species appears to have evolved from a very small founding population,[14,15] but it is arguable from a statistical perspective whether genetic diversity between individuals is indeed as great as that between populations as originally claimed by Richard Lewontin,[16,17] despite the repetition of this claim in many other articles. Recent work shows that it may indeed be possible to separate populations statistically using genetic data, and that these clusters correspond to geographical areas that may offer some epidemiological insight,[18,19] although this claim has also been subject to criticism and reanalysis.[20]

In addition, there is the added social problem of identifying racial origins of individuals who may, for example, look black but have mixed genetic ancestry. In any discussion of racial medicine, then, the genetic makeup of individuals targeted for particular drug treatments should be clearly evaluated. This is, however, often impossible for the same reasons why individually tailored medical treatments are visionary but impractical.

A second major problem is that some drugs have been approved for use with specific ethnic groups following *retrospective* analyses of data from completed clinical trials that happened to have participants that could be identified by self-report as either black or white,[8] or have originated from small-scale trials without adequate cross-group comparisons. The progress of the drug *BiDil* (a combination of two generic drugs called hydralazine and isosorbide dinitrate, a vasodilator and nitric acid donor, respectively), which was approved by the U.S. Food and Drug Administration (FDA) in 2005 specifically to treat heart disease in U.S. blacks, has been described recently by Jonathan Kahn as a "tangled tale of inconclusive studies, regulatory hurdles and commercial motives," thus darkening a complicated picture of genome-targeted drugs.[21]

Here, the retrospective analysis of the Vasodilator Heart Failure Trial that led originally to the patenting of *BiDil* as a racial medicine focused on only forty-nine self-reported black participants who used this drug combination. A more specific trial, however, known as the African-American Heart Failure Trial (A-HeFT), was initiated in 2001 and prematurely stopped in 2004 following findings of significantly more deaths in participants from the placebo arm of the study. A-HeFT enrolled 1,050 self-reported African Americans to examine more closely the effects of *BiDil* on this group. Two striking facts about this trial were that, first, most of the patients in the trial continued to take their normally prescribed heart medications, while no comparative group of whites were enrolled in the trial. It is thus entirely possible, as suggested by Kahn,[21] that *BiDil* is just as effective in white patients as in blacks and that the subsequent A-HeFT trial in blacks only may have been motivated by commercial interests related to the length of patents filed for the drug as a racially targeted one.

But are all racially targeted clinical trials suspect? There are a number of other completed and ongoing clinical trials focusing on specific ethnic groups, or including larger cohorts of different ethnic groups, including the African-American Study of Kidney Disease and Hypertension (AASK), the Antihypertensive and Lipid-Lowering Treatment to Prevent Heart Attack Trial (ALLHAT), and the Losartan Intervention for Endpoint Reduction in Hypertension Trial (LIFE). These include much larger groups for study (for example, >15,100 participants of African descent for ALLHAT and an additional 8,000 Hispanic participants for comparison). The AASK trial examined the effect of different drug interventions among African Americans at risk for cardiovascular problems resulting from existing kidney disease but failed to show an effect of differential treatments on the occurrence of a cardiovascular event,[22] but did find that one particular type of drug known as an angiotensin-converting enzyme inhibitor, or ACE inhibitor, was more effective than other drugs in preventing progression of kidney disease in African Americans who were not diabetic and who suffered from mild to moderate symptoms.[23]

The ALLHAT trial was set up to examine four different drugs used to control blood pressure and their effects on risks of heart disease, stroke, and mortality, and was one of the largest such trials established across the North American continent, including Canada. Although not specifically targeting African Americans, a large proportion (35%) of the 33,357 participants were from this ethnic group. The trial reported that African Americans using an ACE inhibitor suffered significantly higher rates of stroke and combined cardiovascular disease, as shown by a similar trend for heart failure, than those using a diuretic drug instead.[24] In the LIFE trial, African Americans with left ventricular hypertrophy had a lower risk of suffering from death, heart attacks, or stroke when placed on one drug versus another, but the small percentage (6%) of this group overall within the trial prevents definitive confidence in this result.[25]

Many clinicians are of the opinion that "inclusion of subjects by race/ ethnicity in clinical trials is essential to define the reasons for differences in health outcomes, *whether environmental or biological*" (emphasis added).[8] That being the case, what needs to happen is a better consideration of those genetic and environmental contributors to disease profiles.

Human populations do appear to cluster by geographical area. There is also no question that individuals have different susceptibilities to diseases depending on their genetic legacy. But the question remains open whether grouping individuals together on the basis of diseases that have strong environmental components is an effective way of targeting specific drug applications.

The guest for the class on this topic at UCL was Dr. Mark Thomas, a geneticist in the Biology Department at UCL. When asked about the practical relevance for EM that might arise from developing an understanding of how evolution shapes patterns of medically relevant genetic variation, with a particular focus on pharmacogenomics, he replied with the following:

> This is the most practical application in evolutionary medicine without a shadow of a doubt. It's absolutely clear-cut…it's not teleological…evolutionary studies… are impacting medicine in a real and profound way, and it's all from evolution; it's all from understanding molecular evolution that our ability to construct models or appropriate statistical tests comes from, and then it's been applied to medicine. So, it is the least teleological and the most applicable of all areas, I think, in evolutionary medicine.

While this area rightly remains contentious, the implications of how EM can become actively engaged in this debate remain clear, particularly as biologists uncover the evolutionary history of allelic variations within different human groups. In terms of how this new area within medicine might fit within the framework originally designed by Nesse and Williams, the impact of novel environments, historical legacies, and even genetic quirks could each and all be considered relevant.

Epigenetics

Again, this is a relatively new field that has emerged in recent years. While treated initially with scepticism (see Jablonka and Lamb[26] for comments on early resistance to this field), epigenetics is gaining increasing recognition for its importance in determining phenotypes. The term (meaning literally "outside the genome") was coined about 50 years ago by the late Conrad Waddington—a geneticist at the University of Edinburgh—and refers to the expression of traits that do not involve actual changes in the genome. Epigenetics as a topic is perhaps the perfect complement to the discussion of racial medicine, because it clarifies that one's genetic heritage is not deterministic. Rather, particular features of one's environment, or even one's parents' and grandparents', can alter the molecular switches that may either turn on or off gene expression. Such mechanisms often involve a process known as DNA methylation, where a methyl group (one carbon and three hydrogen atoms) is added to particular sections of DNA, or by alterations to the chromatic packaging of DNA.

Numerous animal models have illustrated epigenetic effects *in utero* on off-spring. For example, several researchers have experimented with dietary and other supplements given to agouti mice (mice with a specific gene that makes them susceptible to diabetes and cancer; they are also phenotypically obese and yellow in colour).[27-29] In one such experiment, pregnant agouti mice fed the phytoestrogenic compound genistein—commonly found in soy products and thought to be protective against certain cancers—gave birth to offspring with a different coat colour and with a lowered predisposition toward obesity. Analysis of the genome of these offspring revealed six sites located near the Agouti gene (normally responsible for the specific phenotypic characteristics of the agouti mice) that had been methylated, with additional evidence that this methylation had occurred early in embryonic development.[28]

Fetal Programming

Subsumed under the banner of epigenetics is the topic of fetal programming that has been covered previously in other books on EM.[30] Although earlier work in the field of fetal programming did not use the word *epigenetic* per se, there are now a number of articles suggesting epigenetic mechanisms can explain the link between poor nutrition *in utero* and health in later life,[31-34] although specific identification of genomic mechanisms remains elusive in studies of humans.

Despite the lack of an early link between those researchers studying epi-genetics and those involved in fetal programming, the latter field has overtly embraced evolutionary theory (see chapter 4) such that researchers in this field have embarked on writing introductory and popularised texts about the field of EM.[35-37] Although early researchers such as Elsie Widdowson had discussed fetal development in relation to later health, it was David Barker and his colleagues at Southampton University who fuelled research in this area with their analyses and publications. Initial data derived from a careful search by the Medical Research Council (MRC) in the UK for any historical records of maternal and infant health that could be linked to surviving adults. They found a series of very detailed records begun by the chief health visitor, Miss Ethel Margaret Burnside, for children born in Hertfordshire between 1911 and 1945. These records ran from birth through to the fifth year of these children's lives. The MRC team was able to locate many of these now adult individuals using various national databases and to compile health informa-tion for them; more men were successfully located than women, who tended to change their name on marriage.[35] When these data were analysed, there appeared to be a relationship between low birth weight and an increased risk for CVD and other metabolic diseases, results that have been replicated in many other studies elsewhere. These articles spawned a plethora of research on the association between fetal development and adult health, and led to an increasing interest by medical doctors involved in this field with evolution-ary concepts.

One of the guests in class who came to discuss foetal programming at UCL was Keith Godfrey, a medical doctor located in Southampton in the DOHaD Centre (Developmental Origins of Health and Disease). When asked about the perceived current synergy between those people studying foetal programming and an interest in evolutionary theory, Godfrey replied:

> Absolutely right! We're not as medical docs going to solve this in isolation. The reason is because of the timescales over which developmental effects act—because of these, possible solutions have to encompass life history and life course perspectives, and the complexities are great here. There are developmental effects on maturational processes—that's, for example, things like the timing of puberty—and these have evolutionary implications, and we'll only understand them by going backwards as well as trying to predict, going forwards. Some things that we think of as beneficial in the short term—such as high food intakes in growth restricted kids—actually if you take account of evolutionary perspectives, you might start to see that they could have disadvantageous consequences. So, we've a lot to learn. Epigenetics is starting to provide a mechanism through which aspects of evolutionary theory may operate. This is probably not just about natural selection, but includes a whole new range of processes, coupled with natural selection. So, we've a lot to learn and both camps have got things to learn from each other.

Since the early work by Barker, there has also been increasing recognition that the period of early life when phenotypic traits can be modified probably extends beyond the fetal period to encompass at least critical phases during childhood. In particular, researchers in foetal programming and health across the life course are discovering that the combination of low birth weight with rapid growth in early childhood increases the risk of diseases associated with metabolic disorders, including obesity, hypertension, CVD, and type 2 diabetes.[38–40]

There are other areas of growth, development, and health that might also be affected by epigenetic modifications that occur during childhood. Recently, Nunez and colleagues have shown that Bangladeshi women who grow up in the UK as migrants have significantly higher levels of salivary progesterone and higher rates of ovulation than Bangladeshi women who grow up in their home country, where they are exposed to a greater number of infectious and parasitic diseases.[41] These hormonal levels also appear to be set in women between birth and 8 years of age, with little modification until puberty, and no modifications during adulthood regardless of environmental changes that might occur during this time. Higher levels of progesterone are associated with an increased risk for breast cancer in later life, suggesting there might be a trade-off between early fertility and later health risks. This study contributes to a growing recognition of the importance of plasticity during development in predicting and adjusting the phenotype appropriately to an individual's environment.[42] Thus, epigenetic modifications can occur far more rapidly than modification to the DNA.

There is also a growing body of research in both animal models and humans demonstrating a transgenerational effect of environmental conditions that goes beyond one generation. For example, Michael Anway and colleagues have shown that female rats exposed during early pregnancy to endocrine disruptors produced male offspring with lower sperm counts and higher infertility than normal unexposed male rats.[43] Furthermore, the effects lasted through at least four generations. The effects were believed to be caused by altered methylation in the germline. Despite the importance and publicity that the study of endocrine disruptors has generated, there are as yet no studies showing similar effects in humans.

However, other studies examining nutritional status in one generation have been able to point to effects among grandchildren. For example, Bygren and colleagues used a historical rural Swedish cohort to look at the transgenerational effect of fluctuations in food supplies on grandchildren.[44] They found that excess food documented in the environment of paternal grandfathers when they were aged 9–12 years old (referred to as the slow growth period) resulted in a shorter life span in their male grandchildren, and an increase in risk of cardiovascular disease.[45] This article, however, suffers from low sample sizes and an inability to link general information concerning food availability to actual consumption patterns of individuals or families, raising the possibility of spurious results.

Genomic Imprinting

Another topic that sits squarely within epigenetics is genomic imprinting, also covered in earlier volumes on EM.[46] This subfield first developed during the 1980s with the first "imprinted" gene—insulin-like growth factor 2 (IGF-2), also known as somatomedin A—discovered in 1990, and primarily responsible for embryonic growth. At least one hundred imprinted genes have been discovered up until 2005. It is believed, however, that imprinted genes represent less than 1% of the total genome. Imprinting of genes appears to violate the normal rules of Mendelian inheritance since expression of the genes is determined by only one allele from either the father or mother that can either switch on or off gene function, much as environmental factors now appear to be able to do this in some circumstances as described above, depending on the particular stimulus.[47–49] Imprinting is perceived as a kind of genetic arms race between maternal and paternal genes and may have evolved in eutherian and marsupial mammals because of the potential to manipulate the provision of resources to internally gestated offspring; egg-laying species such as birds do not imprint genes since resources in an egg are finite once it is laid.[50,51]

The importance of imprinted genes was demonstrated early in murine studies of IGF-2. If the normally paternally expressed allele was switched off, then offspring experienced a 40% reduction in growth, whereas if there were mutations in the maternal allele for IGF-2r, then the resulting offspring were oversized, unviable phenotypes.[52] Other early experiments by Azim Surani in Cambridge were conducted to see if mammals could reproduce parthenogenetically (from one sex only). In these experiments, maternally derived DNA was removed from mouse oocytes and replaced with two sets of paternally derived DNA, and conversely paternally derived DNA was removed from oocytes and replaced with two sets of maternally derived DNA. In neither set of experiments were the offspring viable.[53]

Other work suggests that imprinted genes influence various areas of neurological development. For example, Keverne and colleagues used mouse chimeras to explore the influence of maternal versus paternally derived alleles on the development of intelligence and emotion.[54] They found that maternally derived genes clustered in the cortex of the brain, a region associated with intelligence and planning, whereas the paternally derived genes clustered in the hypothalamic region, associated with primitive emotions. Such findings reverse the common male/female cultural stereotypes.

The role of imprinted genes in affecting intelligence is confirmed by specific human disorders that are known to derive from dysfunctions in normally imprinted genes. Two such syndromes are known as Prader-Willi syndrome and Angelman's syndrome. In the former, problems in the expression of paternally imprinted genes (locus 15q11-q13) on chromosome 15 lead to dysfunction associated with activities controlled by the hypothalamus, such as hyperphagia (overeating) and obesity, placidity in temperament, and an underdeveloped adult sex drive. In Angelman's syndrome, problems in the expression of the maternally imprinted gene (UBE3A) on chromosome 15 result in physical defects related to activities controlled by the cortex and striatum, including mental retardation, spasmodic movements, and difficulties with speech.[55,56]

David Haig has contributed significantly to the literature outlining the evolutionary significance of genomic imprinting by explaining its origin in terms of parental conflict as well as maternal-fetal conflict.[46,50,57] A good example of this kind of conflict is illustrated in the respective roles of imprinting genes in fetal development. Paternally expressed alleles in the genome are responsible for factors such as placental development (where it is argued it is in the father's interests to acquire resources for his offspring at the potential expense of the mother), whereas maternally expressed alleles control embryonic growth, therefore potentially conserving maternally derived resources.

Haig has argued that such conflict can explain maternal prenatal conditions such as preeclampsia as well as calcium metabolism.[50,58]

Finally, the two areas of fetal programming and genomic imprinting are beginning to converge. Some researchers have recently suggested that low birth weight, which we know is associated with health problems in later life, might originate from dysregulation of imprinted genes that control fetal development, and that this dysregulation might itself have environmental origins such as differential nutrition.[52,59] The potential marriage of these two areas at present originates from those studying genomic imprinting, but it will probably not be long before those scholars engaged in studying fetal programming begin to look more closely at the kinds of mechanisms that might be responsible for programming the fetus *in utero*, as suggested by one of Godfrey et al.'s recent articles.[31]

Epigenetics is an exciting and growing area of research that has the potential to explain many current aspects of disease and epidemiology with very clear practical applications to medicine. There is even the growing potential to develop epigenetic drugs that could be used to inhibit or enhance DNA methylation or to alter chromatin structure.[60] In relation to Nesse and Williams's framework, the concept of novel environments and their effects in producing epigenetic changes should be considered appropriate here.

Emergency Medicine

In a fascinating and much publicised lunch-hour lecture at UCL in 2004, Melvyn Singer, a professor of Intensive Care Medicine and centre director of the Bloomsbury Institute of Intensive Care Medicine at UCL, who has published widely in the field of emergency medicine, delivered a provocative talk entitled "Are We Ignoring the Lessons of Waterloo at Our (Patients') Peril?"[61] Drawing on historical records during this stimulating hour, Singer discussed the remarkably low mortality rates for severely wounded men in battles such as Waterloo, where most appeared to have survived. For example, of 52 men from the 13th Light Dragoons wounded by cannon, sabre, or gunfire during the battle of Waterloo, only 3 died, while only 6 men died among 102 wounded survivors on board the ship *Victory* during the battle of Trafalgar, this latter number including men with amputations or gangrene.

By pointing out these apparent medical anomalies, Singer speculated whether the absence of modern emergency room (ER) procedures might have helped to save the lives of these wounded soldiers. From a contemporary clinical perspective, this kind of speculation might sound very strange. But, in his academic publications, Singer and colleagues have suggested that multiorgan failure as a sequela of trauma (as in battles), systemic infection, or other life-threatening illnesses might be an *adaptive*, functional response to physiological injury that could protect the body against death.

In one relevant publication, Singer et al. (p. 545) have written:[62]

> Multiorgan failure is an attempt by the body to ensure cell survival in the face of sustained critical illness with affected cells entering a dormant state analogous to hibernation or aestivation. This response enhances the recovery of organ function should the patient survive...and might have evolved as a mechanism that increases the chances of survival in animals in which an external insult is potentially overwhelming.

For the field of EM, it is encouraging to see clinical doctors at the front line of treatment using Darwinian terms such as *evolved* in discussing their work. On the more practical level are the implications of Singer and colleagues' work for medical treatments. For they go on to suggest that many interventions hitherto used fairly routinely in ER contexts, such as mechanical ventilation, blood transfusions, intravenous administration of exogenous hormones, supplementation of nutritional status or other medications, may have adverse effects because they interfere with natural physiological defence processes.

These assertions are substantiated by a growing literature in the field of intensive care and emergency medicine that is examining the pros and cons of such interventions. For example, many recent studies have called into question the previously prevailing liberal strategies surrounding the use of red blood cell transfusions.[63–66] A number of clinical trials have, in fact, found increased rates of mortality or morbidity among critically ill patients who have been given transfusions, while others have found either no effect or increased benefits from lowering the haemoglobin concentration in transfusions. Patients in intensive care appear able to tolerate and survive severe anaemia, contrary to previous prevailing wisdom.[64] Reductions in transfusion rates could also lead to lowering of potential disease transmission caused by contaminated blood supplies as well as lowering immune-related responses to transfusions, and other reactions such as iron overload or electrolyte toxicity.[64]

Similarly, Moloney and Griffiths have suggested from the results of meta-analyses that tidal volumes used in mechanical ventilation could be reduced without affecting patient survival and, in some cases, might significantly improve survivorship by reducing conditions such as pulmonary over-distension, oedema, and infection.[67] They also discuss the possibility that mechanical ventilation might contribute to multiple organ failure due to the release of proinflammatory cytokines that affect cell function, a suggestion confirmed in at least one mouse model.[68]

Similarly, there is growing recognition that the reintroduction of oxygen to patients that have, for example, experienced cardiac arrest may actually induce cell death rather than improve the chances of patient recovery. It is in fact this process of reperfusion that actively kills body cells that have already initiated the protective response of lowering cellular activity.[69] Recent experiments have shown that somatic cells can, in fact, survive for long periods following oxygen deprivation.[70] These findings are revolutionizing the study of intensive care medicine and leading to new research into how to reintroduce oxygen more safely into cells that have effectively moved into a dormant state.[71,72]

These articles also tie in with emerging methods in triage where doctors are experimenting with ways of inducing suspended animation to preserve body function through cooling, aortic flushing, and even administration of toxic gases such as hydrogen sulphide—in a sense mimicking the body's own natural defences when multiple organ failure occurs.[73–76] In one aspect, these artificial methods replicate many of the physiological defences described as adaptive by Singer and colleagues, but they are also in response to an increase in the number of studies that are examining the adverse impact of many interventions used routinely in emergency medicine, and the key role of cellular mitochondria in producing these responses.

Linking both Singer's work and that of intensive care physicians and their teams, such as Lance Becker, Mark Roth, and Patrick Kochanek, is a focus on oxidative phosphorylation by the cell's mitochondria that provide energy and consume the bulk of oxygen. Singer suggested that multiple organ failure resulting from trauma or sepsis results from a decrease in oxidative phosphorylation and thereby a reduction in cellular metabolism by the mitochondria. Similar reactions are in fact produced by the clinical interventions described above that also aim to produce a state of suspended animation. For example, the normally toxic gas hydrogen sulphide interferes with oxidative phosphorylation by binding to cytochrome c oxidase, which normally binds to oxygen to produce adenosine triphosphate, or ATP, the primary cellular fuel.

Findings such as these and their link to innovative research investigating evolved physiological responses to trauma make emergency medicine an exciting clinical area that can be added to the growing corpus of topics in EM that have practical significance. In considering Nesse and Williams's overarching framework, the concept of defence is probably the most important in relation to emergency medicine.

Public Health

Public health professionals study problems that have group-level consequences in terms of significantly increased morbidity or mortality for some population of interest. Within EM, articles dealing with public health have mostly focused on the topic of the apparent mismatch between the environment in which humans evolved and more recent environments where conditions are vastly removed from ancestral ones.[1] Hence, there has been a focus on pathologies connected with contemporary lifestyles such as CHD, obesity, hypertension, and so forth.[42,77] Many of these conditions have behavioural causes. For example, obesity is associated with overeating and lack of exercise; CHD with a sedentary lifestyle, smoking, and an unbalanced diet rich in polysaturated fats; hypertension with stressful lifestyles and atherosclerosis, which is again linked to poor diet and lack of exercise over the lifecourse.

The standard approach in public health to change such "irrational" behaviour has been to appeal to good reasons for engaging in alternative, desirable ones, based on the notion that the irrational behaviours stem from a lack of relevant information: provide the information and people will change their behaviour. Health education professionals have thus attempted to get large numbers of people to come to rational decisions about themselves with respect to often fundamental types of behaviour—how they have sex, what they eat, or how they spend their leisure time. For example, antismoking campaigns have long been based on the notion that smoking kills by causing lung cancer—that is, trying to persuade people to preserve their health with explicit warnings about the disease risk of their behaviour. Such appeals to cognitive-level processes have not been overwhelmingly successful, probably because cognitive-level control is relatively weak, while the behaviours themselves have ancient roots. The appeal of foods rich in sugar and fats lies in evolved preferences for these kinds of substances that were once in short supply in the ancestral environment, but where they offered an important nutritional component of the ancient diet. Hence, humans evolved to have high preferences for these kinds of foods with negative consequences in environments with supermarkets.

Evolutionary psychology—particularly the notion that the brain is an evolved organ—suggests a different kind of approach to that of most public health perspectives. It argues that most of the reasons for widespread, persistently maladaptive behaviours are in fact either habitual or motivated by emotional causes. Further, these motivations—due to ancient mechanisms of control over behaviour—made good sense in the environment in which they evolved. Using this evolutionary perspective, a few public health workers have therefore begun to put forward hypotheses about behaviour change based on recognition of potential mismatches.

For example, a particular health problem that has been well studied from an evolutionary perspective is sun tanning.[78] The mismatch in this case is between contemporary cultural values that induce people with fair skin to become tanned in order to look beautiful, and the consequence of excessive exposure to ultraviolet radiation: skin cancer.[79] The standard approach to studying sun tanning is based on psychosocial models, which examine the relationships among beliefs, attitudes, intentions, personality traits, and normative factors to shed light on an individual's decision to sunbathe. For example, recognition that in one's social group there is a norm to protect oneself against excessive sunlight can predict an individual's expressed intention to use tanning protection.[80] However, public health campaigns designed to reduce sunbathing based on the standard health education approach of trying to change beliefs and attitudes have not proven effective.[81,82] Neither has increasing people's knowledge led to significant changes in the desired behaviour.[83,84] Most sunbathers seem to know that ultraviolet radiation is not good for their skin, but nevertheless persist in the practise.

An evolutionary approach, on the other hand, leads to an understanding of empirical patterns associated with sunbathing—in particular the fact that women are more likely than men to sunbathe, singles more than married people, and younger people more than older people. Women, despite being more aware than men that sunbathing is unhealthy for the skin, sunbathe more than men.[85] "That so many should engage in behavior so seemingly self-destructive can only suggest that powerful motivating psychological forces are at work"[86] (p. 421). The reason appears to be that, for cultural reasons, in northern societies cultural values currently associate tanned skin with beautiful, healthy condition. Health and beauty are, of course, important surrogates for reproductive potential, and so getting tanned skin should be an important goal, particularly for young women in the mating game. As an evolutionary approach would suggest, an intervention focused on a message that emphasised the damaging effects of indoor tanning on appearance (rather than an education message) was effective in changing attitudes, intentions, and behaviours, as well as being more effective on younger respondents.[87] Presumably, an approach designed to counter the evolved causes of sunbathing would be even more effective at getting people to protect themselves from the harmful rays of the sun—perhaps by providing alternative means of appearing tanned (e.g., chemical solutions).

In a similar vein, Val Curtis, director of the Hygiene Centre at the London School of Hygiene and Tropical Medicine, was a guest in the UCL EM class one year and ran a workshop for the students. She was able to demonstrate graphically, using a variety of plastic props of disgusting items (such as faeces), that people usually have innate responses to such objects even when they know they are not real. She pointed out that an important problem, especially in developing countries, is childhood infection, largely associated with contact with excreta. In fact, diarrheoa and respiratory tract infections are the biggest killers of children worldwide. Such contact has become endemic due to urban crowding, coupled with a lack of sanitation facilities that can separate people from their waste products. In such situations, hand washing with soap has been shown to reduce the probability of infection by up to half, and respiratory infections by a similar amount.[88]

As a result of these findings, a public-private partnership composed of international funders (such as the Bill and Melinda Gates Foundation), non-governmental organisations, soap-producing multinational corporations, and academic institutions has been set up by Curtis and others to design hand washing campaigns in the developing world.[89] This partnership supervises the implementation of public health programmes at national levels designed to make more people wash their hands with soap, particularly at the junctures that will have the greatest impact on disease reduction, such as after using toilet facilities. Obviously, such campaigns could follow the health education model by telling people about the dangers of diarrhoea for their

children. However, as in other cases, basing campaigns on fear and information about disease risks has not proven generally persuasive. Inspired by
evolutionary psychology, Curtis, together with colleagues, has shown that
there is an appropriate emotion to target in such cases: disgust.

Although a wide variety of objects in the environment inspire disgust—
faeces, rotten meat, dead bodies, insects, unwanted sexual intercourse—
what all of them share is their capacity to serve as habitats for pathogens or
as agents of disease. Disgust is an evolved response to avoid environmental
disease agents.[90] Contact with infectious agents—and thus diarrhoea and
respiratory infections—can therefore potentially be reduced by inducing
disgust reactions, rather than appealing to the need to avoid germs, which a
rational health education approach would suggest. Mass media campaigns
designed to induce disgust reactions at the important junctures of the day
have proven effective at increasing the awareness of the need to hand wash
in several countries.[91] Presumably, using such inspirations to develop programmes for other kinds of public health problems with a behavioural foundation would also improve their effectiveness.

Curtis and colleagues have summarised the situation this way, claiming
that hand washing promotion[92]

> offers one of the few effective preventive interventions
> against the two biggest, and most neglected, child
> killers; the diarrhoeal diseases and the respiratory
> infections (causing almost 2 million deaths each per
> year, according to WHO). Rates of handwashing with
> soap at key junctures around the world are low; typi-
> cally only 5–15% of mothers do so after cleaning up a
> child, or after using a toilet, but can be substantially
> improved if programmes are built on a solid founda-
> tion of understanding target audiences and their hand-
> wash motivation.

The practical benefits of such public health campaigns rooted in evolutionary principles are thus clear. There is a real ability to save not just thousands
but potentially millions of lives around the world, if only health promoters
take more seriously the insights to be gained from adopting a perspective
from evolutionary medicine. The stakes are high and the rewards potentially enormous. Nesse and Williams's concepts and the idea of mismatches
within novel environments is of relevance here, together with the utility of
using evolved emotions as marketing tools.

Evolutionary Obstetrics

This is a topic that has been covered in other books on EM, but is worth
including here because of the many radical clinical changes that have arisen

in obstetrics, partly in response to findings that birth companions can significantly improve the outcome of the birth experience for both mother and baby. One guest that was interviewed in class at UCL, Yehudi Gordon, was a practicing obstetrician generally unexposed to evolutionary theory, but cognizant of anthropology.

Gordon has become well known in the UK as a kind of celebrity obstetrician because he is sought out by many well-known individuals in the entertainment industry (such as Cate Blanchett, Gwyneth Paltrow, and Heather Mills McCartney) to deliver their babies. He now runs a private natural childbirth clinic at the Hospital of St. John and St. Elizabeth in London but has also worked in other hospitals in London, such as the Royal Free. Dr. Gordon is distinguished from other obstetricians by his very early incorporation of childbirth practises that are advocated by many anthropologists, and particularly the use of birthing companions to assist in the psychological and physical process of childbirth. He introduced the use of water births for labouring women and other innovations to relax mothers during parturition. The increasingly difficult process of birthing for human females that accompanied the evolution of bipedality and encephalisation may have led to the need for a birthing companion who could assist a woman in this process.[93,94] Peter Ellison has suggested that the slowing of labour that arises in contexts where women lack a birth companion may represent an evolved mechanism to delay parturition until help can be obtained. Evidence for this assertion comes from studies of doulas, either lay or professional birth attendants, who provide emotional support during the birthing process even if they are previously unknown to the woman in labour.[93] Statistics from such studies show significantly fewer perinatal and postnatal complications, including oxytocin administration, emergency Caesarean sections, and meconium discharge, and that doulas are also associated with faster labour times and better breastfeeding success in women.[95-99]

In his 20 or so years of work, obstetricians such as Gordon have moved obstetric practise back to the acceptance of birth as a natural process as opposed to an illness.[99,100] Gordon has shunned practises such as shaving of the pubic hairs, enemas, episiotomies, and use of the lithotomy (or prone) position during childbirth, which are practises still used in many countries, including the UK and the United States, and which have been shown, in many cases by the Cochrane Collaboration (a group that studies the results of clinical trials), to have either adverse or neutral impacts on the childbirth process.[101-105] But, at the same time, Yehudi Gordon was explicit in class about the still appropriate clinical aspect of obstetrics, its role in reducing maternal morbidity and mortality, and the concept of natural childbirth in all its meanings:

> Natural versus unnatural? I mean it depends on your perspective. Is it natural to give birth in a tribal hut in Kenya...the active birth pictures from our hospital—they're definitely not natural cause you're in a

> concrete building which has got central heating, it's got fans...it's got facilities downstairs so that if anything goes wrong for mother or baby we can perform a Caesarean or deliver the baby with forceps, so there are degrees of natural....In that context, even on the Birth Centre, almost nothing is natural, but it's a little bit closer to nature than in some other places. If you look at the natural maternal mortality when I worked in South Africa...[it] was 1 in 500, and the maternal mortality in the UK is under 1 in 10,000, so it's 20 times at least as safe to have a baby in the UK. So, actually, we're better than nature, but what's the cost, the cost is a lot of intervention....

Whether or not the field of obstetrics generally recognizes the role that evolutionary anthropology has played in contributing to the literature about more natural methods of childbirth is unclear, but it is undisputed that in recent years, the practical benefits of these realisations are being applied in at least some clinical settings. Again, the concept of novel environments and how these produce mismatches with contemporary clinical practise is important.

Conclusion

In order to counter early criticisms of the field that it had little relevance to medical practitioners, this chapter has attempted to cover some areas in EM that have very clear and practical applications to clinical medicine and that have received less coverage in earlier volumes. Clearly there are many areas in EM that are successfully merging with evolutionary biology while retaining relevance for improving treatment and health for patients, as well as enhancing our understanding of the origins of specific conditions. Increasing attention to such areas in EM will potentially deflect criticisms of this field that it has little relevance to medical doctors, and raise awareness of the practical nature of many subfields within this emerging discipline.

Acknowledgements

The first author sincerely thanks all of the academics, clinicians, and students who participated in the Evolutionary Medicine class at UCL from 2001 to 2006. Almost without exception, the individuals approached accepted the invitation to appear as a guest, and agreed to be videotaped during the interview. They sometimes travelled long distances to do this, and all accepted the invitation without any honorarium. Development of the class at UCL was supported by grants from C-SAP (Centre for Sociology, Anthropology and Politics),[106] the Executive Sub-Committee on Innovations in Teaching, Learning, and Assessment (ESCILTA), UCL,[107] and Media Resources, UCL.[2]

References

1. Williams, G. W., and Nesse, R. M. 1991. The dawn of Darwinian medicine. *Q. Rev. Biol.* 66:1–22.
2. Media Resources UCL. http://www.ucl.ac.uk/mediares/.
3. Tate, S. K., and Goldstein, D. B. 2004. Will tomorrow's medicines work for everyone? *Nat. Genet.* 36:S34–S42.
4. Cohen, J., et al. 2006. Sequence variations in PCSK9, low LDL, and protection against coronary heart disease. *N. Engl. J. Med.* 354:1264–72.
5. Haga, S. B., and Venter, J. C. 2003. FDA races in wrong direction. *Science* 301:466.
6. Collins, F. S. 2004. What we do and don't know about "race," "ethnicity," genetics and health at the dawn of the genome era. *Nat. Genet.* 36:S13–15.
7. Sankar, P., et al. 2004. Genetic research and health disparities. *JAMA* 291:2985–89.
8. Taylor, A. L., and Wright, J. T., Jr. 2005. Importance of race/ethnicity in clinical trials. *Circulation* 112:3654–66.
9. American Anthropological Association. 1998. Statement on "race." http://www.aaanet.org/stmts/racepp.htm. Accessed 20/2/2008.
10. Bamshad, M., et al. 2004. Deconstructing the relationship between genetics and race. *Nat. Rev. Genet.* 5:598–609.
11. Outram, S. M., and Ellison, G. T. 2006. Anthropological insights into the use of race/ethnicity to explore genetic contributions to disparities in health. *J. Biosoc. Sci.* 38:83–102.
12. Marks, J. 1995. *Human biodiversity: Genes, race, and history*. Chicago: Aldine.
13. Dressler, W. W., Oths, K. S., and Gravlee, C. C. 2005. Race and ethnicity in public health research: Models to explain health disparities. *Ann. Rev. Anthropol.* 34:231–52.
14. Sherry, S. T., Batzer, M. A., and Harpending, H. C. 1998. Modeling the genetic architecture of modern populations. *Ann. Rev. Anthropol.* 27:153–69.
15. Takahata, N., Satta, Y., and Klein, J. 1995. Divergence time and population size in the lineage leading to modern humans. *Theor. Popul. Biol.* 48:198–221.
16. Lewontin, R. C. 1972. The apportionment of human diversity. *Evol. Biol.* 6:391–98.
17. Edwards, A. W. F. 2003. Human genetic diversity: Lewontin's fallacy. *Bioessays* 25:798–801.
18. Jorde, L. B., and Wooding, S. P. 2004. Genetic variation, classification and "race." *Nat. Genet.* 36:S28–32.
19. Rosenberg, N. A., et al. 2002. Genetic structure of human populations. *Science* 298:2381–85.
20. Serre, D., and Pääbo, S. 2004. Evidence for gradients of human genetic diversity within and among continents. *Genet. Res.* 14:1679–85.
21. Kahn, J. 2007. Race in a bottle. *Scientific American*, August, pp. 26–31.
22. Norris, K. et al. 2006. Cardiovascular outcomes in the African American Study of Kidney Disease and Hypertension (AASK) trial. *Am. J. Kidney Dis.* 48:739–51.
23. Wright, Jr., J. T., et al. 2002. Effect of blood pressure lowering and antihypertensive drug class on progression of hypertensive kidney disease: Results from the AASK trial. *JAMA* 288:2421–31.

24. The ALLHAT Officers and Co-ordinators for the ALLHAT Collaborative Group. 2002. Major outcomes in high-risk hypertensive patients randomized to angiotensin converting enzyme inhibitor or calcium channel blocker vs diuretic. The Antihypertensive and Lipid-Lowering Treatment to Prevent Heart Attack Trial (ALLHAT). *JAMA* 288:1981–97.

25. Dahlof, B., et al. 2002. Cardiovascular morbidity and mortality in the Losartan Intervention for Endpoint reduction in hypertension study (LIFE): A randomised trial against atenolol. *Lancet* 359:995–1003.

26. Jablonka, E., and Lamb, M. J. 1995. *Epigenetic inheritance and evolution: The Lamarkian dimension.* Oxford: Oxford University Press.

27. Cropley, J. E., et al. 2006. Germ-line epigenetic modification of the murine Avy allele by nutritional supplementation. *Proc. Natl. Acad. Sci. U.S.A.* 103:17308–12.

28. Dolinoy, D. C., et al. 2006. Maternal genistein alters coat color and protects Avy mouse offspring from obesity by modifying the fetal epigenome. *Environ. Health Perspect.* 114:567–72.

29. Wolff, G. L., et al. 1998. Maternal epigenetics and methyl supplements affect agouti gene expression in Avy/a mice. *FASEB J.* 12:949–57.

30. Barker, D. 1999. The fetal origins of coronary heart disease and stroke: Evolutionary implications. In *Evolution in health and disease,* ed. S. C. Stearns, chap. 21. New York: Oxford University Press.

31. Godfrey, K. M., et al. 2007. Epigenetic mechanisms and the mismatch concept of the developmental origins of health and disease. *Pediatr. Res.* 61(Pt. 2):5R–10R.

32. Ozanne, S. E., and Constancia, M. 2007. Mechanisms of disease: The developmental origins of disease and the role of the epigenotype. *Nat. Clin. Pract. Endocrinol. Metab.* 3:539–46.

33. Sinclair, S. K., et al. 2007. The developmental origins of health and disease: Current theories and epigenetic mechanisms. *Soc. Reprod. Fertil. Suppl.* 64:425–43.

34. Waterland, R. A., and Michels, K. B. 2007. Epigenetic epidemiology of the developmental origins hypothesis. *Ann. Rev. Nutr.* 27:363–88.

35. Barker, D. 1998. *Mothers, babies and health in later life.* 2nd ed. London: Churchill Livingstone.

36. Gluckman, P., and Hanson, M. 2005. *The fetal matrix: Evolution, development and disease.* Cambridge: Cambridge University Press.

37. Gluckman, P., and Hanson, M. 2006. *Mismatch: Why the world no longer fits our bodies.* Oxford: Oxford University Press.

38. Eriksson, J. G., et al. 2001. Early growth and coronary heart disease in later life: Longitudinal study. *BMJ* 322:949–53.

39. Kuh, D., et al. 2002. Birth weight, childhood growth and abdominal obesity in adult life. *Int. J. Obes. Relat. Metab. Disord.* 26:40–47.

40. Law, C. M., et al. 2002. Fetal, infant, and childhood growth and adult blood pressure: A longitudinal study from birth to 22 years of age. *Circulation* 105:1088–92.

41. Nunez-de la Mora, A., et al. 2007. Childhood conditions influence adult progesterone levels. *PLoS Med.* 4:e167.

42. Bateson, P., et al. 2004. Developmental plasticity and human health. *Nature* 430:419–21.

43. Anway, M. D., Cupp, A. S., Uzumcu, M., and Skinner, M. K. 2005. Epigenetic transgenerational actions of endocrine disruptors and male fertility. *Science* 308:1466–69.

44. Bygren, L. O., Kaati, G., and Edvinsson, S. 2001. Longevity determined by paternal ancestors' nutrition during their slow growth period. *Acta Biotheor.* 49:53–59.
45. Kaati, G., Bygren, L. O., and Edvinsson, S. 2002. Cardiovascular and diabetes mortality determined by nutrition during parents' and grandparents' slow growth period. *Eur. J. Hum. Genet.* 10:682–88.
46. Haig, D. 1999. Genetic conflicts of pregnancy and childhood. In *Evolution in health and disease*, ed. S. C. Stearns, chap. 7. New York: Oxford University Press.
47. de la Casa-Esperon, E., and Sapienza, C. 2003. Natural selection and the evolution of genome imprinting. *Annu. Rev. Genet.* 37:349–70.
48. Leighton, P. A., et al. 1996. Genomic imprinting in mice: Its function and mechanism. *Biol. Reprod.* 54:273–78.
49. Surani, M. A., et al. 1990. Developmental consequences of imprinting of parental chromosomes by DNA methylation. *Phil. Trans. R. Soc. Lond. B Biol. Sci.* 326:313–27.
50. Haig, D. 1993. Genetic conflicts in human pregnancy. *Q. Rev. Biol.* 68:495–532.
51. O'Neill, M. J., et al. 2000. Allelic expression of IGF2 in marsupials and birds. *Dev. Genes Evol.* 210:18–20.
52. Reik, W., et al. 1993. Adult phenotype in the mouse can be affected by epigenetic events in the early embryo. *Development* 119:933–42.
53. Walsh, C., et al. 1994. The non-viability of uniparental mouse conceptuses correlates with the loss of the products of imprinted genes. *Mech Dev.* 46:55–62.
54. Keverne, E. B., et al. 1997. Genomic imprinting and the differential roles of parental genomes in brain development. *Brain Res. Dev. Brain Res.* 92:91–100.
55. da Rocha, S. T., and Ferguson-Smith, A. C. 2004. Genomic imprinting. *Curr. Biol.* 14:R646–49.
56. Tighlman, S. M. 1999. The sins of the fathers and mothers: Genomic imprinting in mammalian development. *Cell* 96:185–93.
57. Moore, T., and Haig, D. 1991. Genomic imprinting in mammalian development: A parental tug-of-war. *Trend. Genet.* 7:45–49.
58. Haig, D. 2004. Evolutionary conflicts in pregnancy and calcium metabolism— A review. *Placenta* 25 (Suppl.): S10–15.
59. Cetin, I., et al. 2004. Fetal growth restriction: A workshop report. *Placenta* 25:753–57.
60. Sigalotti, L., et al. 2007. Epigenetic drugs as pleiotropic agents in cancer treatment: Biomolecular aspects and clinical applications. *J. Cell Physiol.* 212:330–44.
61. http://news.bbc.co.uk/1/hi/health/4035849.stm and http://www.naturecurekingstonfile.org.uk/file_03/file03_10.html).
62. Singer, M., De Santis, V., Vitale, D., and Jeffcoate, W. 2004. Multiorgan failure is an adaptive, endocrine-mediated, metabolic response to overwhelming systemic inflammation. *Lancet* 364:545–48.
63. Corwin, H. L., et al. 2004. The CRIT study: Anemia and blood transfusion in the critically ill—Current clinical practice in the United States. *Crit. Care Med.* 32:39–52.
64. Alvarez, G., Hebert, P. C., and Szick, S. 2001. Debate: Transfusing to normal haemoglobin levels will not improve outcome. *Crit. Care* 5:56–63.
65. Napolitano, L. M., and Corwin, H. L. 2004. Efficacy of red blood cell transfusion in the critically ill. *Crit. Care Clin.* 20:255–68.
66. Reiles, E., and Van der Linden, P. 2007. Transfusion trigger in critically ill patients: Has the puzzle been completed? *Crit. Care* 11:142.

67. Moloney, E. D., and Griffiths, M. J. D. 2004. Protective ventilation of patients with acute respiratory distress syndrome. *Br. J. Anaesth.* 92:261–70.
68. Vaneker, M., et al. 2007. Mechanical ventilation in healthy mice induces reversible pulmonary and systemic cytokine elevation with preserved alveolar integrity: An in vivo model using clinical relevant ventilation settings. *Anesthesiology* 107:419–26.
69. Kutala, V. K., et al. 2007. Role of oxygen in postischemic myocardial injury. *Antioxid. Redox Signal* 9:1193–206.
70. Idris, A. H., et al. 2005. Oxidant injury occurs rapidly after cardiac arrest, cardiopulmonary resuscitation, and reperfusion. *Crit. Care Med.* 33:2043–48.
71. Anderson, T. C., et al. 2006. Transient and partial mitochondrial inhibition for the treatment of postresuscitation injury: Getting it just right. *Crit. Care Med.* 34: S474–82.
72. Shao, Z. H., et al. 2007. Hypothermia-induced cardioprotection using extended ischemia and early reperfusion cooling. *Am. J. Physiol. Heart Circ. Physiol.* 292: H1995–2003.
73. Behringer, W., et al. 2003. Survival without brain damage after clinical death of 60-120 mins in dogs using suspended animation by profound hypothermia. *Crit. Care Med.* 31:1523–31.
74. Blackstone, E., Morrison, M., and Roth, M. B. 2005. H2S induces a suspended animation-like state in mice. *Science* 308:518.
75. Lampe, J. W., and Becker, L. B. 2007. Rapid cooling for saving lives: A bioengineering opportunity. *Exp. Rev. Med. Devices* 4:441–46.
76. Oddo, M., et al. 2006. From evidence to clinical practice: Effective implementation of therapeutic hypothermia to improve patient outcome after cardiac arrest. *Crit. Care Med.* 34:1865–73.
77. Eaton, S. B., et al. 2002. Evolutionary health promotion. *Prev. Med.* 34:109–118.
78. Saad, G., and Peng, A. 2006. Applying Darwinian principles in designing effective intervention strategies: The case of sun tanning. *Psychol. Marketing* 23:617–38.
79. Armstrong, B. K., and Kricker, A. 2001. The epidemiology of UV induced skin cancer. *J. Photochem. Photobiol. B Biol.* 63:8–18.
80. Jackson, K. M., and Aiken, L. S. 2000. A psychosocial model of sun protection and sunbathing in young women: The impact of health beliefs, attitudes, norms, and self-efficacy for sun protection. *Health Psychol.* 19:469–78.
81. Johnson, E. Y., and Lookingbill, D. P. 1984. Sunscreen use and sun exposure: Trends in a white population. *Arch. Dermatol.* 120:727–31.
82. Robinson, J. K., Rigel, D. S., and Amonette, R. A. 1997. Trends in sun exposure knowledge, attitudes, and behaviors: 1986 to 1996. *J. Am. Acad. Dermatol.* 37:179–86.
83. Leary, M. R., and Jones, J. L. 1993. The social psychology of tanning and sunscreen use: Self-presentational motives as a predictor of health risk. *J. Appl. Soc. Psychol.* 23:1390–1406.
84. Rossi, J. S., et al. 1995. Preventing skin cancer through behavior change: Implications for interventions. *Dermatol. Clinics* 13:613–622.
85. Monfrecola, G., et al. 2000. What do young people think about the dangers of sunbathing, skin cancer and sunbeds? A questionnaire survey among Italians. *Photoderm. Photoimmun. Photomed.* 16:15–18.
86. Koblenzer, C. C. 1998. The psychology of sun-exposure and tanning. *Clinics Dermatol.* 16:421–28.

87. Hillhouse, J. J., and Turrisi, R. 2002. Examination of the efficacy of an appearance-focused intervention to reduce UV exposure. *J. Behav. Med.* 25:395–409.
88. Curtis, V. A., and Cairncross, S. 2003. Effect of washing hands with soap on diarrhoea risk in the community: A systematic review. *Lancet Inf. Dis.* 3:275–81.
89. http://www.globalhandwashing.org/.
90. Curtis, V. A., Aunger, R., and Rabie, T. 2004. Evidence that disgust evolved to protect from risk of disease. *Proc. R. Soc. B* 271:131–33.
91. Curtis, V. A., Garbrah-Aidoo, N., and Scott, B. 2007. Masters of marketing: Bringing private sector skills to public health partnerships. *Am. J. Pub. Health* 97:634–41.
92. Curtis, V. A., et al. 2001. Evidence for behaviour change following a hygiene promotion programme in Burkina Faso. *Bull. WHO* 79:518–26.
93. Ellison, P. T. 2001. *On fertile ground.* Cambridge, MA: Harvard University Press.
94. Trevathan, W. 1987. *Human birth: An evolutionary perspective.* New York: Aldine de Gruyter.
95. Campbell, D. A., et al. 2006. A randomized control trial of continuous support in labor by a lay doula. *J. Obstet. Gynecol. Neonatal Nurs.* 35:456–64.
96. Klaus, M. H., Kennell, J. H., Robertson, S. S., and Sosa, R. 1986. Effects of social support during parturition on maternal and infant morbidity. *BMJ (Clin. Res. Ed.)* 293:585–87.
97. Mosallam, M., et al. 2004. Women's attitude towards psychosocial support in labour in United Arab Emirates. *Arch. Gynecol. Obstet.* 269:181–87.
98. Sosa, R., et al. 1980. The effect of a supportive companion on perinatal problems, length of labor, and mother-infant interaction. *N. Engl. J. Med.* 303:597–600.
99. Davis-Floyd, R. E., and Sargent, C. F., eds. 1997. *Childbirth and authoritative knowledge.* Berkeley: University of California Press.
100. Jordan, B. 1983. *Birth in Four Cultures: A Cross-Cultural Investigation of Childbirth in Yucatan, Holland, Sweden and the United States.* London: Eden Press.
101. Basevi, V., and Lavendar, T. 2001. Routine perineal shaving on admission in labour. *Cochrane Database Syst. Rev.* 2.
102. Boulvain, M., Stan, C., and Irion, O. 2001. Membrane sweeping for induction of labour. *Cochrane Database Syst. Rev.* 2.
103. Clemons, J. L., et al. 2005. Decreased anal sphincter lacerations associated with restrictive episiotomy use. *Am. J. Obstet. Gynecol.* 192:1620–25.
104. Diniz, S. G., and Chacham, A. S. 2004. "The cut above" and "the cut below": The abuse of caesareans and episiotomy in Sao Paulo, Brazil. *Reprod. Health Matters* 12:100–10.
105. Gupta, J. K., and Nikodem, V. C. 2000. Woman's position during second stage of labour. *Cochrane Database Syst. Rev.* 2.
106. http://www.c-sap.bham.ac.uk/
107. http://www.ucl.ac.uk/teaching-learning/committees/escilta.shtml.

11

Comparative and Evolutionary Biology in Medical and Dental Education

M. Christopher Dean
Department of Cell and Developmental Biology, University College London

Contents

Introduction

Despite the fact that we share 98.8% of our DNA with chimpanzees, 54% of adults in the United States prefer to believe that humans did not evolve from some earlier species.[1] Steve Jones, the geneticist from University College London, has famously put this into perspective by reminding us that we also share 30% of our DNA with a banana. With this in mind, one of the many things I find myself doing from year to year is teaching anatomy to increasing numbers of students in less and less time. One might think very little has changed with respect to anatomy teaching, but perhaps not surprisingly in the last few years, given the strong feelings of so many in the United States, just a few students have begun to question why lecturers might choose to use examples of animals other than humans to illustrate functional or morphological points. Some, across a spectrum of faiths, have gone as far as to report that they find it offensive to have evolution mentioned when being taught, for example, the anatomy of the human leg or head. Regardless of this trend, many scientists of many faiths see no conflict whatever between scientific inquiry into the nature of the universe and their strongly held beliefs.

My thoughts and opinions about this are in no way about individual beliefs, but rather, that in our eagerness to devise new ways to teach an ever-diminishing core of basic science, we are in danger of divorcing the facts we teach from the fundamental biological framework that underpins the very way we should be thinking when we try to solve biological and clinical problems. These days, the slightest disquiet from those being taught, combined with inevitable pressure on the timetable, might easily be enough to drive basic biology out of the medical and dental curriculum forever, and thereby change the way future students come to think about human diseases, and indeed human beings. I am especially curious about what actually underlies recurring trends in what we teach and how we teach. Much of the new medical and dental core content probably results from the combined effects of pressure on limited resources, the undoubted increasing importance of molecular biology in clinical practice, and a strong shift in expectations for better communication skills and accountability among health workers generally. The trends in what we teach and how we teach may also swing back and forth over time because they mirror the changing skills, enthusiasms, and interests of the teachers themselves. Nevertheless, anatomy often comes off badly in curriculum reviews, perhaps because reducing what is taught is viewed favourably by many students who feel overburdened by basic science and starved of clinical contact in the early years, and perhaps also because this reduction is often viewed favourably by educationalists and influential clinicians for whom anatomy may have been less important in their careers. Irrespective of this, it seems to me that anatomy and indeed physiology both stand out as unique scientific vehicles for broadening students in both biological and clinical science simultaneously.

Recently, a number of remarkable scientific techniques have brought the way we study and think about our own evolutionary history and that of many diseases closer together. The findings of these studies are relevant to what we do from day to day and to how we design clinical practice for the future. What follows in this chapter is my reaction to the thought that, even now, some students would rather not be taught about things that are actually deeply fascinating, and that soon a whole way of thinking might be dismissed as off limits to future medical and dental students at UK universities. Beyond this, I am particularly curious about the point at which, in what seems to me a logical progression of facts about genetic change over time, some people begin to take offence. My thesis regarding this issue is that we have become expert at teaching, learning, and understanding processes and mechanisms that take place over a comparatively short time span. Only rarely in clinical medicine and dentistry do we expect to study things that develop or change over more than a generation or two. The fact that we may differ from each other and respond differently to clinical treatments because of tens of thousands of years of separation, or indeed that we share uniquely human attributes that are rooted millions of years in the past, is now totally alien to many health workers. I wager that those who, almost by reflex, take offence at the word *evolution* would also have problems believing that as little as 20,000 years ago one could walk across dry land from Norfolk to the Netherlands, or that the Dover Strait was forged by two relatively recent catastrophic megafloods.[2,3] The problem, as I see it, is that we are failing to teach students about mechanisms and processes that occur over very long periods of time. Deep time, so to speak, and its role in modern medicine and dentistry, is therefore completely incomprehensible to them (and also increasingly to many of their teachers) far more, I think, than is the concept of deep space, which gets much more factual and fictional media coverage.

In this chapter I propose to paint a very broad picture to make my point rather than focus on a particular process or disease. First I review, very basically, how changes in genes and gene frequencies come about in populations and show how there are many processes that take place over long periods of time that are relevant to medicine and dentistry, but which are not all appropriately described as evolution. In so doing, I hope to show that examples of each can be identified across a spectrum of living organisms. I then want to demonstrate how being able to reconstruct past histories of humans, viruses, and bacteria, and in some cases being able to identify examples of coevolution, has helped us define the course and pattern of many human diseases. Throughout, wherever possible, I have tried to draw examples from prominent papers published recently in highly ranked journals but which I suspect are not often flagged as relevant or significant in the medical and dental literature.

Genes and Populations and the Processes Involved in Changing Genetic Structure

Genes and Gene Frequencies

Human populations, like those of any living organisms, come in all shapes and sizes. Crucial to an understanding of how populations work, in the sense of how their genetic structure either remains the same or changes, be they populations of viruses, bacteria, or humans, is population genetics. Few people would even contemplate that they are genetically identical to their parents or grandparents, or indeed to any of their relatives or ancestors. We all know that we inherit genes from both parents and that this makes us genetically unique (unless we happen to be an identical twin or a clone). In medicine and dentistry we often construct pedigrees to track family history. It is routine to be able to predict for worried parents whose first child has a cleft lip, for example, the probability that a second child will also have a cleft lip (~4% if both parents are unaffected but ~17% if one parent also has a cleft lip). Yet at a population level where there is completely random mating (panmixis) and where no other factors act on that population to either introduce new genes or remove existing genes, the frequencies of genes in a population would remain the same over time. This is so even when inbreeding (which increases homozygosis) or outbreeding (which increases heterozygosis) changes genotype frequencies, which in reality they constantly do. There are, however, two things that do bring about a change in gene frequency in a population and replace one set of genes with another set: migration of individuals into that population and gene mutation.

Migration

Modern humans are characterised by high levels of mobility, and the resulting gene flow has a great homogenising effect on our population genetics.[4] The cell-surface carbohydrates that underlie the ABO blood group system are well studied for obvious reasons and distribute across the world in a way that reflects past human migrations. For example, the O group reaches close to 100% in some South American populations that have avoided intermixture.[4] The B gene map[4] is to some extent the inverse of the O. The highest frequencies in Himalayan regions gradually reduce across central Asia westward in an irregular cline or gradient that reflects the origins of European populations from an ancient central Asian migration.[4] Tooth morphology is also a powerful way of tracking past population history. Polymorphic trait frequencies primarily reflect population movement and admixture over time.[5] In the Far East (Japan, China, Mongolia, and Tibet, for example), shovel-shaped incisors occur with a frequency of ~80%, while in Europe, Australia, and Africa frequencies are closer to 10%. Carabelli's cusp occurs on upper molars with a frequency of ~22% in European populations, but frequencies are as low as 10% in Far Eastern populations and as little as 2% in some Eskimo/Aleut populations. Similarly, while around 90% of Europeans have four-cusped rather than five-cusped lower second molars, the frequency in southern Africa and among aboriginal Australian populations is about half this.[5]

Mutation

Mutations, on the other hand, are comparatively rare events but are actually the sole source of new genes. Rates of nuclear DNA mutation are relatively slow but occur at a fairly consistent rate. We inherit our nuclear DNA from both our mother and our father. However, mitochrondria, the cytoplasmic organelles where many energy-related biochemical pathways take place, have their own DNA (mtDNA). This is because they were at one time independent unicellular organisms[6] that became engulfed and incorporated into the cells of eukaryotes (animals, plants, fungi, and protists). Rates of mtDNA mutation are approximately ten times faster than those in nuclear DNA.[4]

Mutations in bacterial and viral DNA and RNA are crucially important in the context of human infectious diseases. Seventy percent of animal viruses are RNA viruses.[7] They have smaller genomes than DNA viruses and their replication is more error-prone. Because of this and because their mutation rates are higher than those of DNA viruses, their evolution is much faster. RNA viruses have a frightening ability to evolve and adapt to new hosts and increase in virulence. They are less species specific and readily jump species barriers. DNA viruses tend to be more species specific, have narrower host ranges, are more stable, and are usually older than RNA viruses. Slow-evolving DNA viruses are often of ancient origin and many have coevolved with their hosts over long time spans. Fast-changing RNA viruses, on the other hand, have often only been acquired recently, usually at the time when human populations became large enough to support them.[7] The hepatitis B virus (HVB) is a DNA virus but replicates with an RNA intermediate and reverse transcriptase, giving it the same error-prone transcription and capacity for rapid evolution as the retroviruses to which it is related.

Genetic Drift

Once introduced into a population there are just two things that determine what happens to genes. The first is genetic drift, a random or stochastic process that can change gene frequencies relatively quickly just by chance. But gene flow (migration) tends to cancel out the effects of genetic drift, so that the most rapid changes resulting from drift occur in small populations where migration levels are low.[4] At a local level, for example, the distribution of blood group frequencies or dental morphological traits is likely to be the result of genetic drift.[5] Third molar agenesis is clearly multifactorial but is another good example of the effect of genetic drift. Its complex genetic aetiology is clear from the observations that missing third molars run in families and are associated with the late initiation of other teeth, with other missing teeth in the same mouth, and with generally smaller teeth in a dentition.[8] However, there is little evidence that worldwide the incidence of third molar agenesis is anything other than the result of random genetic drift and the so-called founder effect establishing high or low frequencies in newly established populations.[8–10] The incidence of third molar agenesis in Europe hovers around 25% but is reported to be as little as 2% or less in some African populations and as great as 40% in some Eskimo and Aleut populations,[8,9] although a standardised way of recording it, and describing it statistically, might in fact iron out some of these apparently considerable differences.[8] Nonetheless, even 1.6 million years ago in the human fossil record there are examples of individuals with missing third molars.

Selection

The second process that can change gene frequencies in a population is selection. Selection either eliminates mutant genes (balancing selection) or favours them (directional selection). Most people are aware of the classic example of how a deleterious point mutation in the β-globin chain of haemoglobin results in sickle cell anaemia, and that this can actually confer an advantage against malarial infections when present in the heterozygous state. The deleterious *HB^s* gene has therefore, since this time, been maintained in some populations by balancing selection.

Another illustration of how balancing selection maintains an obviously advantageous character is revealed in the red eye of the manatee.[11,12] Clarity of vision is no longer the imperative in manatees that it is in other animals since manatees live in fresh but turbid waters. They rely heavily on their sensory bristles (vibrissae) to navigate and protect themselves against the environment. The cornea is one of the few tissues that contains no blood vessels and is normally crystal clear because a soluble receptor called sflt-1, or vascular endothelial growth factor receptor-1 (VEGFR-1), traps vascular endothelial growth factor-A (VEGF-A), stopping it from directing the formation of blood vessels.[11,12] For all animals there is an obvious advantage to being able to see clearly, and the presence of sflt-1 is evolutionarily conserved. An exception appears to be the manatee cornea, where slft-1 is not present and which therefore becomes vascularised. This recent insight has suggested ways of preventing vascularisation of other tissues, and for treating neovascularisation in tumour angiogenesis.[11,12]

Evolution (L. *evolutio* = unrolling), in general terms and put simply, is no more than a change in gene frequency from one generation to the next. This of course may come about through migration, mutation, genetic drift, or selection. Charles Darwin is famous, not for noticing that plants and animals evolved or changed through time, but for describing a mechanism, natural selection, whereby individuals with favourable traits are systematically favoured for reproduction. Natural selection, or descent with modification, as Darwin put it, remains the single primary explanation for adaptive evolution. No better example of directional selection exists than methicillin-resistant *Staphylococcus aureus* (MRSA). In the 1940s most strains of *Staphylococcus aureus* were sensitive to penicillin, but some that were able to produce the enzyme penicillinase survived short periods of treatment by penicillin and quickly became more common. By the 1960s, between 90 to 95% of *S. aureus* strains were resistant. Methicillin was developed to treat these strains by counteracting the action of penicillinase, but within a year strains of MRSA were already known. Now approximately 40% of the *S. aureus* that cause bacteraemia are resistant to methicillin. I suspect because the timescale of this change is within living memory and because it relates to a bacterium and not a human, few people would have a problem with either the mechanism or this sequence of historical events. Exactly the same processes and mechanisms, but over longer periods of time, are apparently incomprehensible to some because they have never been taught human or comparative biology in an evolutionary context.

Directional selection is rarely as fast as with MRSA, but the last surviving mammoths, the majority of which became extinct ~10,000 years ago, survived on islands north of Siberia for another 6,000 years by reducing their body size by at least 30% in a very short period of time in response to diminishing food supplies.[13,14] Red deer on the Channel Island of Jersey also became dwarfed to a sixth of their body weight in just 6,000 years.[15] Some human fossils recently discovered on the island of Flores in Indonesia, and dated to just ~18,000 years ago, also appear to have become island dwarfs that were around a metre tall before their extinction as little as ~12,000 years ago.[16] It is often the case that tooth size reduces at a slower rate than body size in dwarfed forms that become small very quickly.[13,15,17] This is especially interesting in the context of the rapid reduction in body size, jaw size, and tooth size that occurred among modern humans during the Early and Late Upper Palaeolithic. This trend appears to have been the result of strong directional selection; males were affected more than females, and lower anterior teeth and upper cheek teeth more so than the rest of the dentition.[18] Much of modern orthodontic practice is designed to deal with the disparity in size between teeth and jaws that results in crowded teeth, but despite the long-term trends in tooth size and jaw size reduction that took place during the Early and Late Upper Palaeolithic, crowded dentitions are rare finds in the archaeological record.

Reconstructing the Past

The Evolutionary History of Viruses

Every living virus belongs to a group or family of viruses that share a large amount of their DNA in common. Small changes (mutations) acquired by a virus in the past persist in all its descendants. More recent mutations only exist in the terminal species of a lineage, and some mutations will be unique to an individual virus in the family. Reconstructing genetic relatedness depends upon being able to sort every virus in a family (1) by the DNA sequences they all share in common, (2) by those they share only with a few viruses, and (3) by those that are unique to a single virus only. When the rate of mutation is fairly constant, a timescale can be applied to these genetic changes based on the number of mutations accumulated. A phylogeny, or phylogenetic tree, reconstructed in this way reflects the likely evolution of a viral family through time. The origin and evolution of human immunodeficiency viruses (HIV) has recently been tracked back to wild gorillas, rather than chimpanzees, using similar principles.[19]

Large DNA herpesviruses have a mutation rate similar to cellular DNA and a fairly robust phylogeny.[7] Their evolutionary history has been reconstructed from ~48 herpesviruses known today. They are an ancient group of viruses with representatives in all vertebrate groups (and at least one invertebrate). Between 200 to 100 million years ago the herpesviruses split into three tissue-trophic groups (the alpha neurotrophic and beta and gamma lymphotrophic groups). Each of these groups split again into two or more sublineages sometime before the mammalian radiation about 60 to 80 million years ago.[7] After this time they continued to co-speciate with their hosts. Alpha herpesviruses show an especially tight correlation with phylogenies of primates, ungulates, and carnivores.[7] Interestingly, oral HSV-1 and genital HSV-2 herpesviruses are quite specific to each region in modern humans. Monkey and great ape type 1 herpesviruses, however, infect both oral and genital regions equally. The split between these two distinct human herpesviruses appears to have occurred ~8 million years ago, which suggests these anatomical regions of the body became microbiologically isolated from each other around this time. This begs the interesting question why.[7] Perhaps it has to do with the evolution of bipedalism at that time among the earliest human ancestors, or perhaps with an interesting shift in sociosexual behaviour, or even both,[7] which may have evolved hand in hand.

Eliminating the *Herpes* virus in the same way as the smallpox virus or developing a vaccine effective against it may at first sight seem to be something that could only be advantageous.[20] All humans become infected with multiple *Herpes* viruses during childhood, and after the acute infection the virus enters a dormant state known as latency, which persists for the life of the host. However, *Herpes* virus latency confers symbiotic protection from bacterial infections through prolonged production of the antiviral cytokine interferon-gamma (IFN-gamma) and systemic activation of macrophages.[20] Latency upregulates the basal activation state of innate immunity against subsequent infections, and therefore has symbiotic immune benefits to the host that may be evolutionarily conserved.[20] These are likely to be especially important at mucosal and epithelial sites where *Herpes* reactivation occurs and where most pathogenic challenges initiate.[20] Evaluation of vaccines obviously needs to be carefully considered in a broad comparative biological and evolutionary context, as decreased infection rates may actually be associated with unintended negative consequences for vaccinated individuals. Only with some knowledge of the long evolutionary history of viruses and their more recent coevolution with humans can we expect to understand how to live alongside them.

The Origin of Modern Humans

The DNA of all living people contains a history of our genetic origins. Genetic mutations tend to occur at a fairly constant rate over time, and so a broad timescale of past events can be reconstructed in the same way as for viruses. People that share the same genetic mutation, or marker, share the same common ancestor. By comparing markers in many different populations, it is possible to trace ancestral connections. Mitochondria exist both in the cytoplasm of the egg and in the middle piece of the sperm. At fertilisation, however, the neck of the sperm, with the middle piece and flagellum, fails to penetrate the egg, such that we only inherit our mtDNA from our mother, which is then passed down intact to her children. The Y chromosome, on the other hand, is only passed on intact from fathers to sons. Comparing both mtDNA and Y chromosome distributions in many modern human populations gives us a good idea of when groups of humans parted ways during the great migrations around the planet.[21] Modern humans first appeared between 200,000 and 150,000 years ago in Africa. Today, women of African decent have twice as much genetic diversity in their mtDNA as any others worldwide, which indicates that modern humans must have lived in Africa for twice as long as anywhere else.[21] Many ancient ancestral marker genes are most common in the San and the Biaka. Even simple measurements of skull shape made on 4,666 skulls from 105 living populations show that phenotypic variance decreases over increasing distance from central and eastern Africa today, again indicating an African origin for all living modern humans.[22]

The fossil evidence suggests that some modern humans must have left Africa ~90,000 years ago. They appear to have gone no further than the Middle East before vanishing from the record.[23-26] The genetic evidence shows that a second wave of modern humans, maybe a group as small as only one thousand people, left Africa for Eurasia between 70,000 and 50,000 years ago. All non-Africans today are descendants of these people.[21,23] They carried only a few of the African marker genes with them. It seems through an initial wave of migration along coastal routes that some of these people reached the Andaman Islands, then Papua New Guinea, and finally Australia between 50,000 and 40,000 years ago. All western Eurasians appear to have emerged from a central Asian population that migrated northwest between 40,000 and 30,000 years ago, and a further major migration out of central Asia finally reached Southeast Asia, China, Japan, and eventually the Americas approximately 20,000 to 15,000 years ago.

Coevolution of Humans and Bacteria

About 50% of humans are infected with *Helicobacter pylori*, the gram-negative bacteria that lives happily in the acid environment of the stomach and which can cause peptic ulcers and constitutes a risk for stomach cancer. The pattern of genetic diversity of *H. pylori* decreases with geographic distance from

East Africa, and simulations indicate it must have spread from East Africa ~58,000 years ago.[27] In other words, this mirrors the genetic diversity among modern humans today. This old association between *H. pylori* and *H. sapiens* demonstrates that humans were already infected before their final migrations out of Africa, and that there has been coevolution of both ever since.

Another example of ancient coevolution in humans and a bacterium comes from the mouth and a bacterium involved in caries and periodontal disease. By 2 years of age infants have an oral flora that closely reflects that of their mother. Analysis of many samples of the *Streptococcus mutans* bacteria from diverse modern human populations shows that *S. mutans* also has its roots in Africa,[28] with a main branch extending to Asia, a smaller branch from that leading back to Europe, and additional branches from Asia to the Americas. *S. mutans* is another bacterium that has evolved alongside humans and where a clear line can be traced back to a single common ancestor that lived in Africa between 200,000 and 100,000 years ago.[28]

The Evolutionary Origins of Human Disease Patterns

Some of the most important human infectious diseases include those that could only have emerged within the last 11,000 years following the origins of agriculture.[29] For much of their evolutionary history, humans survived in low-density groups by foraging, hunting, and gathering. The domestication of wild animals and plants first occurred in and spread from the Levant along a fertile arc running along the base of the Taurus and Zagros mountains. Agriculture, followed by pastoralism, arose independently in many regions from China to the Americas, but while cereals, pulses, and tubers were domesticated in many places, animals were only really domesticated in large numbers in western Asia. Very few were domesticated in the Americas, and probably none in tropical Africa or Australia. The large human populations necessary to sustain many infectious diseases did not exist before this time. As more people became increasingly dependent on food production, villages, towns, and then cities arose. Acute crowd infections require at least 500,000 people to maintain them, something that only happened during the Neolithic.[7]

The human smallpox DNA virus, now extinct in the wild, has one of the largest and most complex viral genomes known. Infection with it followed an acute course in humans, which was short-lived and virulent.[7] It probably appeared in human populations around the time of the Neolithic, when one theory holds it may have evolved from an African gerbilpox virus.[7] Those who historically were lucky enough to survive the various plagues and pandemics that have repeatedly swept the world through large populations of humans may have passed on to their descendants some degree of immunity to those diseases. Not so lucky were those in the New World with no previous exposure to diseases of Old World origin when Europeans first arrived in the Americas.[29]

To this day different disease patterns exist in temperate and tropical regions.[29] Temperate diseases tend to be crowd epidemic diseases. Several hundred thousand people in a population are required to sustain, for example, measles, rubella, and whooping cough. Lots of diseases reached humans from animals, and large numbers of domestic animals were in close contact with humans from the beginnings of agriculture and domestication.[29] The greatest concentrations of domestic animals still occur in temperate zones, and this was especially the case early on. Many pathogens that once exclusively infected animals have in a comparatively short time become transformed into pathogens that now exclusively infect only humans.[29] Tropical diseases show a different pattern.[29] Fewer diseases originated from domestic animals because there were and still are fewer of them, but more are derived from primates, which only live in tropical zones. Animals act as reservoirs for disease, but the weakest species barriers exist between humans and non-human primates. Populations of humans in tropical zones were also traditionally smaller, and many diseases show a different pattern with a slow chronic course and with infection of others taking place over many months or years.[29]

The emergence of pastoralism has also shaped our recent evolution. Most people, with the exception of those of African or European origin, find the sugar in milk (lactose) indigestible after the time of childhood weaning.[30] The ability to drink milk after this time is a useful trait with a variety of origins among different pastoralist and semipastoralist populations. Lactase persistence results when a single gene fails to switch off. In northern European populations, this results from a small change in the "enhancer" region upstream from the lactase gene, but in other populations, for example, of African origin, a different mutation seems to keep the gene from being switched off.[30] It seems these small random mutations evolved relatively recently and spread quickly over the last 7,000 to 3,000 years, which implies there must have been strong directional selection that conferred a particular advantage. Lactase persistence may in the past have allowed people to stay alive by drinking milk in times of drought without the risk of diarrhoea, which would otherwise simply exacerbate dehydration.[30] Interestingly, this is a rare example of convergent evolution in modern humans, one that has apparently arisen for the same reason several times but through different genetic mutations.[30]

What Makes Modern Humans So Special?

For those of us inclined to want to lose weight, it comes as depressing news to hear that one has to walk 15 km to work off the calories in just one McDonald's Big Mac. Even if we choose to run that distance, it only takes the accompanying portion of fries to supply the necessary calories to do this.[31] Walking on two feet (bipedalism) is an extremely efficient way of getting around largely because of the way we do it. No other primates can lift one leg off the ground and keep their pelvis, trunk, and head level without tilting strongly to the other side to counterbalance this. The small gluteal muscles that lie over the hip joint allow us to walk with minimal displacement of our centre of gravity, and therefore minimal loss of energy.[32] When this goes wrong, so to speak, we become seriously disabled and cannot walk at all without showing a positive Trendelenburg sign, where one hip slumps awkwardly downward on the opposite side of the lesion as we try to raise a foot off the ground.

Energetically efficient bipedalism is the hallmark of all hominins (members of the zoological tribe to which modern humans belong[33]), and it evolved as long as 6 million years ago, perhaps as the earliest hominins started to travel over long distances. Endurance running appears to have evolved at a later stage in human evolution, ~2 million years ago, and yet again our anatomy transformed from a short-legged round-bellied ape-like form to our modern leaner, longer-legged body plan.[31] In particular, we have an elongated lumbar region, larger articular joint surfaces to deal with high impact loads, and long spring-like collagen-rich tendons in the legs, as well as some specialised anatomy designed to stabilise the head and neck and counteract the strong rotational torques of running.[31] Our physiology also ensures we are able to lose heat by being relatively hairless and through sweating and mouth breathing on exhalation.[31] We also employ other cranial cooling systems to keep the brain from overheating.[34] In short, surprising as it may seem to some of us, we are all by virtue of our evolutionary past pretty well adapted anatomically and physiologically to run marathons.

Quite late in human evolution, bigger brains relative to body size reached modern human proportions. Growing a big brain, and learning to use one, takes a long time but is a quintessentially human characteristic. The other hallmark of being a modern human that goes hand in hand with a big brain is a prolonged period of growth and development and a long life span.[35] Investing time and energy in raising offspring over a long time comes at a cost to any animal. Modern humans appear to have evolved ways to offset some of this cost that are reflected in our reproductive biology. Growing very slowly between the end of weaning and the onset of puberty costs families less in terms of food than if we grew faster over this period as, for example, chimpanzees do. Were we to grow like chimpanzees, it would cost traditional hunter-gatherers an extra 1,267–1,500 kcal per day to support the same number of offspring.[36] Our modern human adolescent growth spurt is the final catch-up phase that makes up for our slow period of growth. Growing up slowly is just one mechanism whereby we are able to produce more offspring. Another is grandparenting.[37] Modern humans are the only primates with a menopause after which females do not bear more children. However, the investment of older generations toward the rearing of children allows parents more time to provision and may be part of a package of modern human life-history attributes that ensures a high reproductive output despite the costs of growing and living so long.[35,37,38] Acknowledging the ways in which we have become unique as a species ought surely to frame the way we think about young and old people as well as the way we plan to treat and care for the young and the old.

Comparative and Evolutionary Biology and Problem-Based Teaching

I think there is now nowhere in the UK where the dental or medical curriculum contains a taught and examined component of comparative anatomy, physiology, or human biology. The human biology course and accompanying textbook written for first-year medical and dental students by J. Z. Young at University College London[39] was once a compulsory component of the core curriculum. Years ago, in many medical and dental schools students were also taught physiology and anatomy with a generous component of illustrative comparative material that probably stretched their curiosity and problem-solving skills to the limit, but may at the same time have sown seeds of inventiveness. Solving comparative problems was routine in tutorials, for example, with respect to respiratory physiology and anaesthetics. Questions such as how frogs breathe without a diaphragm or how swans can breathe at all with a dead space in the trachea more than three times greater than that expected for their body size were used as problem-solving vehicles to fix facts that were clinically crucial.[39] The whole animal kingdom seemed fair game in physiology tutorials,[40,41] and hours in my own preclinical years were spent debating how diving mammals occasionally acquire dental caries,[42,43] why camels do not burn their feet on the hot desert sand, and why it is advantageous for salmon to swim hundreds of miles to spawn upriver but then immediately die en masse.

There are lots of good things about new medical and dental undergraduate curricula, but there were also lots of good things about some of the old curricula. In our eagerness to move with the times, it would be a pity to dwell only on humans when we teach anatomy, physiology, and the pathology of diseases, and indeed a tragedy for future students to deny comparative and evolutionary biology a place in what we teach and how we teach it.[44] In concluding, I come back to my thesis, that by denying students an education about the many biological processes and mechanisms that take place over great spans of geological time, we are actually denying them an education in fundamental biological theory. Not only are we neglecting to provide them with an important framework for thinking through and solving biological and clinical problems, we are, I think, also contributing to the rising disbelief in evolutionary biology. What one has never been taught in context can never have any context. Today, this will have increasingly serious consequences for the way we try to understand modern humans. In the future, it will change the way we define and treat diseases.

References

1. Pagel, M. 2007. Selling evolution. *Nature* 447:533.
2. Gupta, S., et al. 2007. Catastrophic flooding origin of shelf valley systems in the English Channel. *Nature* 448:342.
3. Gibbard, P. 2007. Europe cut adrift. *Nature* 448:259.
4. Harrison, G. A., et al. 1988. *Human biology.* Oxford: Oxford Science Publications.
5. Scott, R. G., and Turner, C. G. 1997. *The anthropology of modern human teeth, dental morphology and its variation in recent human populations.* Cambridge Studies in Biological Anthropology. Cambridge: Cambridge University Press.
6. Poole, A., and Penny, D. 2007. Eukaryote evolution: Engulfed by speculation. *Nature* 447:913.
7. Van Blerkom, L. M. 2003. Role of viruses in human evolution. *Yrbk. Phys. Anthropol.* 46:14.
8. Garn, S. M., Lewis, A. B., and Vicinus, J. H. 1963. Third molar polymorphism and its significance to dental genetics. *J. Dent. Res.* 42 (Suppl.):1344.
9. Brothwell, D. R., Carbonell, V. M., and Goose, D. H. 1963. Congential absence of teeth in human populations. In *Dental anthropology,* ed. D. R. Browthwell, 179. Oxford: Pergamon Press.
10. Rozkovcova, E., et al. 2004. Agenesis of third molars in young Czech populations. *Prague Medical Report* 1:35.
11. Ambati, B. K., et al. 2006. Corneal avascularity is due to soluble VEGF receptor-1. *Nature* 443:993.
12. Marte, B. 2006. Developmental biology: Red-eye redirected. *Nature* 443:928.
13. Vartanyan, S. L., Garutt, V. E., and Sher, A. V. 1993. Holocene dwarf mammoths from Wrangel Island in the Siberian Artic. *Nature* 362:337.
14. Lister, A. M. 1993. Mammoths in miniature. *Nature* 362:288.
15. Lister, A. M. 1989. Rapid dwarfing of red deer on Jersey in the Last Intergalcial. *Nature* 342:539.
16. Brown, P., et al. 2004. New small-bodied hominin from the Late Pleistocene of Flores, Indonesia. *Nature* 431:1055.
17. Lister, A. M. 2001. Gradual evolution and molar scaling in the evolution of the mammoth. In *The world of elephants,* 648. Rome: International Congress.
18. Hillson, S. M. 1996. *Dental anthropology.* Cambridge: Cambridge University Press.
19. Van Heuverswyn, F., et al. 2006. Human immunodeficiency viruses: SIV infection in wild gorillas. *Nature* 444:164.
20. Barton, E. S., et al. 2007. Herpesvirus latency confers symbiotic protection from bacterial infection. *Nature* 446:326.
21. Horai, S., et al. 1995. Recent African origin of modern humans revealed by complete sequences of hominoid mitochondrial DNAs. *Proc. Natl. Acad. Sci. U.S.A.* 92:532.
22. Manica, A., et al. 2007. The effect of ancient population bottlenecks on human phenotypic variation. *Nature* 448:346.
23. Stringer, C. B. 2003. Out of Ethiopia. *Nature* 423:692.
24. Mellars, P. 2004. Neanderthals and the modern human colonization of Europe. *Nature* 432:461.
25. Mellars, P. 2006. New radiocarbon revolution and the dispersal of modern humans in Eurasia. *Nature* 439:931.
26. Templeton, M. 2002. Out of Africa again and again. *Nature* 416:45.
27. Linz, B., et al. 2007. An African origin for the intimate association between humans and *Helicobacter pylori. Nature* 445:915.

28. Caufield, P. W., et al. 2007. Population structure of plasmid-containing strains of *Streptococcus mutans*, a member of the human indigenous biota. *J. Bacteriol.* 189:1238.

29. Wolf, N. D., Dunavan, C. P., and Diamond, J. 2007. Origins of major human infectious diseases. *Nature* 447:279.

30. Check, E. 2006. How Africa learned to love the cow: The development of lactose tolerance in sub-Saharan Africa, a fascinating tale of genetic convergence. *Nature* 444:993.

31. Bramble, D. M., and Lieberman, D. E. 2004. Endurance running and the evolution of *Homo*. *Nature* 432:345.

32. Aiello, L., and Dean, C. 1990. *Human evolutionary anatomy.* London: Academic Press.

33. Wood, B. A. 2005. *Human evolution, a very short introduction.* Oxford: Oxford University Press.

34. Dean, M. C. 1988. Another look at the nose and the functional significance of the face and nasal mucous membrane for cooling the brain in fossil hominids. *J. Hum. Evol.* 17:715.

35. Dean, M. C. 2006. Tooth microstructure tracks the pace of human life-history evolution. *Proc. R. Soc. B* 273:2799.

36. Gurven, M., and Walker, R. 2006. Energetic demand and the evolution of slow human growth. *Proc. R. Soc. B* 273:835.

37. Hawkes, K., et al. 1998. Grandmothering, menopause and the evolution of human life histories. *Proc. Natl. Acad. Sci. U.S.A.* 95:1336.

38. Dean, C. 2007. Growing up slowly 160,000 years ago. *Proc. Natl. Acad. Sci. U.S.A.* 104:6093.

39. Young, J. Z. 1971. *An introduction to the study of man.* Oxford: Oxford at the Clarendon Press.

40. Schmidt-Nielsen, K. 1972. *How animals work.* Cambridge: Cambridge University Press.

41. Schmidt-Nielsen, K. 1979. *Animal physiology: Adaptation and environment.* 2nd ed. Cambridge: Cambridge University Press.

42. Ness, A. R. 1966. Dental caries in the platanistid whale *Ina geoffrensis. J. Comp. Pathol.* 76:271–78.

43. Miles, A. E. W., and Grigson, C. 1990. *Colyer's variations and diseases of the teeth of animals.* Cambridge: Cambridge University Press.

44. Harris, E. E., and Malyango, A. A. 2005. Evolutionary explanations in medical and health profession courses: Are you answering your students' "why" questions? *BMC Med. Educ.* 5:1.

12

Is There a Place for Evolutionary Medicine in UK Medical Education?

Sarah Elton
Functional Morphology and Evolution Unit, Hull
York Medical School, University of Hull

Paul O'Higgins
Functional Morphology and Evolution Unit, Hull
York Medical School, University of York

Contents

Introduction

The new synthesis of evolution[1] and recent developments in evolutionary developmental biology[2] underpin modern biological sciences. However, there is relatively little emphasis on evolution in UK medical curricula,[3,4] despite the fundamental role of evolutionary theories and cvoncepts in the basic medical sciences of anatomy, physiology, and biochemistry. The limited use of evolutionary ideas in medical teaching is even more surprising when one considers their power in explaining and understanding some of the major issues faced by clinicians in their work on a day-to-day basis, such as antibiotic resistance, the obesity epidemic, and subfertility. If medicine is to recognise the relevance of evolutionary principles—the benefits and disadvantages of which are outlined in a number of chapters in this volume—it is logical that the foundations of evolutionary thinking are established through the initial education and training received by medical students.

Medical education in the UK, in which students follow a 5-year undergraduate degree, often (but by no means exclusively) straight after secondary school, is very different from the graduate entry model of the United States. We focus on the UK system in this chapter, although many of the general arguments and ideas presented are applicable to medical education in the United States, mainland Europe, and elsewhere. In the past decade, UK medical schools have experienced significant shifts in the emphasis and content of their curricula. Few now adhere rigidly to the preclinical and clinical divisions that once characterised undergraduate medical degrees. The General Medical Council (GMC), who regulates UK medicine and medical education, advises that clinical contact should occur early in the undergraduate course and that science should be covered in the context of clinical cases.[5,6] Inevitably, this has led to some traditional areas of medical education, such as biochemistry, anatomy, and physiology, being de-emphasised, while others, like communication skills, have become more prominent. As always, the 5-year medical curriculum has to resolve many competing demands on the timetable and student workloads to ensure that graduates are sufficiently knowledgeable and skilled to pursue a career in medicine and adapt to sometimes rapidly changing theories and practice. The net result of this is that there is less didactic, classroom-based teaching, less factual information, more emphasis on enhancing skills of independent learning, and greater focus on the development of clinical skills and attitudes.

Evolutionary biology was never a cornerstone of medical education, although where relevant it was introduced by some, if not all, teachers. Despite its increasingly obvious applicability in some areas of modern medicine (see Chapters 4, 5, 7, and 10), the continuing absence of evolutionary biology can be linked in part to the profound shifts in UK medical education that have occurred over the past decade. Specifically, in making the basic medical sciences more peripheral, underpinned by clinical issues rather than scientific method, and with less overall depth of content, the natural homes for discussions of selection, adaptation, descent with modification, and stochasticity have largely been removed. This makes the task of introducing evolutionary principles even more difficult than before.

Downplaying the role and importance of evolution is not restricted to UK medical education, however, and reorganisation of the curriculum does not comprehensively explain why evolution is often disregarded. Indeed, a recent study has demonstrated that the word *evolution* is used only infrequently in studies of antibiotic resistance published in international biomedical journals.[7] This may well indicate that evolution is seen as being tangential to medicine.[7,8] Evolution is also undoubtedly controversial, with the passage of time since Darwin's initial work apparently failing to dull scepticism about natural selection in certain sectors of the population. For example, it has been reported that between 4 and 11% of biology undergraduates and 10% of medical students at the University of Glasgow in the late 1980s and 1990s did not accept the occurrence of evolution in biology.[3,4,9] The situation in the United States is potentially even more serious, with relatively widespread resistance to evolutionary theory, either through total rejection of evolutionary concepts or through a lack of understanding about their general importance.[10] The well-publicised battles over the place of evolution in U.S. public school syllabuses and the small but significant movement toward teaching so-called alternatives, such as Intelligent Design, in UK schools show that educating people about evolution is not straightforward or likely to be universally well received. In an already crowded undergraduate medical curriculum, which satisfactorily trains medics from a proximate perspective, potential for the negative reception of evolutionarily focused courses is likely to be another reason why there has been no widespread integration of evolution into medical education in either Europe or the United States.

In this chapter we consider if and how evolutionary thinking about health and disease could play a role in medical education in the UK. We discuss how the UK curriculum has changed and what effects this might have on opportunities for bringing evolutionary thinking to the understanding of human variation and disease. Subsequently, we consider the constraints that arise because of the clash between some religious beliefs and evolution itself. Finally, we suggest ways in which evolution might be made more prominent in the UK curriculum, notwithstanding the problems and challenges faced in doing so.

Education versus Training in UK Medical Schools: Prospects for the Integration of Evolutionary Approaches into the Current Medical Curriculum

Traditionally, the undergraduate medical degree in the UK was fact and workload heavy. Students followed a curriculum that comprised 2 or 3 years of basic science education followed by a further 3 years of mainly hospital-based clinical training. They often complained that the initial years were too focused on basic science and demanded a more clinically relevant curriculum. In *Tomorrow's Doctors*,[5,6] a landmark document that heralded massive changes to the way medical students in the UK are educated and trained, this burden of factual information was highlighted, the implication being that students were bombarded with too much information that was not deeply learned and hence rapidly forgotten. The response was to embed what are termed basic sciences within a much more clinically orientated curriculum, as well as to promote skills of self-directed learning. Educators came to agree with and support the widespread student opinion that a medical course should focus principally on a core of knowledge, skills, and attitudes related directly to medical practice.[11] In so doing, the emphasis moved toward the skills and tasks relevant to clinical practice and away from in-depth study of scientific methodology and the scientific underpinnings of medicine. Inevitably, these changes have caused a shift from an educational process that exposed students to scientific concepts and their development toward a learning environment in which training for tasks has a central role, justi-fied on the basis that too many facts are forgotten and often serve no useful purpose in and of themselves. This ignores the possibility that the valuable part of the study of science was never the memorising of facts, but rather the educational experience of having to understand how scientific ideas arise and are tested and how we come to know things. We argue that this latter element, fundamental to a curriculum in which evolutionary approaches to medicine could play a central role, is a key part of the education as opposed to the training of doctors.

It can be extraordinarily difficult to truly demarcate education and train-ing, and the two terms are often conflated in common usage. Generally, edu-cation refers to gaining knowledge and understanding concepts, whereas training equips people with specific skills. In the health care arena, training has been defined as "a fixed process with a defined beginning and end," whereas education is "an holistic continuous and ongoing development of the individual"[12] (p. 379). From these different but not mutually exclusive definitions, it could be argued that the medical school curriculum comprises a patchwork of education and training, in which the acquisition of specific skills (for example, clinical examination or taking a history) that usually take a finite amount of time to learn and perfect are coupled with explor-ing and understanding theories and underlying concepts through a process that should involve deep learning with no fixed endpoint. It is commonly perceived that there is a conflict between the ideals of education and the pur-pose of training, with training often seen to be authoritarian (i.e., not driven by the interests and curiosities of the learner or the trainee) and narrow in scope.[13] However, training does not necessarily result in a lack of in-depth understanding of a particular topic, and a strong case can be made for the benefits of integrating training and education.[13]

Encapsulated within *Tomorrow's Doctors* is the fundamental idea that medi-cal education should provide a future foundation for learning in practice and beyond. Modern doctors require the ability to evaluate ideas and data, and must develop skills in critical thinking and reasoning that are generic rather than specifically targeted to particular tasks. In this aspect, undergraduate medical degrees do (or should) not differ from most other degrees. In addi-tion to the educative part of the medical curriculum, which equips doctors with generic skills of reasoning and understanding, the training component provides the technical skills that could not be acquired by, for example, an undergraduate in history or even bioscience. This theoretically results in the integration of education and training in medical schools, with increasing attention on continued professional development in postregistration medical practice potentially acting to reinforce the importance of both aspects at the undergraduate as well as the postgraduate level. However, our feeling is that the main purpose of medical schools as seen by the public, some medical students, most practitioners, and a large proportion of people working within them is to provide for the basic training of doctor-practioners who will be functional within the medical economy rather than to produce educated doctor-thinkers who can challenge it. We are not arguing here that training is less valuable in medical schools than education. Indeed, each approach is necessary and the two should be integrated. Nonetheless, we believe that the main emphasis is on training ("I'm training to be a doctor" rather than "I'm being educated to practice medicine"), and as a result, the value and importance of the educative aspect of the medical curriculum is diminished in the eyes of many participants. Those interested in bringing evolutionary thinking into medical education are therefore swimming against the tide of current opinion.

Playing devil's advocate, it is possible that within UK undergraduate medical courses training is actually more important than education. On graduating from medical school, most doctors undergo a very tightly defined specialist medical training and then spend the rest of their careers diagnosing and treating diseases within a single area. Thus, one might seriously suggest that an undergraduate training programme that equips doctors to communicate with patients and to follow diagnostic and therapeutic algorithms informed by evidence-based medicine is actually more appropriate than a broader medical education in the liberal tradition. Marks (chapter 13) equates doctors to plumbers or car mechanics, whose job it is to "diagnose a problem and to effect, or at least to recommend, a solution to it." Plumbers and car mechanics are usually trained in specific skills rather than educated within a broader intellectual tradition of household or vehicle maintenance. We admit to being deliberately provocative by taking Marks's analogy one step further and observing that if doctors are skilled technicians or "body mechanics," they do not need to be educated.

There is also an economic argument for the training rather than the education of doctors. It is clearly (and we argue unfortunately) possible to produce graduates simply by training them in sets of economically and practically useful skills. Indeed, there is real concern in the academic community that this is becoming more prevalent.[14] The National Health Service (NHS) requires large numbers of practitioners able to follow appropriate algorithms and conduct their work within its clinical and administrative structures. It thus might make sense to produce these practitioners as "efficiently" as possible, efficiency here being framed in terms of lower cost for the state rather than on the basis of any benefit to the individual student. Following this factory model, it is not necessary to bring anything to the attention of medical students that does not directly address the needs of the working clinic or hospital. Understanding evolutionary approaches to medicine would thus fall outside the remit of medical training colleges. Such a model may be desirable from an economic perspective that focuses entirely on the provision of services, including health care. It may not be desirable for the personal development of the individual. A liberal education emphasises the benefits of curiosity and developing human knowledge[15] as well as instilling and reinforcing values of autonomy, responsibility, and citizenship.[16–18] A training programme does not require these ideals and, unlike university education, has no ultimate aim to prepare individuals in ways that may progress knowledge. Thus, training alone is not desirable if medical doctors are to challenge and advance medical-scientific understanding of health and disease at a fundamental level that may eventually impact on practice.

The picture we draw gives a jaundiced, extreme perspective on current trends in medical education that have fortunately not yet been fully realised. It is probably fair to say that modern UK medical curricula still encourage the deep learning of some science as well as acquisition of skills. However, it is increasingly common for medical students to study science in the same way that a beginner in a foreign language class starts to become familiar with novel words and grammatical constructs. The superficial learning of vocabulary, analogous to discrete medical facts such as functional cascades of enzymes or the structure of DNA, is not the same as the development of deeper knowledge that can be applied to a range of topics in a variety of circumstances. This "shopping list" approach might be reinforced by problem-based learning (PBL), popular in several UK medical schools, which may fail to promote the understanding of basic science.[19] At our own institution, the PBL process hinges on a number of virtual patients that provide triggers for the week's learning. The intention is that basic medical sciences are covered in the context of clinical issues, and PBL plus clinical placements are the hub of the curriculum, integrating its various aspects. While this approach is effective in preparing students for the complexities of clinical problem solving and practice, it largely fails to provide deep engagement with scientific method or theory. Indeed, a recent meta-analysis of PBL outcomes[20] showed a slight negative effect for PBL in relation to the acquisition of concepts.

Nonetheless, integration has widespread support in UK medical schools, following the strong suggestion by the GMC that medical courses should integrate disciplines and bring clinical aspects of training into the whole curriculum. The key here is that the GMC recommends that clinical practice and experience be the integrating force, rather than any basic scientific principle. This is not at all surprising given the current focus of undergraduate medical education and training in the UK. However, one consequence of this is that there are no closely argued and tested bodies of knowledge (disciplines), such as biochemistry, physiology, or anatomy, that underpin the curriculum. Instead, there is a mish-mash of facts and ideas, brought together and structured only by the clinical case used to integrate them. Thus, the historical and intellectual arguments (which are often evolutionary) that led to the construction of certain ideas and which now form the basis of practice are neglected. An evolutionary framework could be used to fill this intellectual void: evolutionary medicine introduces the underlying principles of many of the basic science topics that doctors must understand,[21] allows integration of several aspects of the medical curriculum, and hence could provide a means by which deep, as opposed to surface, learning is promoted.

Making space in the first 2 years of medical school for the clinical work and patient contact that integrates the curriculum has been achieved almost entirely by the removal of the basic science that was once regarded a necessary part of the educational experience for future doctors. At this time, therefore, there is no home in the core UK curriculum for evolutionary medicine. This situation is exacerbated by time constraints that mean the acquisition of "necessary facts" and competences in "essential skills" for practice take priority over deepening and broadening educational experiences such as the holistic study of medicine from the perspective of individuals, populations, and societies woven together by their common evolutionary thread. Surface learning, although not a new phenomenon when faced with the challenges of a medical course,[22] remains a mainstay of the student experience. However, current trends in this direction are mitigated to some extent by the emphasis on developing skills of lifelong and self-directed learning, which have been shown to promote deep learning.[22]

Another compensation for the move away from basic science education toward training in skills that have clinical relevance is the incorporation of student-selected components (SSCs) within the curriculum. SSCs (also termed special study modules [SSMs] in some medical schools) currently comprise one-quarter to one-third of timetabled and self-directed learning time and "allow students to study in depth areas of particular interest to them"[6] (p. 3). Two-thirds of SSCs should be related to medicine, broadly defined, so theoretically students have considerable scope not only to broaden and deepen their clinical learning, but also to take options from a wide range of disciplines. At our own institution, for example, students can study topics ranging from languages and literature to human evolution and computer programming. Many of these are run by tutors from outside the medical school, and the vast majority have an educational focus rather than being designed to equip students with particular clinical skills. It is in this area of the curriculum that modern UK medical students are most likely to get a taste of a liberal education.

SSCs (and potentially, by extension, intercalated degrees) offer a clear opportunity to bring evolutionary thinking about health and disease into undergraduate medical education. In a recent survey of UK medical schools, fewer than half of the respondents incorporated evolution into their curricula, and it has thus been recommended that evolutionary biologists work to develop courses in evolutionary medicine for undergraduate medical students.[3] Downie[3,4] suggests that the core medical curriculum should include the basic principles and concepts of evolutionary theory, with SSCs providing additional in-depth courses for those students who are especially interested. This approach is laudable but may not result in the widespread integration of evolutionary thinking and medical education. Despite their huge potential,[23] in practice SSCs are frequently piecemeal bites of learning undertaken in short bursts that can fail to encourage the long sequence of learning that leads to deep understanding. Some medical schools[24] recognise this and make the acquisition of skills the main focus of SSCs, thus rendering them functional extensions of the core. Despite being a convenient vehicle for topics that are not addressed adequately in the core curriculum, sequestering the bulk of concepts relating to evolutionary medicine into SSCs would send out the message that this way of thinking is marginal and can be contained, separate from the main body of medical studies. This is clearly not the case, since evolutionary explanations facilitate the understanding of many important medical phenomena ranging through body structure and function, reproduction, infection, cancer, mental illness, and more. Indeed, evolutionary phenomena are so pervasive that, as argued above, they offer the opportunity to build links between and integrate many aspects of medical training.

Evolutionary Anatomy in the Medical Course

Confronted with the reality of most modern UK medical curricula, those advocating a move toward a medical education in which evolutionary thinking is a key element have a dilemma. Evolutionary theory and concepts have to compete with other areas of basic science for time. In arguing for the incorporation of evolutionary aspects of medicine into the core curriculum it is necessary to make a case that these are more essential than the science components that have been trimmed from previous curricula. The extent to which they are eventually incorporated depends to a large degree on their perceived relevance, the expertise and interests of faculty, and whether they are introduced as stand-alone SSCs or woven into core science teaching. However, it is uncertain whether this is achievable at all—although it seems logical that neo-Darwinism and evolutionary developmental biology, as basic foundations of modern biological science, could truly integrate the multiple strands of medical education, current students, staff, and policy makers seem more than content with the focus on the clinic as the integrating factor.

So would it be more realistic to attempt an evolutionary approach in a single discipline within medicine? Although many advances in genetics might provide a suitable vehicle for the teaching of evolutionary principles to medical students, we will focus here on our own discipline, anatomy. Anatomy is still a significant component of the undergraduate medical curriculum, and anatomical structure readily lends itself to the concept of adaptation and hence evolution. In the UK as well as the United States there is a long tradition of evolutionary research in anatomy departments. Biological or physical anthropology is a relatively young discipline, and for many years anatomists in medical schools undertook the comparative anatomical and functional morphological studies that were necessary to make sense of our anatomy, the human fossil record, and our own evolutionary origins. Thus, there remains a ready-made expertise in evolutionary theory and concepts in anatomy departments that is not always as obvious in some other disciplines, such as modern physiology and biochemistry. Two of the criteria we mention above as being necessary for the incorporation of evolutionary approaches—perceived relevance and interest and expertise of staff —are therefore in place. However, the third factor we mention—whether evolutionary approaches can be introduced as part of the core—is not as easy to resolve.

In older UK medical curricula, anatomy often took the central position in the first 2 years of medical school. It was taught as a series of information points about the topography and relations of structures within the body. This in itself is a fairly trivial but interesting pursuit: the naming of parts, the statement of where they are in relation to each other, and a description of their blood supply and nerve supply can all be compared to the learning of the vocabulary of a language without the literature associated with it. In some medical schools a limited amount of literature was included in the curriculum, in the form of relationships between structure and function in clinical practice. This makes for better training and ultimately more effective application in clinical practice but does not necessarily inform students about fundamental organising principles. Rote learning does not encourage students to synthesise ideas, which has been identified as a general problem in the teaching of evolutionary principles to many undergraduates,[25] not just medical students. For undergraduate medics learning anatomy, such difficulties might be heightened, given that it is not only expected but actually encouraged, for example through the use of mnemonics, that they memorise facts, thus precluding true understanding of anatomy in a way that might allow future advances or stimulate novel insights.

Clinical application is not the "literature of anatomy"—the literature is the study of how anatomical structure comes to be. By understanding the evolution and development of anatomical structures it is possible to work out from first principles why things are arranged in certain ways and therefore avoid a large amount of rote learning. The negative aspect of this is that compared to rote learning, substantially more effort is required to understand the nature of and influences on anatomical structure, both proximate and ultimate. The proximate influence is the ontogeny of the individual from the fertilised ovum, through adulthood to senescence and death. Of particular interest to anatomists is how and why a single fertilised ovum undergoes the transformation that results in complexity of body form and neural structure. This science of developmental biology has made enormous strides in the last 25 years with the discovery of genetic mechanisms that regulate the patterning and subsequent growth of the embryo and foetus that eventually generate the mature body form. The explanations for this patterning lie in deep understanding and knowledge of genetics and proteins in a world of biochemistry and cellular and molecular biology. However, these explanations in and of themselves are still insufficient, given that ontogenies are evolved. Natural selection over many generations favours particular ontogenetic trajectories leading to particular adult morphologies.

Anatomists wanting to educate students on the structure of, for instance, the upper limb are faced with several alternative ways of approaching the topic. The first is to simply take the students through a Hayne's manual* of anatomy, a method that has much in common with most anatomy textbooks. A second option is to take a deeper approach; rather than teach students that the median nerve innervates the muscles of the flexor forearm compartment and ventro-lateral hand, educators place the structure and function of the forelimb in the context of the transformations that have occurred during phylogeny and ontogeny. This empowers students by providing an explanation of the structures they observe, and allowing them to work out key aspects of anatomy from first principles. This is a deeply satisfying pursuit that encourages deep learning. However, it is questionable whether this approach is economically sensible and efficient, given the pressures on the curriculum and the actual requirements of medical practice. Hence, the evolutionary developmental basis of anatomy is marginalised in favour of a more functional, economically justifiable, and practical rote learning of facts. These issues were sufficient to inhibit deep study of anatomy in most medical schools before the curricular revisions engendered by *Tomorrow's Doctors* and are even more pressing now. As palaeoanthropologists and comparative morphologists, we would love to teach anatomy from an evolutionary perspective. However, as teachers in a UK medical school (and as described in this chapter), the practical opportunities for this are slim, despite our involvement in curriculum design as well as delivery from the inception of our institution.

Religion and Evolutionary Medicine

The emphasis on clinical issues as the integrating factor in UK medical curricula and time pressures on students, staff, and learning alike are not the only reasons why there is little support for a medical education based on evolutionary ideas. Although less pervasive in public life than in the United States, religious beliefs do influence what occurs in UK society; recent examples include governmental policy encouraging state-funded faith schools, religiously inspired terrorist attacks allegedly carried out by medical doctors, and the heated opposition to adoption rights for homosexual couples. Several studies[3,4,9,26,27] have indicated that religious views are a key factor in the individual's dismissal of evolution or belief in creationist ideas. In a sur-

---

* Hayne's manuals are popular in the UK and give a simple step-by-step pictorial guide to the maintenance and repair of motor cars without providing any explanation of how vehicles actually work.

vey conducted at the University of Glasgow in 1999, 10% of medical students did not accept evolution.[9] A creationist mind-set, in which humans are seen as separate from nature, has also been argued to be common among practicing medics.[28] We argue that religious beliefs affect what is taught in UK medical schools and how topics are presented, with evolutionary theory and concepts being obvious casualties of this.

In our experience, students over the last few years have tended to become more vocal in expressing their religious views where they conflict with what is taught (see chapter 11). Indeed, there are suggestions that medical students are increasingly withdrawing from parts of their courses that conflict with religious teachings.[29] However, data from the University of Glasgow indicate that the proportion of first-year students rejecting the idea of Darwinian evolution remains relatively constant from year to year.[3,4,9] It thus appears, albeit anecdotally, that rather than there being an increase in the number of students with religious views that lead them to reject evolution, there is a greater willingness to challenge whether evolution should be taught at all. This is not to say that most people of faith reject evolution or, even if they do, object to it being a part of the curriculum. However, an increasingly prevalent argument among those that design and deliver medical curricula, and one that appears to persuade many medical teachers, is that since evolution is not a necessary part of the training of doctors, there is no reason to "rock the boat" by ramming it down students' throats. This worryingly echoes experiences of biology teachers in the United States, who avoid teaching evolutionary theory and concepts because they are afraid of negative responses from parents or administrators.[27] It also further encourages religion-inspired censorship of the single underlying unifying principle of biology and medicine—evolution.

Should we even be raising the subject of religion in a discussion of the role of evolutionary medicine in UK medical curricula? After all, UK society (for the greater part) has travelled a long way from the Victorian outcry against Darwin's writings, and religious belief is not incompatible with balanced, objective scientific inquiry and does not automatically imply rejection of evolution. However, it has been noted that many academics have appeared unusually reticent in exploring the attitudes of the student body to religion, evolution, and creationism.[9] This suggests that academics are imposing upon themselves a type of censorship—possibly due to concern about accusations of intolerance, or alternatively because of apathy about how students interact with their discipline—over a topic that is actually fundamental to the understanding of numerous biological processes. It has been very strongly argued that medical researchers have a duty to promote the fact that far from being esoteric and irrelevant to normal life, evolution is a process that impinges upon the lives and health of people on a daily basis.[7] It has been equally strongly argued that in medical students the rejection of evolutionary theory on the basis of religious belief is not only at odds with the pursuit of science (and medical schools still emphasise the importance of science in their application processes even if basic science is encased within a clinical wrapper in the new curriculum) but also worrying in a supposedly evidence-based discipline.[3,4,9] Thus, as academics who work within the paradigm of neo-Darwinism, we should not be afraid to teach evolution in a medical context or medicine in an evolutionary context. We should also challenge contentions that evolution is peripheral and therefore expendable in an atmosphere of increasingly vocal student representation concerning the design, content, and delivery of university syllabuses and curricula.

Conclusion: Use of Evolutionary Medicine in Medical Education

The chapter has outlined some of the problems and challenges faced by those who would like to see evolutionary theory as the integrating factor in modern UK medical curricula. It is probably not overstating the case to say that this is unlikely to become a reality, now or in the future, however much we need to emphasise the importance of evolution. But this does not automatically mean that there is no place for evolutionary medicine within medical schools. Day,[30] an anatomist clinician who was also one of the foremost palaeoanthropologists of his generation, has argued that to be truly effective, the medics of the future will need to be fully conversant with evolutionary ideas. It is encouraging that evolutionary medicine has been embraced by at least some practicing clinicians not only as a framework for generating research and audit questions, but also as a way to engage with patients (see, for example, chapters 5, 6, and 10). The potential utility and applications of evolutionary medicine have even been highlighted in veterinary practice.[31] Nonetheless, it has been recognized for several years that there is a real lack of engagement with evolutionary ideas in medical teaching as well as research.[7,8,32] So how might we go about introducing (or reintroducing) evolution into medical curricula? The "foot in the door" principle may be useful here. Rather than opting for the top-down, integrative approach that is likely to prove frustrating and fruitless, it could be more rewarding and productive to introduce evolutionary ideas into the core curriculum as and when opportunities present themselves. Harris and Malyango[33] and Dean (chapter 11) give useful examples and case studies of where and how this can be achieved—obvious ones include malaria and heterozygote advantage for the sickle cell gene, and antibiotic resistance, both of which could easily be incorporated into PBL sessions, tutorials, or lectures. In our own teaching, we introduce evolutionary ideas into the core curriculum through plenary (lecture) sessions on human locomotion and obesity. This opportunistic strategy should also avoid the pitfall of stretching evolutionary interpretations of some conditions beyond the realms of what is testable or plausible, a common criticism of evolutionary medicine.[28,34]

In addition, a real strength of UK medical curricula as outlined in *Tomorrow's Doctors* is the opportunity for students to broaden their educational horizons and augment their intellectual skills through SSCs. Although relying on SSCs for the teaching of evolution is not necessarily the optimal or even a sustainable strategy, they could be exploited much more by basic scientists as well as interested clinicians to encourage students to think about the evolutionary context of biology in general and medicine in particular. Our own experiences, as well as those reported elsewhere,[3,4,9] indicate that SSCs on evolution are chosen by students, and those who undertake them engage with the material and enjoy the experience. There is also the option for students who are very interested in evolution to intercalate in a subject concerned with evolutionary processes, such as evolutionary anatomy, medical anthropology, or biological anthropology.

To conclude, making evolution a more prominent part of UK medical curricula is a challenge but not an insurmountable one. Resistance may be felt from at least two quarters, the vocal minority (both students and staff) who either do not accept evolution or who think it is irrelevant to modern humans, and the die-hard medical educators who believe that the only meaningful way to educate and train doctors is through clinical experience and examples. However, there is a reasonable proportion of academics and teachers in medical schools who, although largely silent to date, would support a greater emphasis on evolution and who would also be very much in favour of encouraging deeper learning and understanding. We accept that a combination of sporadic evolutionary teaching in the core curriculum plus preaching to the converted (or at least the interested and open-minded) through SSCs and intercalated degrees is not likely to result in the wholesale adoption of evolutionary medicine by the UK medical profession. However, judging by the experiences of educators elsewhere,[10] it will keep evolution on the educational agenda, will help to nurture and encourage those with a true interest in the subject, and although it may not spawn a generation of Darwinian clinicians, will certainly help to breed some medics with a theoretical and applied interest in evolutionary medicine that informs their work and research.

Acknowledgements

We are grateful to the three reviewers for their thoughtful comments and suggestions.

References

1. Huxley, J. S. 1942. *Evolution: The modern synthesis.* London: Allen and Unwin.
2. O'Higgins, P., and Cohn, M. 2000. *Development, growth and evolution: Implications for the study of hominid skeletal evolution.* London: Academic Press.

3. Downie, J. R. 2004. Evolution in health and disease: The role of evolutionary biology in the medical curriculum. *Bee-j* 4. http://www.bioscience.heacademy.ac.uk/journal/vol4/beej-4-3.pdf. Accessed July 23, 2007.

4. Downie, J. R. 2004. Evolutionary biology. *Lancet* 363:1168.

5. General Medical Council. 1993. *Tomorrow's doctors*. London: GMC.

6. General Medical Council. 2003. *Tomorrow's doctors*. London: GMC.

7. Antonovics, J., et al. 2007. Evolution by any other name: Antibiotic resistance and avoidance of the E-word. *PloS Biol.* 5:e30.

8. MacCallum, C. J. 2007. Does medicine without evolution make sense? *PloS Biol.* 5:e112.

9. Downie, J. R., and Barron, N. J. 2000. Evolution and religion: Attitudes of Scottish first year biology and medical students to the teaching of evolutionary biology. *J. Biol. Educ.* 34:139–46.

10. Sloan Wilson, D. 2005. Evolution for everyone: How to increase acceptance of, interest in, and knowledge about evolution. *PloS Biol.* 3:e364.

11. Harden, R. M., Laidlaw, J. M., Ker, J. S., and Mitchell, H. E. 1996. AMEE medical eductaion guide number 7: An education strategy for undergraduate, postgraduate and continuing medical education 2. *Med. Teach.* 18:91–98.

12. Cooper, S. 2005. Contemporary UK paramedical training and education. How do we train? How should we educate? *Emerg. Med. J.* 22:375–79.

13. Winch, C. 2000. Education and training. In *Educational issues in the learning age*, ed. C. Matheson and D. Matheson, 116–28. London: Continuum.

14. Lomas, L. 1997. The decline of liberal education and the emergence of a new model of education and training. *Educ. Train.* 39:111–15.

15. Association of American Colleges and Universities (AACU). 1998. Statement on liberal learning. http://www.aacu.org/About/statements/liberal_learning.cfm (accessed July 23, 2007).

16. Strike, K. A. 1999. Liberalism, citizenship and the private interest in schooling. In *The aims of education*, ed. R. Marples, 50–60. London: Routledge.

17. Winch, C. 1999. Autonomy as an educational aim. In *The aims of education*, ed. R. Marples, 74–84. London: Routledge.

18. Hirst, P. H. 1999. The nature of educational aims. In *The aims of education*, ed. R. Marples, 124–32. London: Routledge.

19. Vernon, D. T. A., and Blake, R. L. 1993. Does problem-based learning work? A meta-analysis of evaluative research. *Acad. Med.* 68:550–63.

20. Gijbels, D., et al. 2005. Effects of problem-based learning: A meta-analysis from the angle of assessment. *Rev. Educ. Res.* 75:27–61.

21. Nesse, R. M., and Williams, G. C. 1995. *Evolution and healing: The new science of Darwinian medicine*. London: Weidenfeld and Nicholson.

22. Dacre, J. E., and Fox, R. A. 2000. How should we be teaching our undergraduates? *Ann. Rheum. Dis.* 59:662–67.

23. MacNaughton, R. J. 1997. Special study modules: An opportunity not to be missed. *Med. Educ.* 31:49–51.

24. Stark, P., et al. 2005. Student-selected components in the undergraduate medical curriculum: A multi-institutional consensus on assessable key tasks. *Med. Teach.* 27:720–25.

25. Moore, R. et al. 2002. Undergraduates' understanding of evolution: Ascriptions of agency as a problem for student learning. *J. Biol. Educ.* 36:65–71.

26. Short, R. V. 1994. Darwin, have I failed you? *Lancet* 343:528–29.

27. Moore, R. 2000. The revival of creationism in the United States. *J. Biol. Educ.* 35:17–21.

28. Ebert, D., and Sokolova, N. V. 2001. Morning has broken: Ten years after the dawn of evolutionary medicine. *J. Evol. Biol.* 14:194–96.

29. Foggo, D., and Taher, A. 2007. Muslim medical students get picky. *Sunday Times,* October 7. http://www.timesonline.co.uk/tol/news/uk/health/article2603966.ece (accessed October 8, 2007).

30. Day, M. 1999. Foreword. In *Evolutionary medicine,* ed. W. R. Trevathan, E. O. Smith, and J. J. McKenna, eds. Oxford: Oxford University Press.

31. LeGrand, E. K., and Brown, C. C. 2002. Darwinian medicine: Applications of evolutionary biology for veterinarians. *Can. Vet. J.* 43:556–59.

32. Lee, M. 1995. Evolution and healing. *Lancet* 346:686.

33. Harris, E. E., and Malyango, A. A. 2005. Evolutionary explanations in medical and health profession courses: Are you answering your students' 'why' questions? *BMC Med. Educ.* 5:16. DOI: 10.1186/1472-6920-5-16.

34. Bull, J. J. 1995. (R)evolutionary medicine. *Evolution* 49:1296–98.

13

Would Darwin Recognize Himself Here?

Jonathan Marks
Department of Anthropology,
University of North Carolina at Charlotte

Contents

Introduction

One of the things you can never escape in talking about Darwinism is that it comes in so many forms. And it is their underlying assumptions, or epistemes, that not only differentiate those forms of Darwinism from one another, but which make them interesting as bodies of knowledge. For the present context, then, which evolutionary theory should I invoke in order to make medical sense?

Dr. Harry Haiselden, producer and star of the 1917 movie *The Black Stork*, actually did facilitate the deaths of babies born crippled, and always claimed to be speaking for evolutionary medicine,[1] and who is to say he wasn't? Philippe Rushton claims to speak for Darwin today when he argues that Africans are r-selected and Asians K-selected, which he measures with not only IQ, but skull size, penis size, law abidingness, and degree of civilization.[2] Richard Dawkins claims to speak for Darwin today when he argues that whatever you do is to facilitate the replication of your genes, or alternatively of your memes, but either way, it is definitely not God's will.[3,4]

The point is that Darwinism itself partakes of multiple meanings, not necessarily connected at all to the ideas actually maintained by the celebrated hypochondriac of Down House, or even to the ideas that may be normative today in evolutionary biology or anthropology.

Science and Culture

We begin with Madison Grant, a wealthy New York lawyer, bearer of the names of two American presidents, nature enthusiast, and author of the 1916 best-seller *The Passing of the Great Race*.[5] He claimed to speak for Darwin in arguing that the Nordics were responsible for all civilizations, and that the lesser peoples entering the United States in the early twentieth century would bring it to biological ruin. The book's preface is by his friend Henry Fairfield Osborn, the leading evolutionary biologist in America. Grant articulates a problem, the influx of immigrants to America from the inferior stocks of southern and eastern Europe, and proposes a recognizably Darwinian solution: "a rigid system of selection through the elimination of those who are weak or unfit" (p. 146). We could haggle over just what some of the words mean, but Grant goes on to elaborate for us: it is about sterilizing the social discards—"the criminal, the diseased, and the insane, and ultimately...worthless race types" (p. 146).

Madison Grant was one of the leaders of the American Eugenics Society, and he was a major advocate for its scientific goals—both of which had been successfully achieved by 1927, in a congressional bill selectively restricting immigration (the Johnson-Reed National Origins Act, 1924) and a Supreme Court decision (*Buck v. Bell*, 1927) allowing states to sterilize poor citizens against their will.[6]

It can hardly be surprising, then, that as soon as the Nazis gained control in Germany, they turned explicitly to Madison Grant for inspiration, as a squib from the *New York Times* reported on August 2, 1933. Indeed, the Germans found one metaphor particularly resonant, that the race or nation is a super-organism, whose literal health is at risk by the proliferation of inferior peoples. Consequently, what we called eugenics, they tended to call race hygiene. And in 1933, a course called Race Hygiene for Physicians was offered to medical students, and the professor who taught it bragged that what other nations (i.e., the Anglophone ones) had only proposed, the new Germany was actually implementing.

And when Madison Grant died a few years later, the *New York Times* observed (May 31, 1937) *not* that his life's work resulted in tens of thousands of Americans being sterilized against their will, or in the deaths of thousands of refugees from the Nazis who could no longer find asylum in America. His public legacy is that he saved the California redwoods!

To say that eugenics was pseudoscience is simply false—the major scientists in America all kowtowed to Madison Grant. *The Passing of the Great Race* was favorably reviewed in *Science* by a geneticist from MIT,[7] although Sir

Arthur Keith was a bit more stinting in his praise for the book in *Nature*.[8] The Harvard physical anthropologist Earnest Hooton ridiculed Grant after his death,[9] but Hooton had served faithfully under him on the advisory board of the American Eugenics Society in the 1920s, and had written to him politely after receiving an inscribed copy of Grant's last book: "I don't expect that I shall agree with you at every point, but you are probably aware that I have a basic sympathy for you in your opposition to the flooding of this country with alien scum."[10]

There is no "pseudo" about it: leading scientists tended, by and large, to agree with what Grant had to say. And if I sound like I am taking this personally, well, they are talking about my grandparents here.

Neither can we call this bad science. British eugenicists thought themselves superior to American eugenicists, but that is hardly surprising. Eugenics was, in fact, a reasonable scientific solution to a set of social problems, given three cultural facts: (1) no income tax and hence no significant federal budget or venues for assisting the poor, (2) the belief that the needs of the state or race take precedence over the rights of the individual, and (3) the idea that large groups of people can be linearly ranked. Like any other cultural understanding of the world, eugenics was sensible and scientific for its time and place, given a set of prevalent assumptions about biology and society.

The problem of eugenics is not solved, however, by brushing it aside and calling it bad, as if that was then and this is now, as if they were blinded by culture, but we are smarter than that. Geneticist Steve Jones[11] suggested drawing just such a lesson: "As is often the case in science, geneticists have become much more humble about their understanding of their subject as they realise how little they really know. Eugenics was based on ignorance and prejudice rather than on fact; a science with these at its centre was bound to die" (p. 446). But this fails to address how scientists at the time were unable to see that it was based on ignorance and prejudice, rather than on fact, much less that it was fated to die. Rather, the problem of eugenics is solved in appreciating that science is inevitably a cultural property and, as such, does not stand outside of the universe of values and morality.

What I am trying to set up here is that a classic idea—that there are good uses of evolution and bad uses of evolution—is simply a strategic illusion. So, for a recent example, Rick Bribiescas writes:[12] "Shoddy and irresponsible research tainted by political agendas has made its way into the academic mainstream…. But to totally discount evolutionary theory because of its previous misuse would be to outlaw matches in response to the crimes of arsonists" (p. 12). (The ellipsis is standing in for references to Stephen Jay Gould and myself, by which I am immensely flattered, although I would point out that neither of us has even remotely suggested discounting evolutionary theory for its misuse.)

Arguing about its *misuses* at all, however, presupposes that you can readily distinguish them from its *uses*, which the eugenics movement shows that you simply cannot. This passage presumes a classic categorical distinction between scientific ideas and their applications, or between the world of facts (on the one hand) and values (on the other). That distinction is self-serving, because it permits scientists not to have to worry about the implications or the cultural context of what they say and do, for they are now absolved of any responsibility for it. That distinction began to be untenable with the famous 1947 epigram by the father of the atomic bomb, J. Robert Oppenheimer, to the effect that "physicists have known sin." That sin was not plagiarism or avarice or heresy; it was developing and releasing an astonishing power of destruction upon the human race and the world. And neither was it an evil application—it was their job, it was their responsibility to the war effort.

The older view is a philosophical fossil from the Enlightenment, that humans are rational beings and animals are not, and that science is merely heightened rationality, free of interest conflicts or prejudice—positions that no competent philosopher would hold today. By the mid-twentieth century it had come to be accepted that it is not so much rationality that makes us human, but culture (in the anthropological and philosophical, rather than in the newer, ethological sense)—something locally rational and coherent, yet to a large extent also irrational and capricious, but always ultimately meaningful. Moreover, the idea that science stands apart from culture has never really held up. Science has always required patronage, and that can come from business, or from the state, or simply from the support of the masses. Science is always connected to other interests—nationalist, capitalist, populist—science is cultural. As a human activity, science must be cultural; to see it as disconnected from culture is to see it as somehow nonhuman, and as standing apart from the set of values, codes, and stories through which our humanity is actually constituted. It is not that culture gets in the way of science, then, it is that everything human, including science, is *ipso facto* cultural.

To maintain that science takes place among humans interacting socially in a network of meanings and interests, and then to set science apart from that very network of meanings and interests, is to make scientific behavior nonhuman, which is paradoxical, given that the only known scientific activity is in fact carried out by humans. So if the wall between science and culture or facts and values is porous, and that distinction from the Enlightenment is unsustainable, then you can no longer think of science as having products or uses that are either good or bad. Science itself is both good and bad,[13,14] and consequently it is the responsibility of scientists, as it is the responsibility of anybody, to appreciate the significance of that distinction and presumably to side with the good, or else to bear the consequences of failing to do so. And that, I might add, is precisely the point of the Garden of Eden story, as its authors actually intended it to be understood.

Right and wrong are very real and very human, even if they are learned and locally specific. The development of categories of behavior, of codes and standards—some being special, or sacred or taboo—is an autapomorphic feature, a uniquely derived trait of our species, and is as effective a zoological identifier of humans among the living primates as bipedalism is. You can train a dog, and you can train a chimpanzee, but you cannot get them to remember the Sabbath Day and to keep it holy. It just makes no sense to them: crickets chirp, kangaroos hop, and humans make, follow, and break rules. It is what we evolved doing. Obviously primates strategize and modulate their behavior, but this is different: it comprises the distinction between not doing something because you do not think you can get away with it, and not doing something because you *just should not*. And it does not matter why—*that is just not the way we do things around here, and it just is not right.*

So anyone who says "I am a scientist, so I don't have to worry about right and wrong—that's somebody else's problem" is actually denying, in his or her own behavior, what is arguably a fundamental product of human evolution, for any activity not connected to the realm of morality and values is not human, pure and simple. For science to be such an activity, it would have to be undertaken by intelligent chimpanzees, or perhaps by Vulcans or androids—but not by humans.

I am dwelling on this point for a reason: culture is not the frosting on the human cake, which can and perhaps ought to be scraped off and set aside for separate enjoyment; it is the eggs in the cake. It is inseparable from being human; it is the "tradition of all the dead generations [that] weighs like a nightmare on the brain of the living"[15] (p. 15). And yet many scientists still get it backwards, believing that they can separate a deep human nature from a superficial human culture, and often in the name of Darwin.[16] But of course culture is also the product of millions of years of coevolution with the human species and cannot be so casually separated from it.

Darwin's Ventriloquists

A famous philosopher reputedly said, "For many shall come in my name, saying, I am Christ; and shall deceive many" (Matthew 24:5). Some millennia later, another famous philosopher reputedly said, "Je ne suis pas Marxiste" (Frederick Engels to Edouard Bernstein, November 3, 1882). Marx denies being a French Marxist, and Jesus warns against impersonators. The message is the same: don't be fooled! Both men were heads of movements and embodied ideologies reaching far beyond anything they had ever actually said or implied.

Darwin is no different, although perhaps less prescient. Within just a few years after his death, the British Eugenics Society would be founded by his cousin Francis Galton, and later headed by his son Leonard. It would be hard not to make the association between Darwinism and eugenics. Likewise, Herbert Spencer's phrase "survival of the fittest" was widely being adopted as a template for exploitative colonial and domestic economic practices. The phrase was adopted by Darwin himself in the fourth edition of *On the Origin of Species*, and frankly, who ever heard of "social Spencerism"?

In America, the eugenicists also had Darwin to back them up. Indeed, it was the succession of ideas in Hunter's *A Civic Biology*[17] (out of which John

T. Scopes had taught evolution, and was thereby arrested), in which evolution led directly into racism and eugenics, that induced the lawyer Clarence Darrow to evolve from biology's most vociferous champion to its most vociferous critic in the space of just a few months.[18–20] Darwin becomes an icon not just of evolution, but of any naturalistic theory trying to gain a foothold on the social and ideological landscape. It is consequently necessary to look carefully at the claims made in his name: to what extent might an evolutionary medicine be based on *assumptions* about evolution, rather than on *conclusions* about it?

A basic tension runs through the field of evolutionary biology. Every generation has tried to grapple with it, and it remains largely unresolved, except for the methodological approaches and unqualified assertions made by different scholars, and the periodic calls to question their assumptions. I refer to the assumption of adaptationism: that things have a purpose, or else they would not be there, and that our job as "Darwinians" is to identify that purpose. The problem can be seen up front, in Nesse and Williams's manifesto of "Darwinian medicine," *Why We Get Sick*:[21] "Studies of the functional reasons for human attributes are based on a method of investigation recently named the *adaptationist program*," they observe, italicizing the key phrase for didactic purposes. "By suggesting the functional significance of some known aspect of human biology, you may logically be able to predict some other, unknown aspects."

This sounds normative, and indeed uncontroversial, unless one looks at the endnotes. There, the diligent reader learns, "The term adaptationist program was first used, disparagingly, by S. J. Gould and R. C. Lewontin[22] in their much-cited article 'The Spandrels of San Marco.'"[21] In other words, Nesse and Williams have adopted a term coined in derision of the very views they hold, without engaging the nature of the criticisms themselves, or even acknowledging in the body of the text that their views on evolution are not quite universal.

The controversy is largely independent of Darwin. If you assume things are there for a purpose, then a reasonable scientific goal would be to identify that purpose. But if, on the other hand, things may or may not be there for a purpose, then the enterprise of identifying that purpose becomes less scientific and more tendentious. Gould and Lewontin were specifically calling attention to the falsity of the assumption that everything is there for a purpose. To devote one's attention to identifying a purpose that might not be there in the first place is scholasticism, hardly science. Indeed, Gould and Lewontin's point is that the adaptationist program is not an intellectual leap forward at all, but a throwback to natural theology, the biology that Darwin himself learned in college and rejected.

This tension reflects to some extent a methodological divide in evolutionary biology. Anatomists tend to see adaptation in organisms, tend to expect stability through time, and tend to explain unexpected change or difference by recourse to Darwinian (directional) selection. Geneticists, on the other hand, tend to see slop in genomes, tend to expect persistent change through time via mutation, and tend to explain unexpected stability or uniformity by recourse to conservative stabilizing selection. The expectation of slop and noise is so great in genomics that any evidence at all of adaptation via directional selection is considered publishable and newsworthy. An entire Japanese school of molecular evolution came to be centered on the study of neutral variation in the 1960s, and by the 1970s the French molecular biologist François Jacob could famously declare that evolution is like a tinkerer, not like an engineer.[23,24]

Jacob's choice of metaphor is itself noteworthy. "Tinkerer" is familiar to any anthropologist as the English equivalent of Claude Lévi-Strauss's *bricoleur*, who constructs myths out of whatever motifs are handy, and shapes them into a coherent and meaningful story.[25] Genomes are not engineered by selection, argues Jacob (indeed, citing Lévi-Strauss), they are cobbled together, and somehow manage to get the job done. This is not the adapationist program; this is a view of evolution in which history and chance have a large say in the final product—and, most importantly, in which that final product is very hard to tell from one that has been precisely engineered.

Again, this reflects a tension inscribed into the very core of evolutionary theory. Genetic drift, the gene pool's random deviations from mathematical expectations, was devised and mathematically explored by Sewall Wright, and adopted around mid-century by major American evolutionary biologists. Sewall Wright modeled evolution in subdivided, structured populations (where nonadaptive change could play a significant role), while in England, Ronald Fisher modeled it in large, randomly mating populations (where it would not). The American school of evolutionary genetics consequently came to see a greater role for genetic drift than the British school.[26,27*]

What, then, of the adaptationist program? In fact, it has been criticized in every generation. William Bateson criticized it shortly after the turn of the last century. Earnest Hooton[28] criticized it in the 1930s ("It is impossible to distinguish between a non-adaptive variation and one which is environmentally actuated and directed"), and so did Sherwood Washburn[29] in the 1960s: "When I read the descriptions of the importance of adaptive characteristics, I am not sure that there has been any progress since the nineteenth century." "What I am protesting against strongly is the notion that one can simply take a factor, such as a high cheekbone, think that it might be related to climate, and then jump to this conclusion without any kind of connecting link between the two elements."

---

* Motoo Kimura, founder of the Japanese school of molecular population genetics, earned his doctorate under Wright.

Yes, there is a problem with the adaptationist program: if you do not know that a feature is adaptive, then speculating on *how* is it adaptive is grossly unscientific. In fact, at least according to Gould and Lewontin, it is un-Darwinian.

Three Diseases

A contrast from human genetics may be instructive. On one end of the spectrum, one in thirteen African Americans is a carrier of the sickle cell allele (OMIM* 603903), and we have a good idea why. Populations of tropical Africa have adapted genetically to endemic malaria in the last 10,000 years, and this adaptation involves a trade-off between the bad effects of having sickle cell anemia and the worse effects of having malaria. Their descendants may carry the genetic adaptation even into malaria-free places. On the other end of the spectrum, 1 in 250 Afrikaners (white South Africans of Dutch descent) have the allele for porphyria variegata (OMIM 176200), and we also have a good idea why. One of the original seventeenth-century Dutch settlers had the gene, and virtually everyone who has it today in South Africa is simply descended from her. It is not an adaptation to anything, just a passive demographic consequence of population expansion—in this case, with an unfortunate attribute in the genome of one founder.

Neither of these origin narratives is contested. We have a lot of physiology in support of the sickle cell story, and a lot of genealogical history in support of the porphyria story. But suppose we did not. How would we know whether any particular feature of the gene pool was likely to be an adaptation (like sickle cell in Africans) or not an adaptation (like porphyria in Afrikaners)?

In fact, it is commonly difficult to discern whether particular diseases typical in particular populations are there as the products of drift or selection. Cystic fibrosis, for example, has been linked (unconvincingly) to cholera, typhus, lactose intolerance, bubonic plague, tuberculosis, asthma, and genetic drift.

Tay-Sachs disease, or TSD (OMIM 272800), is a recessive lethal allele carried by one in thirty Ashkenazi Jews. We might expect that, if its frequency were the result of founder effect, then all the copies of the defective allele would contain the same lesion. On the other hand, if selection were elevating its frequency, we might expect to find a high degree of heterogeneity in the structure of the defective allele across affected individuals.

As early as the 1960s, the sickle cell model was providing inspiration for a disease that TSD might confer protection against. The leading candidates were tuberculosis and hyposalinity. But molecular studies showed that Ashkenazi Jews have two common variants of the TSD allele, one present in about 70% of copies and one present in about 20% of copies. Could that be two different founder effects independently elevating the frequencies of

---

* Online Mendelian Inheritance in Man, the standard reference for human genetics.

two different versions of a very rare disease? Or could it be selection elevating the proportions of diverse mutations for TSD, but doing so grossly unequally? Advocates of drift are plentiful, and a new hypothesis, connected to old prejudices, harnesses the prevalence of TSD in Jews to heterozygous protection against stupidity.

The point is that, once again, TSD is a fairly well-studied and well-known disease. But the demographics of the target population are complex, and the patterns do not easily sort themselves out into an unambiguous story. How valuable, then, would the adaptationist program be, if it would give you a choice between tuberculosis and IQ as explanatory hypotheses, but not even consider random genetic drift as an alternative?

And if the situation is at best ambiguously adaptive for a comparatively well-known aspect of the gene pool like TSD (and cystic fibrosis, which is comparably ambiguous), then how appropriate will the adaptationist program be for the examples invoked by Nesse and Williams, which include morning sickness and jealousy?

Once again, my purpose is not to discredit the endeavor of introducing evolution into medicine, but to highlight the fact that there are multiple coexisting theories of evolution, and the one that is presented as normative may actually be an idiosyncratic mutation—or at least a variant accompanied by gratuitous pleiotropic assumptions.

Nothing-butism

I do not mean to suggest that evolutionary medicine is alone in its Darwinian ventriloquism. Periodically, ethologists "realiz[e] that man is descended from a primitive ancestor, [and] say that he is only a developed monkey," a situation derided by an earlier generation of evolutionary biologists as "nothing-butism"[30,31](p. 20). Since then, the mould has produced titles like *The Territorial Imperative*,[32] *The Naked Ape*,[33] *The Descent of Woman*,[34] and *Demonic Males*.[35]

Sir Edmund Leach cogently dispatched Ardrey's work decades ago, calling attention to the false assumption about human social relations invented by Thomas Hobbes, but now being associated with Darwin.[36] The most recent theory to do this is the demonic males theory,[35] which holds that studying the behavior of chimpanzees is unproblematically continuous with studying the behavior of humans, and the former illuminates the latter—for we are, obviously, very close relatives. Following in the lineage of naked apes and hairy dolphins, we are now just—and of course that is a critical "just"—brainy chimps. In this vein of understating the differences between humans and apes, primatologist Craig Stanford writes, "In their emotions, cognition, linguistic ability, homicidal brutality and erotic sexuality, the apes and we are far more alike than we are different"[37](p. xi). But what can this likeness in, for example, erotic sexuality mean if a chimpanzee male is stimulated by purple estrus swellings and copulates for 15 seconds? Is it possible that we have produced a generation of primatologists that have come to know more about apes than they do about people?

It is an interesting idea that differences between ourselves and apes are either trivial or illusory, and that evolutionary thought might somehow imply a methodological need to merely trim them away or push them aside. Darwin and Huxley thought no such thing—they were hard-pressed just to argue for any anatomical and emotional continuity at all, where the differences between humans and apes were so obvious as to be taken for granted. And it is precisely this unrealistic interpretation of Darwinism, that humans are not really different from apes in any important ways, that is perennially revived in nothing-butism.

Nothing-butism was most trenchantly described to me by the late Sherry Washburn as the study of human evolution that begins by assuming we are not human. What he meant was this: consider the distal hindlimb of chimpanzee and human—one adapted for grasping, the other for weight bearing. They look rather alike; they are made of pretty much the same parts in pretty much the same relations. But if one is interested in the human foot—how it works, what it means, even where it came from—there is very little that studying a chimp foot can tell you, except by way of contrast, that you cannot learn better from studying human feet. A chimp foot can be trained to bear weight to some extent, and a human foot can be trained to grasp to some extent. But what is interesting about the two feet, from the standpoint of evolution, is how they *differ*. That is also why, in spite of having similar forelimb parts in similar relations to those of a sparrow, you still cannot get off the ground by flapping, while the sparrow can.

Returning to the ape, then, if locomotion is so different between us, and the structures have been reworked so as to alter the basic function, then what about cogitation? Is it possible that the threefold growth in size, extensive cortical convolutions, and neurological reorganizations facilitating speech[38,39] make human thought and behavior different from—not bigger than, not more complex than, not a variant of—chimp behavior? I believe it does. The value of chimp feet for understanding human feet lies in their contrast, not in their sameness. Likewise, the value of chimp brains and behaviors lies in their contrast to humans. Once again, this is not about piety or humanism, it is about scientific methodology. Labeling things that look different and do different things as "the same"—because their parts roughly correspond, and 6 or 7 million years ago their ancestors were the same—is not only weird but perverse. It is antievolutionary.

Who but a creationist would deny the very thing that Darwin was trying to demonstrate in *On the Origin of Species*? The purpose of that book was to exclude the miraculous from the history of life, as Lyell had in the history of the earth, and people like Morgan, Tylor, and Marx had in the history of society. By 1859 the antiquity of the earth had already been established, and the succession of life had already been established. The book is called *On the Origin of Species* because literally that—the origin of species—was the site at which natural theologians were still invoking miracles. Darwin successfully relocated the miracles to the origin of life, rather than of species themselves, and reduced them in quantity to "a few...or...one" in the famous last line of the book.

The origin of species, said Darwin, lies not in theology, but in adaptive divergence, and it is exactly that divergence that interests us as post-Darwinian biologists and anthropologists. In other words, it is evolutionary to acknowledge the difference of humans, and an evolutionary theory that fails to come to grips with that is not going to be of much use as an analysis of behavior, or a representation of nature.

Representing Nature

Cladistics is a school of evolutionary thought that is obsessed with proximity of descent—how to establish it, catalog it, manage it—at the expense of any reciprocal interest in adaptive divergence, which is (in the argot) autapomorphy and therefore not of intellectual value.[40] Privileging descent over divergence, the cladists have been reclassifying life as if there were no divergence.

Thus, in this scheme, reptiles cannot contrast against birds. Reptiles must be obliterated, because they are not close relatives of one another. Cladists give us, then, Squamata (incorporating birds, dinosaurs, and crocodiles) and Archosauria (incorporating lizards). We gain a consistent criterion for classifying (descent). What we lose in the process is the significance of a feathered, flying reptile on the ecological history of life on earth.

As early as 1962, Morris Goodman was proposing a reduction in the rank at which humans were separated from apes, and the obliteration of the distinction between humans and great apes, on the grounds that genetically humans are so very similar to apes that you can hardly tell them apart, and that phylogenetically humans fall within apes. He failed to persuade the principal systematists of that generation, for the simple reason that he really did not know what he was talking about. As Simpson[41] put it, in as close to monosyllables as he could make it,

> It is abundantly established that anatomically, behaviorally, and in other ways controlled or influenced by total genetic makeup *Homo* is very much more distant from either "*Pan*" or "*Gorilla*" than they are from each other. That fact is not overbalanced by the failure of just one kind of data to reflect that distinction clearly or in equal degree. The distinction is real, and it still justifies the classical separation of Pongidae and Hominidae in classification. (p. 369)

The dispute was not over data-driven scientific conclusions about evolution; it was epistemic—it was about whether we should privilege genetic relationships over all other relationships (such as anatomical or ecological), and whether we should privilege ancestry over divergence. Simpson was not a creationist, he simply did not accept the primacy of genetics over paleontology.[42] If you cannot tell a human from a chimp by looking at its hemoglobin, then why not just look at something else? And certainly, we are all tailless, social, arm-rotating species, but only one of us has driven the others to the brink of extinction, and is threatening the very planet itself, and that seems to be a significant chunk of biological knowledge.

Somewhat extraordinarily, in the last few years, the biological anthropology community has actually caught up with Goodman's error, which is why we now have to talk about hominins instead of hominids.[43] It is what follows from assuming that (1) our genetic similarity to the apes is so transcendent as to necessitate a distinction at a lower taxonomic level than the family (humans, chimps, gorillas, and orangutans are now all in the family Hominidae) and (2) descent is taxonomically significant while divergence is not (humans, chimps, and gorillas are separated from orangutans in the subfamily Homininae, and humans are not alone until we descend to the Tribe Hominini).

The new classification again has the advantage of applying a single criterion—descent—consistently, and the disadvantage of emphasizing the temporal divergence of the orangutan over the ecological divergence of the human. It is thus not out of a misplaced sense of piety that one can question that decision. I agree with Simpson; I just do not think that a classification that willfully ignores the major highlights in the history of life is necessarily a good one. And those highlights are constituted by adaptive divergences—people from apes, anthropoids from prosimians, tetrapods from fish, birds and mammals from reptiles, eukaryotes from prokaryotes, and many more—all of which are missing from a taxonomic system like this, as Simpson of course noted many years ago. And more importantly, one loses the evolution itself when encoding into a classification the message that we did not really diverge from apes, or that even if we did, it is somehow not very important.

The point I am trying to get at here is that speaking for evolution is a tough job, and overstating our relationship to the apes can paradoxically produce as antievolutionary a result as understating it.

An Intersection of Darwinism and Medicine

Just as there are weird theories of human behavior and systematics, with highly contestable assumptions, deriving legitimacy by claiming to speak in Darwin's name, there are already programs of "racialized medicine" doing the same. *BiDil*, the first racial pharmaceutical (targeted specifically at black hypertensives), would seem to violate our knowledge of the patterns of human diversity (which are principally cultural, polymorphic, clinal, and local) as readily as a perpetual motion machine would violate our knowledge of physics. We already know that much in the way of racial difference in health is due to differences in quality of health care—in other words, to sociocultural causes.[44-46] We also know that most genetic differences in the human species are from person to person, not from group to group. And finally, we know that racial groupings subsume tremendous amounts of biological heterogeneity. To the extent, then, that we encounter genetic differences between large groups of people, they are expressed as different allele frequencies, and to the extent that any group differs in health-related issues, these are expressed as risk probabilities—hardly the situation most amenable to the development of a racial pharmacopoeia.

Since race is a very poor surrogate for genotype (structured as polymorphism and clines), and since genetics is a minor component (compared to social and economic factors) of group differences in health-related issues, racial medicine cannot even be a useful first step toward individualized medicine. Nevertheless, racial medicine is powered by a more powerful force than science—capitalism. *BiDil*, although itself a financial flop, is the vanguard of the racial "niche market."[47-52]

In 2002, the *New York Times Magazine* ran an essay[53] by a psychiatrist at the American Enterprise Institute—a "right-wing think tank" that is home to Charles Murray, author of *The Bell Curve*.[54] The author of "I am a racially profiling doctor," Sally Satel, drew together three themes: (1) biologically reifying race (explicitly decrying the politically correct fluff put out by anthropologists and sociologists), (2) speaking for Darwin ("Skin color itself is not what is at issue—it's the evolutionary history indicated by skin color"), and (3) touting *BiDil* and the forthcoming racial pharmacopoeia.[53]

Three years later, the *New York Times* published an op-ed[55] by an "evolutionary developmental biologist," Armand Marie Leroi, who told its readers:

> The recognition of race may improve medical care. Different races are prone to different diseases....Such differences could be due to socioeconomic factors. Even so, geneticists have started searching for racial differences in the frequencies of genetic variants that cause diseases. They seem to be finding them.
>
> Race can also affect treatment. African-Americans respond poorly to some of the main drugs used to treat heart conditions—notably beta blockers and angiotensin-converting enzyme inhibitors. Pharmaceutical corporations are paying attention. Many new drugs now come labeled with warnings that they may not work in some ethnic or racial groups....
>
> Such differences are, of course, just differences in average. Everyone agrees that race is a crude way of predicting who gets some disease or responds to some treatment. Ideally, we would all have our genomes

sequenced before swallowing so much as an aspirin. Yet until that is technically feasible, we can expect racial classifications to play an increasing part in health care.*

Conclusion

Darwin, race, and Big Pharma—not exactly an association that evolutionary medicine can take much pride in. How, then, do we identify the good evolutionary medicine, and distinguish it from the bad? History shows that it is not very easily done, except in retrospect, when it is too late.

I regard health care professionals, as I suspect most Americans do, with the same esteem I accord my plumber and computer technician. Their job is to diagnose a problem and to effect, or at least to recommend, a solution to it. Their ideas upon the descent of species, or for that matter upon the heliocentric solar system, matter little to me as long as they take care of my stomach ache, toilet bowl, and laptop, respectively. Add to that the prospect that a random, well-read practitioner of evolutionary medicine is as likely as not to be mistaking it for racial essentialism, or to be an unwitting shill for the pharmaceutical industry, and I hope I can be forgiven for approaching the subject with some degree of caution.

The role of evolution in human affairs, or in medicine itself, is not really at issue. The issues are: (1) What are the specifics of evolution that we are being asked to apply? (2) Where does the authority for producing those instantiations of evolutionary theory reside? In other words, when medicine and evolution overlap, who really does speak for Darwin, and how can we tell?

Another way of framing the issue is to combine Theodosius Dobzhansky's famous aphorism, that nothing in biology makes sense except in the light of evolution, with Emile Durkheim's, that social facts can only be explained in terms of prior social facts. Perhaps there is something biological that does make sense without evolution, but the functioning human body is assuredly not it. Yet this is precisely the context in which one needs to acknowledge the construction of the boundary between functioning and malfunctioning, by which Foucault begins to intrude upon Darwin.[56,57] After all, a strong case can be made for medicine needing to become more cultural, with perhaps even greater urgency than it needs to become more evolutionary. And part of that culturalizing must include the recognition that as long as Darwin's name confers scientific legitimacy, it can and will be exploited.

---

* In spite of a storm of scholarly protest from anthropologists and geneticists, the *New York Times* would not publish any substantive letters, and the Social Science Research Council established a web page for the criticisms of Leroi's essay: raceandgenomics.ssrc. org.

With political ideologies and financial profits increasingly at stake, Darwin's name is in constant jeopardy, and somehow the onus always seems to fall upon us anthropologists to keep it as unsullied as possible.

References

1. Pernick, M. 1996. *The black stork*. New York: Oxford University Press.
2. Rushton, J. P. 1995. *Race, evolution, and behavior: A life-history approach*. New Brunswick, NJ: Transaction.
3. Dawkins, R. 1976. *The selfish gene*. New York: Oxford University Press.
4. Dawkins, R. 2006. *The God delusion*. New York: Bantam.
5. Grant, M. 1916. *The passing of the great race*. New York: C. Scribner.
6. Kevles, D. J. 1985. *In the name of eugenics*. Berkeley: University of California Press.
7. Woods, F. A. 1918. *The Passing of the Great Race*, 2d ed. [review]. *Science* 48:419.
8. Keith, A. 1917. Is the Anglo-Saxon doomed? *Nature* 99:502.
9. Hooton, E. A. 1940. *Why men behave like apes and vice versa*. Princeton, NJ: Princeton University Press.
10. E. A. Hooton to Grant, November 3, 1933. Earnest A. Hooton Papers. Peabody Museum, Harvard University.
11. Jones, J. S. 1992. Eugenics. In *The Cambridge encyclopedia of human evolution,* ed. S. Jones, R. Martin, and D. Pilbeam, 442. New York: Cambridge University Press.
12. Bribiescas, R. 2006. *Men*, 12. Cambridge, MA: Harvard University Press.
13. Proctor, R. N. 1991. *Value-free science? Purity and power in modern knowledge*. Cambridge, MA: Harvard University Press.
14. Collins, H., and Pinch, T. 1993. *The Golem: What everyone should know about science*. New York: Cambridge University Press.
15. Marx, K. 1852. *The 18th Brumaire of Louis Bonaparte*, 15. New York: International Publishers, 1981.
16. Pinker, S. 2002. *The blank slate: The modern denial of human nature*. New York: Viking Penguin.
17. Hunter, G. 1914. *A civic biology presented in problems*. New York: American Book Co.
18. Darrow, C. 1925. The Edwardses and the Jukeses. *The American Mercury* 6:147.
19. Darrow, C. 1926. The eugenics cult. *The American Mercury* 8:129.
20. Marks, J. 2006. The scientific and cultural meaning of the odious ape-human comparison. In *The nature of difference: Science, society and human biology*, ed. G. Ellison and A. Goodman, 35. Baton Rouge: Taylor & Francis.
21. Nesse, R. M., and Williams, G. C. 1995. *Why we get sick: The new science of Darwinian medicine*. New York: Times Books.
22. Gould, S. J., and Lewontin, R. C. 1979. The Spandrels of San Marco and the Panglossian Paradigm: A critique of the adaptationist programme. *Proc. R. Soc. B* 1161:581–98.
23. Kimura, M. 1983. *The neutral theory of molecular evolution*. New York: Cambridge University Press.
24. Jacob, F. 1977. Evolution and tinkering. *Science* 196:1161–66.
25. Levi-Strauss, C. 1962. *The savage mind*. Chicago: University of Chicago Press.
26. Fisher, R. 1930. *The genetical theory of natural selection*. Cambridge: Cambridge University Press.
27. Wright, S. 1931. Evolution in Mendelian populations. *Genetics* 16:97–159.
28. Hooton, E. A. 1930. Doubts and suspicions concerning certain functional theories of primate evolution. *Hum. Biol.* 2:223–49.

29. Washburn, S. L. 1963. The study of race. *Am. Anthropol.* 65:521–31.
30. Huxley, J. 1947. *Touchstone for ethics*, 20. London: Harper.
31. Simpson, G. G. 1949. *The meaning of evolution*. New Haven, CT: Yale University Press.
32. Ardrey, R. 1966. *The territorial imperative*. New York: Atheneum.
33. Morris, D. 1967. *The naked ape*. New York: Random House.
34. Morgan, E. 1997. *The descent of woman*. 4th ed. London: Souvenir Press.
35. Wrangham, R., and Peterson, D. 1996. *Demonic males: Apes and the origins of human violence*. Boston: Houghton Mifflin.
36. Leach, E. 1966. Don't say "boo" to a goose. *New York Review of Books*, December 15.
37. Stanford, C. 2001. *Significant others: The ape-human continuum and the quest for human nature*, xi. New York: Basic Books.
38. Preuss, T. 2001. The discovery of cerebral diversity: An unwelcome scientific revolution. In *Evolutionary anatomy of the primate cerebral cortex*, ed. D. Falk and K. Gibson, 138–64. Cambridge: Cambridge University Press.
39. Balter, M. 2007. Neuroanatomy: Brain evolution studies go micro. *Science* 315:1208.
40. Eldredge, N., and Cracraft, J. 1980. *Phylogenetic patterns and the evolutionary process: Method and theory in comparative biology*. New York: Columbia University Press.
41. Simpson, G. G. 1971. Remarks on immunology and catarrhine classification. *Syst. Zool.* 20:369.
42. Simpson, G. G. 1964. Organisms and molecules in evolution. *Science* 146:1535.
43. Marks, J. 2005. Phylogenetic trees and evolutionary forests. *Evol. Anthropol.* 14:49.
44. Martins, D., and Norris, K. 2004. Hypertension treatment in African Americans: Physiology is less important than sociology. *Cleve. Clin. J. Med.* 71:735–43.
45. Sankar, P., et al. 2004. Genetic research and health disparities. *JAMA* 291:2985–89.
46. Dressler, W., Oths, K., and Gravlee, C. 2005. Race and ethnicity in public health research: Models to explain health disparities. *Annu. Rev. Anthropol.* 34:231–52.
47. Duster, T. 2005. Medicine. Race and reification in science. *Science* 307:1050–51.
48. Kahn, J. 2005. Misreading race and genomics after BiDil. *Nat. Genet.* 37:655–56.
49. Kahn, J., and Sankar, P. 2006. Being specific about race-specific medicine. *Health Affairs* 25:375–77.
50. Duster, T. 2006. The molecular reinscription of race: Unanticipated issues in biotechnology and forensic science. *Patterns Prejudice* 40:427–41.
51. Kahn, J. 2006. Patenting race. *Nat. Biotechnol.* 24:1349–51.
52. Ellison, G. 2006. Medicine in black and white: BiDil®: Race and the limits of evidence-based medicine. *Significance* 3:118–21.
53. Satel, S. 2002. I am a racially profiling doctor. *New York Times Magazine*, May 5.
54. Herrnstein, R. J., and Murray, C. 1996. *The Bell curve: Intelligence and class structure in American life*. New York: Touchstone Books.
55. Leroi, A. M. 2005. A family tree in every gene. *New York Times*, March 14, p. A23.
56. Foucault, M. 1973. *The birth of the clinic: An archaeology of medical perception*. New York: Vintage.
57. Roughgarden, J. 2005. *Evolution's rainbow: Diversity, gender, and sexuality in nature and people*. Berkeley: University of California Press.

Index

Milton Keynes UK
Ingram Content Group UK Ltd.
UKHW020315111024
449327UK00040B/1123